Laser Welding

edited by
Dr. Xiaodong Na

Laser Welding
Edited by Dr. Xiaodong Na

CBS, Edition **2017**

Published by InTech
Janeza Trdine 9, 51000 Rijeka, Croatia

Publishing Process Manager Jelena Marusic

Technical Editor InTech DTP team

Cover Designer InTech Design Team

Image Copyright Olegusk, 2010. Used under license from Shutterstock.com

Additional hard copies can be obtained from orders@intechopen.com

Laser Welding, Edited by Dr. Xiaodong Na
p. cm.
ISBN 978-953-307-129-9

INTECH

open science | open minds

Contents

Preface

For so many years ever since the invention of the laser technology in 1960s, laser welding/processing started its debut in various industries and has progressed to be one of the most useful techniques in high-speed, automated welding.

This book is entitled to laser welding processes. The objective is to introduce relatively established methodologies and techniques which have been studied, developed and applied either in industries or researches. State-of-the art developments aimed at improving or next generation technologies will be presented covering topics such as monitoring, modelling, control, and industrial application. This book is to provide effective solutions to various applications for field engineers and researchers who are interested in laser material processing. This book is divided into 10 independent chapters corresponding to recent advances in the field.

Editor

Dr. Xiaodong Na
Cummins Fuel System Specific Control
United States of America

1

Intraoral laser welding

Carlo Fornaini, MD, DDS, MSc
University of Parma, Faculty of Medicine, Dental School
Italy

1. Introduction

Just after the introduction of the first laser by Maiman, in 1960, there has been a very fast evolution of this new technology characterized by the constant progression about techniques and applications, devices ever more efficient, smaller and cheaper, and the introduction of ever-new wavelengths.

One interesting application of this new technology was the possibility to weld many kinds of metals and, in industrial fields, this procedure spread in a very short time.

Laser welding was firstly introduced in jewellery during years 70 and, just after, it was successfully used also by dental technicians (Maddox, 1970).

Initially, CO2 and Nd:YAG were used but, finally, the second one rapidly conquered the market due to the results obtained (Shinoda et al, 1991- Yamagishi et al, 1993)

Laser welding, in fact, gives a greater number of advantages than traditional welding.

First of all laser device saves time in commercial laboratory because welding is completely done directly on the master cast. Inaccuracies of assembly caused by transfers from the master cast along with investment are reduced. (Berg et al, 1995)

Then, the heat source is a concentrated and high power light beam able to minimize distortion problems on the prosthetic pieces. (Santos et al, 2003)

Another interesting aspect is the possibility to weld very close to acrylic resin or ceramic parts without physical (cracking) or colour damage (Bertrand et al, 1995)): this means to save time and money during the restoration of broken prosthetics or orthodontics appliances because of the possibility to avoid the remaking of the non-metallic portions.

This welding technique may be used on every kind of metal but the property to be very active on titanium makes it very interesting and specific for the prosthetics over endosseous implants. (Walter et al, 1999)

Many laboratory tests have demonstrated laser welding joints have a high reproducible strength for all metals, consistent with that of the substrate alloy. (Bertrand et al, 2004)

All these advantages gave to this procedure a great diffusion in the technician laboratories and stimulated the companies to put in the market more and more upgraded appliances.

Some aspects, such great dimensions, high costs and delivery system by fixed lenses today still characterize these machines, strictly limiting their use only to technician laboratories.

Moreover, the management of this appliance is very difficult, due to the number of the parameters involved and the factors related to the welding process. For these reasons, it

results strictly dependent by operator and influenced by the duration of the training period (Bertrand & Poulon-Quintin, 2009).
The first aim of this study was to value the possibility to utilize, for laser welding, the same device normally used in dental office. This gives the advantage to be used by dentist himself, being easier and with few parameters to adjust, and to avoid costs for the appliance, because it is the same of dental cares.
Moreover the dentist may avoid to send prosthetics to the lab and, sometimes, to take impressions, with patient receiving his repaired prosthetics after a very short waiting.
The second aim was to reach the result to weld directly into the mouth by means of the utilisation of a fibber-delivered laser after a careful valuation of the biologically compatibility of the procedure.
The advantages of this technique, defined Intraoral Laser Welding (ILW), consist in the possibility to fix the position of the different parts of the prosthetics without utilisation of acrylic resins and/or silicon impressions, and to repair damaged fixed prosthetics without their removal from the mouth.

2. Materials and Methods

The first step of our research was to determine what wavelength, among these normally utilised by dentist in his office and also in industrial field to make a real welding process (CO_2 -10600 nm, Diode laser -810 nm and Nd:YAG -1064 nm.) was able for our work. Some tests have been realized with metallic plates showing that proper wavelength was the Nd:YAG laser.
In fact, in dental CO_2 laser the pulse durations are too short (microseconds) and cannot give the thermal elevation necessary to obtain a fusion of metal while in Diode dental laser output power is too low (from 5 to 10 watt) and cannot give the Energy necessary to make a real welding process.
We decided so to use the appliance FIDELIS PLUS III (FOTONA, Slovenia) (Fig. 1) which is a combination of two different wavelengths, Er:YAG (λ=2940 nm) and Nd:YAG (λ=1064 nm).

Fig. 1. The utilized appliance Fidelis Plus III

The first allows to the dentist to treat hard tissues (enamel, dentin and bone) by a mechanism, which, utilizing the affinity of this laser with water and hydroxyapatite, induces the explosion of intracellular water molecules and so causes the ablation of the tissues. (Keller & Hibst, 1989)
Its utilisation may be extended also to the dermatology, where it can be employed, in addition to the elimination, by vaporizing them, of lesions such condyloma, naevi, warts, mollusca contagiosa, to the treatment of cheloid scars and wrinkles with the so-called "resurfacing". (Khatri, 2003)
Nd:YAG laser allows to the dentist to make surgery with complete emosthasis, thanks to the affinity of this wavelength with haemoglobin, and so to avoid the use of sutures. (White et al, 1992).
The delivery system, in this laser, consists of optical fibres of different sizes, chosen in conformity with the kind of application; the dimension go from 200 µm (endodontics) to 900 µm (bleaching).
The peculiarity of the appliance FIDELIS PLUS III is given by the possibility to have, in addition to pulse duration of microsecond which are necessary during dental interventions, even pulse durations of millisecond (15 or 25). This gives the possibility to use it also in flebology, in the treatment of inestethisms of vascular origin, thanks to the affinity of this wavelength for haemoglobin. (Scherer & Waner, 2007)
The optical fibre delivery system is a very important advantage of this device, by the point of view of the intraoral welding, because it is a very flexible and ergonomic, able to penetrate into the oral cavity.
We decided to use a fibre of 900 µm of diameter, normally used for bleaching and biostimulation.
Initially a handpiece with a 2 mm-spot (Fotona R 30), normally used in dermatology, was chosen and, by reducing the working distance, a spot of 1mm was obtained. Manufacturer took part in the first experimental step of our work by the realization of an handpiece able to generate a 0.6 mm spot. The aim was to increase the Fluence (J/cm2) which is the most important parameter determining the quantity of energy delivered to a surface, by a factor of 10, while also utilizing the device's maximum energy output (9.90J).

Fig. 2. The metallic support used in order to put firmly handpiece

Before each test we evaluated the output power with a powermeter (Ophir Nova II with thermal head F150A, Ophir, Jerusalem, Israel) in order to verify the stability of the laser's energy delivered.
A metal support, in which the handpiece was securely placed, was developed to maintain the correct working distance and to obtain a better management of the welding process (Fig.2).
To determine the proper parameters of the device, in order to obtain a real welding process causing the minimal thermal damage to the matter, several tests were performed.
A CrCoMo plate (25X50X10 mm) (Fig.3) was cast, sandblasted with alumina powder (50-mm), rinsed and degreased with acetone.

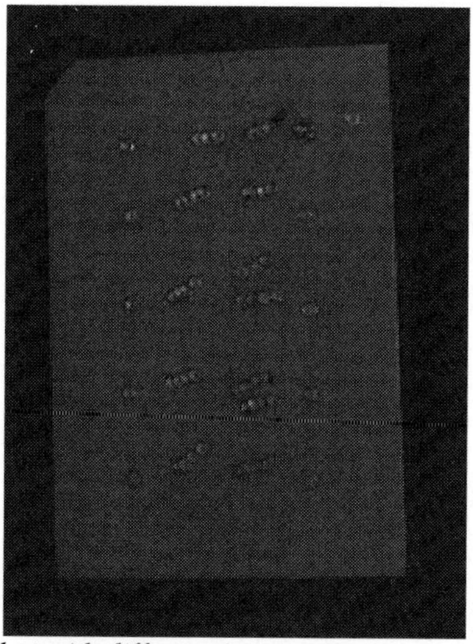

Fig. 3. The CrCoMo plate shot with different parameters.

Various combinations of welding parameters were tested on this alloy plate. The tests consisted of shooting the plate's surface with the laser: the spot's configuration was then analyzed using an interferometric technique. Interferometry is a non-contact optical technique for measuring surface height and shape with great speed and accuracy. Interferometry makes it possible to precisely measure, in three dimensions, shape and size of the laser's crater in the metal surface and allowed us to choose the laser parameters that well welded with minimal collateral damage to the surrounding area.
Interferometric analysis were performed by Prof. Caroline Bertrand at CNRS Laboratory of Bordeaux (France).
The following parameters were selected:
Output Power= 9.85 W, Frequency= 1 Hz, Pulse Duration= 150 msec, Working Distance 40 mm, Energy= 9.85 J, Fluence= 3300 J/cm2.
Best results were obtained by using the maximum power output and the minimum frequency of the device and at a pulse duration of 25 msec we noted cracks and fissurations (Fig 4) while, at a pulse duration of 15 msec, we did not observe these features (Fig. 5).

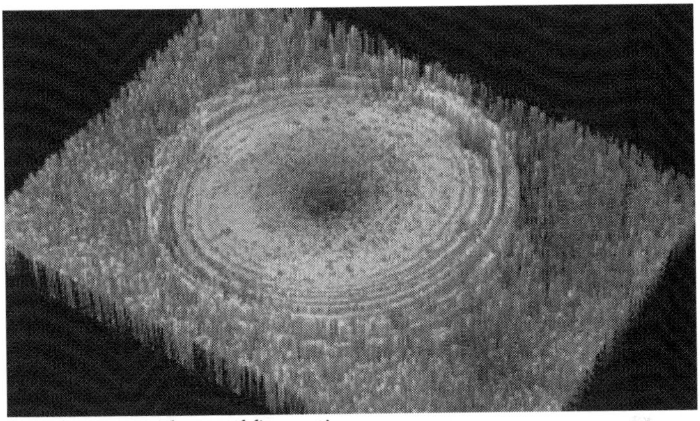

Fig. 4. 15 msec shot: no evidence of fissurations.

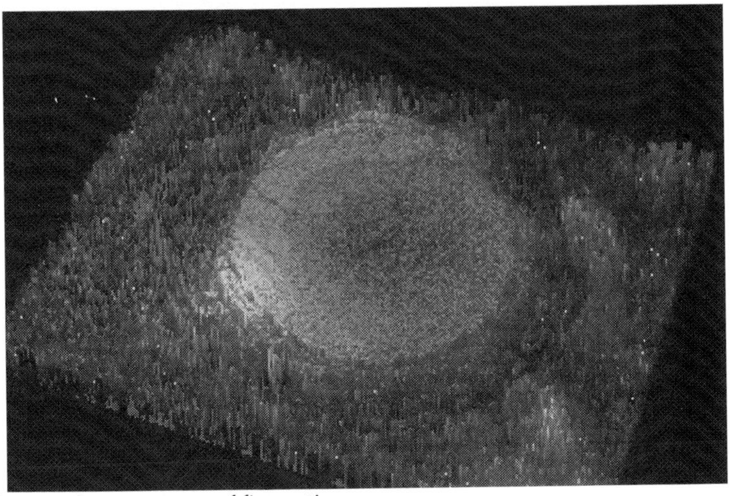

Fig. 5. 25 msec shot: presence of fissurations.

Other tests were then realized in order to compare the quality of welding process obtained with the office laser vs. that obtained by a technician laboratory welding laser.
We compared, by optical microscope, using laser beam over different Co-Cr-Mo plates, the welding process obtained by dental office Nd:YAG laser (Fidelis Plus III, Fotona), with the parameters previously described, and dental technician laboratory Nd:YAG laser (Rofin, Germany). This device was used with these parameters:
Volt 310, Energy/Pulse 3.0 J, 4.5 Hz Frequency, Pulse Duration
1.9 msec, Output Power 2.6 KW, Fluence 1516 J/cm2
We appreciated that they were very similar, except for the dimension of the welded bead, that was smaller in dental laser tested plates, due to the different spots used. (Figure 6)

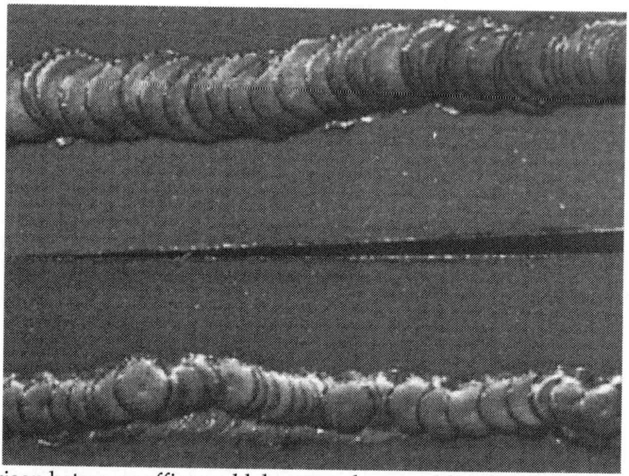

Fig. 6. Comparison between office and laboratory laser welding beads

Then we made other tests to see ultrastructural aspects and strength of the laser welding joints.

Sixteen Argeloy NP Special® (Argen Corporation, Dusseldorf, Germany) (composition: 31.5%Cr, 5%Mo, 59.5%Co, 2%Si, 1%Mn, 1%Other) plates of the dimension of 15mm x 15mm x 0.15mm were divided in four groups.

In each group the plates were welded to obtain two samples of two plates welded in the median portion on the two sides.

The plates of the first group were welded by a Nd:Yag laser device normally used in dental office (Fidelis Plus III, Fotona, Ljubljana, Slovenia) without metal filler.

The plates of the second group were welded by the same device but with the apposition of a metal filler (CoCr- Schweissdraht, Dentaurum, Ispringen, Germany) (Co 65%, Cr 28%, Mo 5.5%).

The plates of the third group were welded by a Nd:YAG laser device normally used in dental technician laboratory (Rofin, Hamburg, Germany) without metal filler.

The plates of the fourth group were welded by the same device but with the apposition of a metal filler (CoCr- Schweissdraht, Dentaurum, Ispringen, Germany) (Co 65%, Cr 28%, Mo 5.5%).

The parameters used were the same as previously described.

In samples where filler was added, three passages were done: the first passage was realized without metal apposition to fix the position, the second one with filler and the third one without metal to regularize the welded fillet. In the other samples only the first and the third passes were done.

We used the described parameters, even if very different, because they are the ones currently used in dental laboratory by Rofin and in dental office by Fidelis.

Every plate was marked with an alphanumeric code and sent to the laboratory for the analysis with optical microscope, SEM and EDS: the analyst knew only the code but not the type of the samples in order to realize a blind study.

The samples were firstly englobed in epoxydic resin and then polished by abrasive papers and diamond pastes to 1 µm and observed by optical microscope, then chemical attack by $HCl + H_2O_2$ solution, and observations by SEM were realized.
SEM and EDS analysis was performed by Prof. Francesca Passaretti at CNR-IENI Laboratory in Lecco (ITALY) and mechanical tests were performed by Dr.Elena Villa in the same Laboratory.
In scanning electron microscopy (SEM), an electron beam is scanned across a sample's surface. When the electrons strike the sample, a variety of signals are generated and it is the detection of specific signals which produces an image or a sample's elemental composition. The three signals which provide the greatest amount of informations in SEM are the secondary electrons, backscattered electrons, and X-rays. (Goodhew et al, 2001)
Secondary electrons are emitted from the atoms occupying the top surface and produce a readily interpretable image of the surface. The contrast in the image is determined by the sample morphology. A high resolution image can be obtained because of the small diameter of the primary electron beam (Bozzola et al, 1999).
Backscattered electrons are primary beam electrons which are 'reflected' from atoms in the solid. The contrast in the image produced is determined by the atomic number of the elements in the sample. The image will therefore show the distribution of different chemical phases in the sample. As these electrons are emitted from a depth in the sample, the resolution in the image is not as good as for secondary electrons. (Julian, 2005)
Interaction of the primary beam with atoms in the sample causes shell transitions which result in the emission of an X-ray. The emitted X-ray has an energy characteristic of the parent element. Detection and measurement of the energy permit elemental analysis (Energy Dispersive X-ray Spectroscopy or EDS). EDS can provide rapid qualitative, with adequate standards, and quantitative analysis of elemental composition with a sampling depth of 1-2 microns. X-rays may also be used to form maps or line profiles, showing the elemental distribution in a sample surface. (Goldstein et al, 2003)
Twenty steel round orthodontic wires (Filo Tondo Duro Leone .030 C8080-30, Leone, Firenze, Italia and 14" Straight Wire, ortho Organizer Inc, San Marcos, Ca, USA) were also prepared as reported:
2 wires 0.75 mm not welded
2 wires 0.50 mm not welded
2 wires 0.75 mm welded by Rofin without metal filler
2 wires 0.50 mm welded by Rofin without metal filler
2 wires 0.75 mm welded by Rofin with metal filler
2 wires 0.50 mm welded by Rofin with metal filler
2 wires 0.75 mm welded by Fidelis without metal filler
2 wires 0.50 mm welded by Fidelis without metal filler
2 wires 0.75 mm welded by Fidelis with metal filler
2 wires 0.50 mm welded by Fidelis with metal filler
Metal filler, parameters and steps of the welding process were the same of these used for the plates.
The samples were analyzed by the Dynamic Mechanical Analysis (DMA) which can be simply described as the application of an oscillating force (stress) to a sample and the analysis of the material's response to that force (Menard KP, 2008).

The appliance used was the Dynamic Mechanical Analyzer Q800 (TA Instruments, New Castle, Delaware, USA) which makes the tests in single cantilever configuration, under static or dynamic conditions and up to a maximum force of 18 N.
The limitation of this device is that it can analyse only samples of little dimensions and this was the reason why we used orthodontic wires.

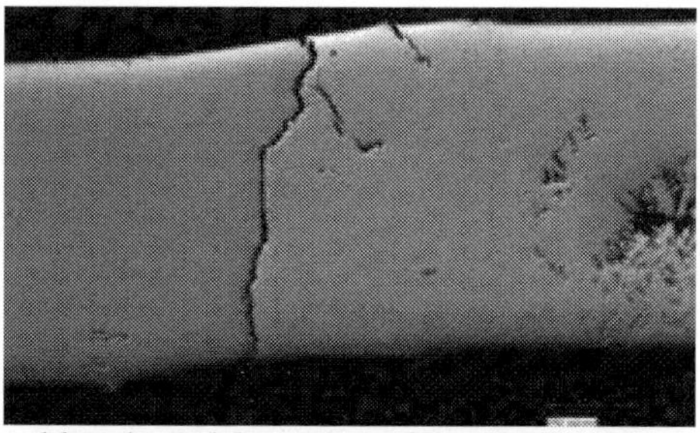

Fig. 7. Sample of laser welding by Fidelis without filler observed by optical microscope.

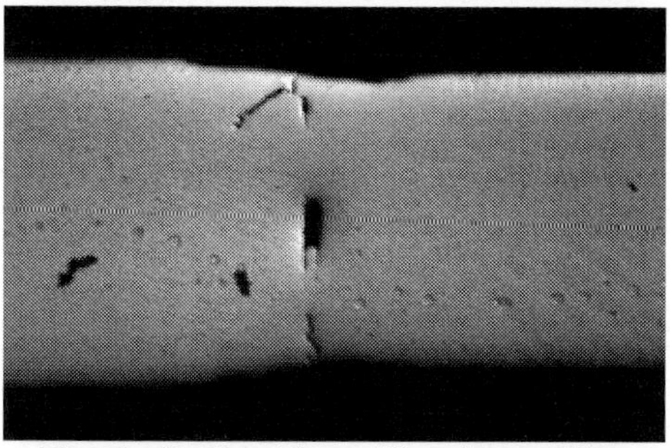

Fig. 8. Sample of laser welding by Rofin without filler observed by optical microscope.

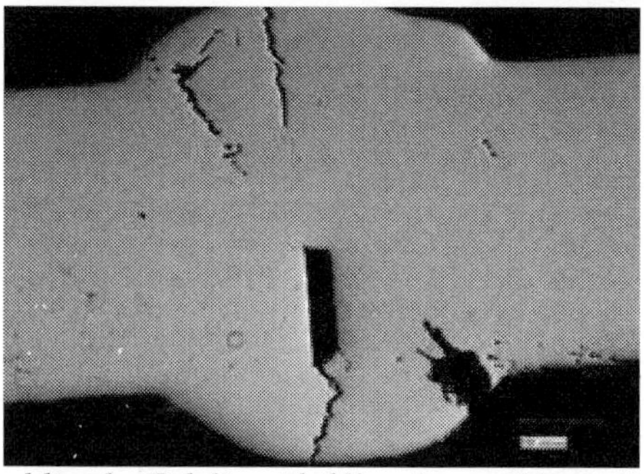

Fig. 9. Sample of laser welding by Fidelis with filler observed by optical microscope.

By optical microscope observation, the only significant difference between the groups regarded those welded without filler metal, in which it was seen a greater number of

fissuration in the plates welded by Fidelis plus III (Figg. 7 and 8) while in the groups with filler the differences were minimal (Figg. 9 and 10).
These aspects were confirmed by the observations of the samples after chemical attack by HCl and H_2O_2 and SEM observation (Figg. . 11-12-13 and 14).

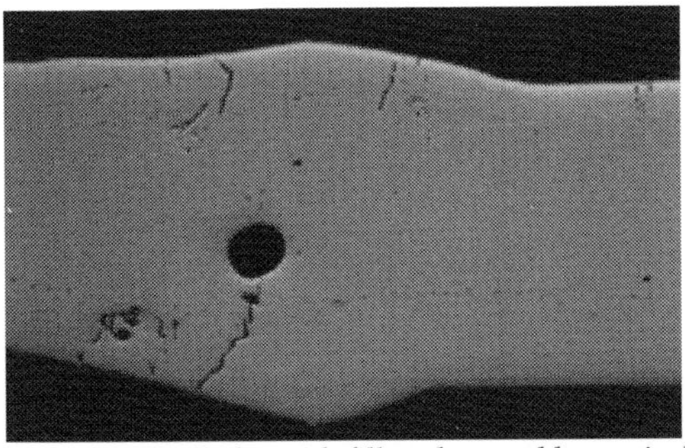

Fig. 10. Sample of laser welding by Rofin with filler observed by optical microscope.

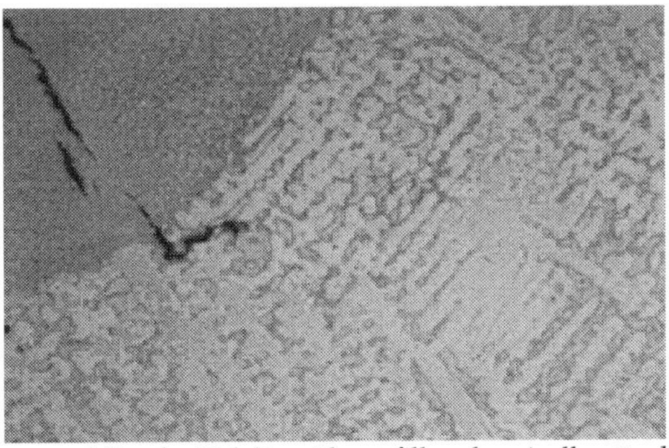

Fig. 11. Sample of laser welding by Fidelis without filler chemically attacked

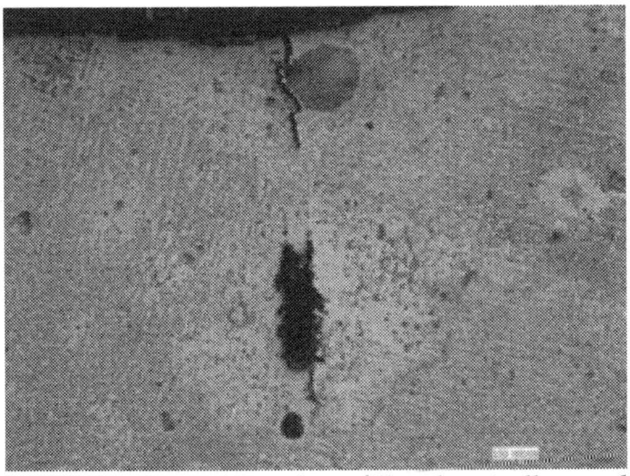

Fig. 12. Sample of laser welding by Rofin without filler chemically attacked

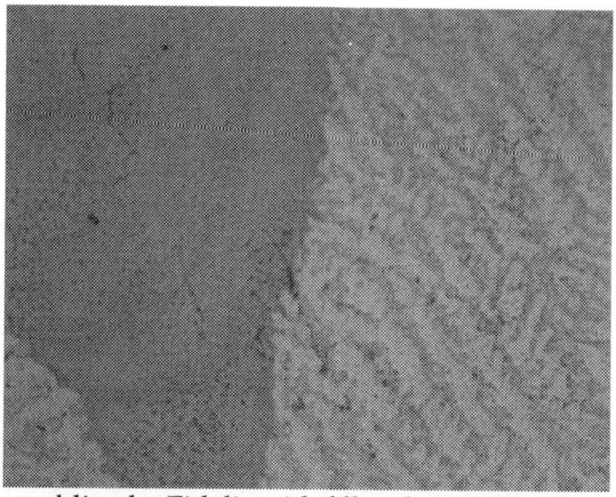

Fig. 13. Sample of laser welding by Fidelis with filler chemically attacked

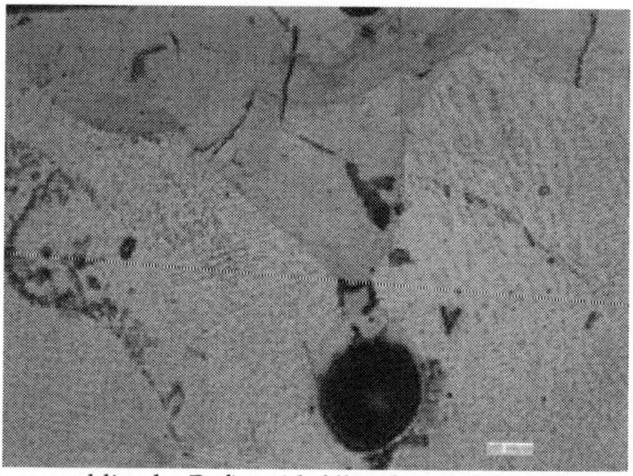

Fig. 14. Sample of laser welding by Rofin with filler chemically attacked

The EDS analysis in the welding zone showed an homogeneous composition of the CoCrMo alloy with no significant differences in the groups (Tabb. 1-2-3-4, results in %).

Spectrum	Co	Cr	Mo	W	C	Ga
1	57,67	23,33	5,57	7,57	2,86	3
2	58,64	22,18	5,71	8,28	2,82	2,38
3	58,98	22,41	5,57	7,77	2,8	2,47
4	58,64	22,32	5,97	7,91	3,11	2,06
5	58,43	23,42	4,7	7,43	2,98	3,05
6	60,26	22,45	4,29	7,06	3,01	2,91
7	57,26	23,03	5,32	7,58	3,62	3,19

Table 1. EDS analysis of plates welded by Fidelis without metal apposition

Spectrum	C	Cr	Co	Ga	Mo	W
1	3,86	22,93	57,71	2,68	5,2	7,61
2	4,57	22,34	57,31	2,32	5,6	7,85

Table 2. EDS analysis of plates welded by Rofin without metal apposition

Spectrum	Co	Cr	Mo	W	C	Ga
Spectrum 1	57,15	22,94	5,34	7,81	3,82	2,94
Spectrum 2	58,87	24,05	6,1	4,69	4,8	1,5

Table 3. EDS analysis of plates welded by Fidelis with metal apposition

Spectrum	C	Cr	Co	Ga	Mo	W
1	4,06	22,74	40,89	1,37	15,21	15,73
2	3,4	23,05	58,76	3,03	4,72	7,05
3	3,96	22,77	56,49	2,45	6,12	8,21
4	3,73	22,49	60,07	2,84	4,04	6,83

Table 4. EDS analysis of plates welded by Rofin with metal apposition

The mechanical tests done on orthodontic wires showed an elastic behaviour of the samples very similar (Tabb.5-6-7-8: green=not welded, red=Fidelis, Blue=Rofin)) with minimal differences between the samples.

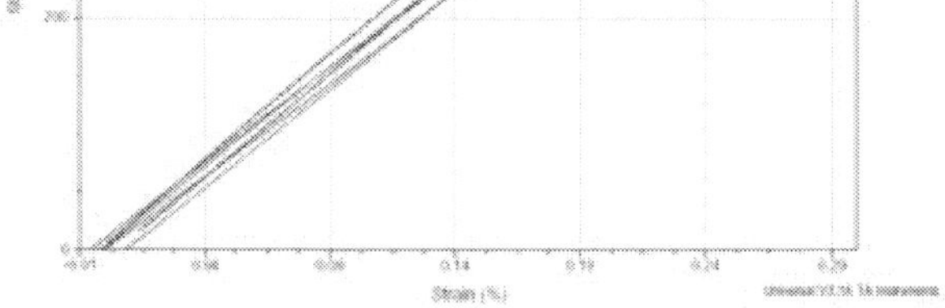

Table 5. Mechanical tests of wires up to 2 N

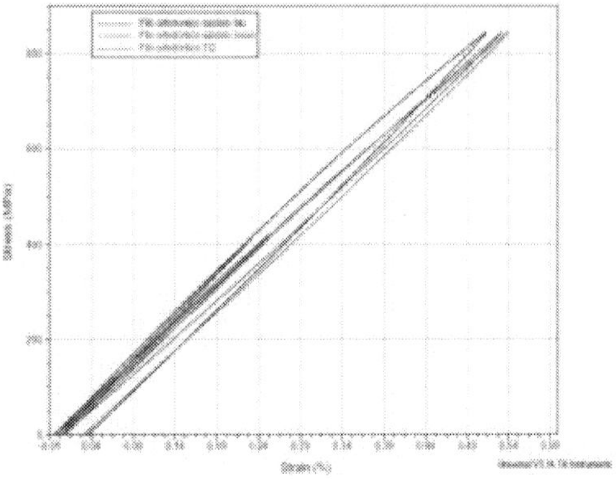

Table 6. Mechanical tests of wires up to 4 N

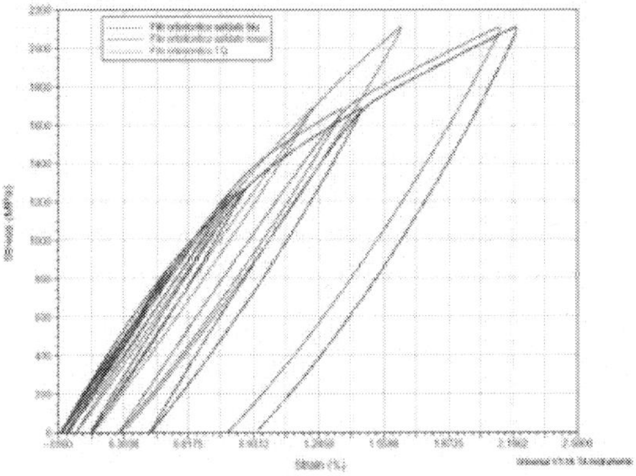

Table 7. Mechanical tests of wires up to 10 N

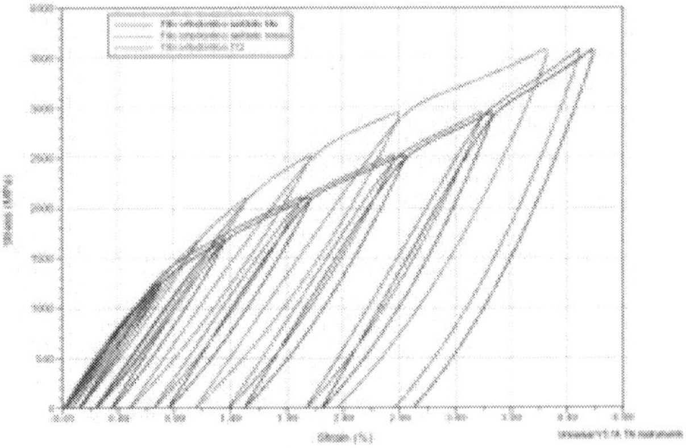

Table 8. Mechanical tests of wires up to 17 N

The residual deformation seen in the arches of 0.5 diameter was greater than in the arches 0.75 and in the arches welded with filler was greater than in those without and, in every case, no wire was broken even under the maximum strength.
The optical microscope observation of the wires after the mechanical tests didn't show significant differences between the samples. (Figg. 15 and 16)

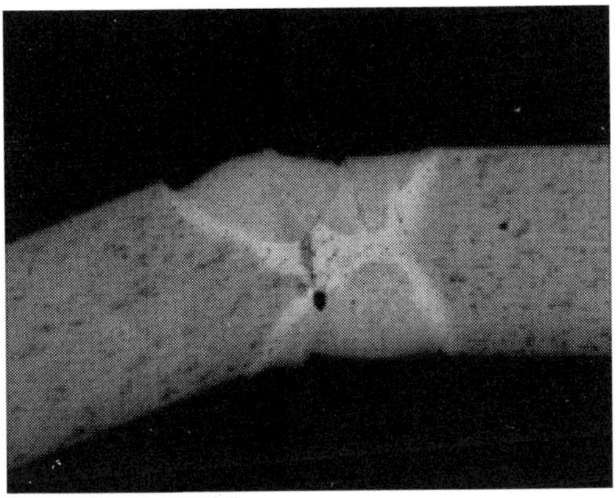

Fig. 15. Sample of laser welding by Fidelis

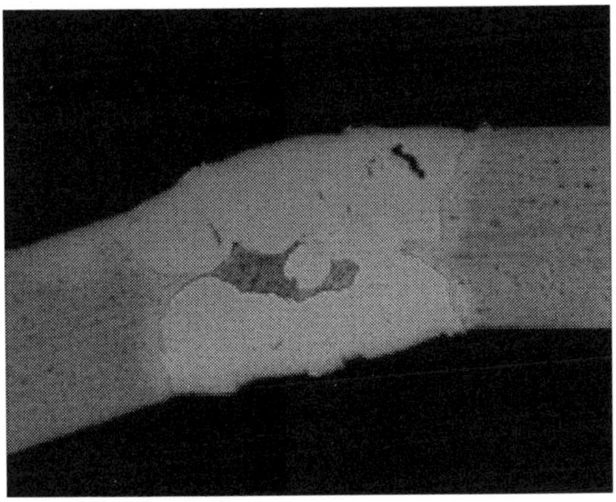

Fig. 16. Sample of laser welding by Rofin

Firstly it is necessary to do some general considerations about laser welding process to help the interpretation of these results
Laser technology is the most efficient method for applying thermal energy to small areas and, according to many Authors (Daves, 1992 and Ream, 1988), it is one of the best fusion welding techniques for dissimilar metals. This depends from the possibilty, by modern laser appliances to focus ligth beam to a fine focal point. This beam imparts energy into the metal causing to heat up locally to a temperature above the liquidus. The metal evaporates and a cavity is formed immediately under the heat source and a reservoir of molten metal is

produced around it. As the heat source moves forward, the hole is filled with the molten metal from the reservoir and this solidifies to form the weld bead. (Baba et al, 2004)
The presence of cracks on the fracture surfaces of the specimans were evident in several studies made on laser-welded Co-Cr-Co, while they were not reported on the fracture surface of the laser-welded titanium.(Watanabe et al, 2002)
This is due to the different values of thermal conductivity among metals: that of CoCrMo is higher than of the titanium.
After welding, the solidification of the molten CoCrMo alloy may occur more quickly, if compared with titanium: in fact the heat generated by the laser energy rapidly diffuses to the surrounding solid metal.
This rapid solidification causes a drastic constriction of the welded region at the same time.
Thus, a concentration of stress occurs at the laser-welded region , and so cracks are created. (Duhamel and Banas, 1983)
Also in the samples analysed in this study, fissurations and cracks were observed into the welded zone.
The greater differences between the plates welded by laboratory laser and these welded by dental office laser regarded the group without filler where these belonging to the second group had a greater number of cracks than these of the first group.
This difference was significantly reduced by the use of filler metal, and this is an important indication for our clinical practice where we decided to use always a filler metal.
The mechanical tests of this study hold an importance by the clinical point of view; in fact in all the tests with all the samples, the elastic module of the orthodontic wires was very high, and this means that our technique may be used intraorally as an aid to the orthodontic therapy.
We must emphasize that dental laboratory laser operates under an atmosphere of shielding gas, in this particular case argon, and this might be an important aspect regarding to the differences of the "without filler welded samples" two groups.
It'll be interesting to observe if, comparing the same appliances utilised in this work, by using both of them under an atmosphere of shielding gas, the differences between the groups will be the same.
In order to make these tests and also to have the possibility to weld titanium we added to the appliance a gas cylinder connected to a pipe spreading to the laser impact beam.

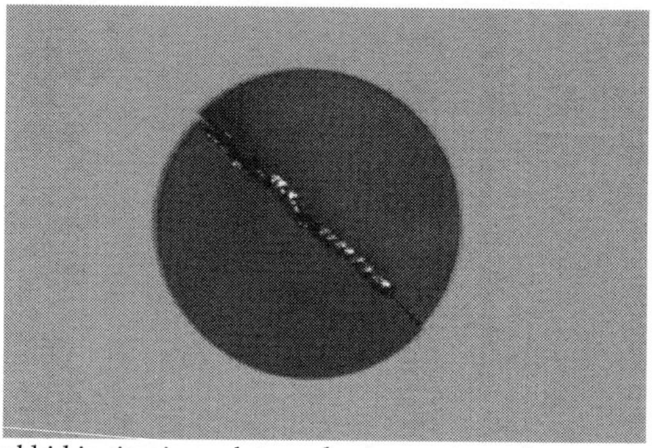

Fig. 17. Laser welded bid in titanium plate under argon atmosphere

In fact, shielding gas, is necessary to protect the weld area from atmospheric gases, such as oxygen, nitrogen, carbon dioxide, and water vapour that can reduce the quality of the weld or make the welding process more difficult to be used. (Watanabe & Topham, 2006)
Argon provides greater cleaning action than other gases, and, because it is heavier than air, it blankets the weld from contamination. For these reasons we decided to use argon, according to most of the Authors. (Yamagishi et al, 1993- Taylor et al, 1998)
So, we made some tests with Fidelis Plus III under an argon gas atmosphere on titanium plates, with and without apport metal, and we appreciated the absence of oxidation in the welded area. (Fig. 17)
In order to define thermal increase in the biological structures (sulcus, pulp chamber, bone and root) close to the thermal affected zone during the welding process, we made another in region and vitro study.
Two calf jaws freshly sacrificed were kept at room temperature and in six molars of each one a hole was made by a micromotor drill, in the distal- labial area.
Then, with the drill inserted into the tooth, its exact location into pulp chamber was checked by X-Rays (Fig. 18 and 19).

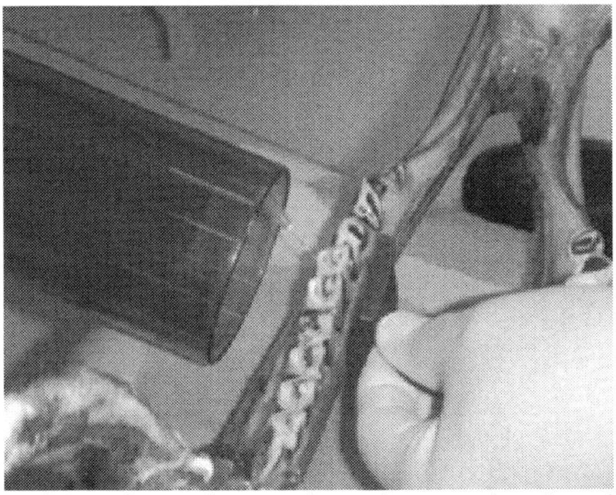

Fig. 18. Drill inserted into the tooth to evaluate the exact location of the drill in pulp chamber.

In each tooth, two further holes were made by a micromotor, one into the bone and one into the root.
Then, four thermocouples k-type were connected to every tooth and fixed with thermoplastic paste (Impression Compound Red Sticks, Kerr) into pulp chamber, sulcus, bone and root.
The thermocouples were then connected to a 4 channels Thermometer (LUTRON TM-946) PC-integrated in order to record and save data. (Fig. 20)

Fig. 19. Rx Check to evaluate the exact location of the drill in pulp chamber.

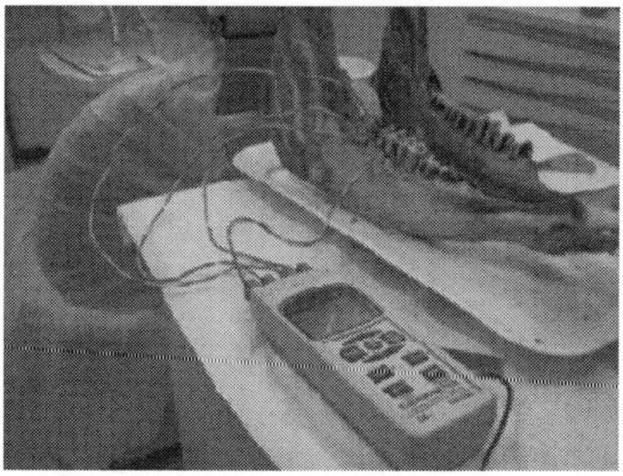

Fig. 20. 4-channel Thermometer (LUTRON TM-946) device.

Twenty-four metallic plates in ARGELOY NP SPECIAL (31.5% Cr, 5% Mo, 59.5 Cr, Co 2%, Si 1%, Mn 1%, others 1%), sized 5x35 mm and 0.5 mm thick, were curved to hemispherical shape (15 mm ray) and a couple of them was placed on each tooth previously prepared. (Fig. 21 and 22)

Fig. 21. Metallic plates in ARGELOY NP SPECIAL curved to hemispherical shape

Fig. 22. The couple of metallic plates placed on every previously prepared tooth

Every plate couple was welded, by Nd:YAG Laser Fidelis Plus III, in three points (occlusal, vestibular and lingual) to fix the position and the thermal rise was recorded by the four thermocouples. (Fig.23)

Fig. 23. Metallic plates welded, by Nd:YAG Laser, in three points (occlusal, vestibular and lingual).

The parameters were:
OUTPUT POWER 9.85 W, FREQUENCY 1 Hz, PULSE DURATION 15 msec, SPOT DIAMETER 0.6 mm, WORKING DISTANCE 40 mm, ENERGY 9.85 J, FLUENCE 3480 J/cm2.
We have not taken into consideration the Rayleigh length which is the distance from the focus at which the cross-sectional area of the beam is doubled. That is very important in laser welding procedure and we believe this might be the spotlight of a next in vitro study

even if we think that it's very difficult to keep a constant focal distance during intraoral welding procedure.
The recorded values were integrated by PC software (LUTRON SW-U801-WIN) in order to have an image of the highest, lowest and mean values .
The higher thermal elevation was pointed out by thermocouples inserted into the pulp chamber while the lowest regarded the bone.
The standard deviation (SD) was under 0.5.
The Mean values of temperature increase were as follows:
Pulp chamber: Mean value: 0.714 ± 0.45 °C (Max: 1.5 °C; Min. 0.1 °C).
Bone: Mean value: 0.100 ± 0.1 °C (Max: 0.3 °C; Min. 0.0 °C).
Sulcus: Mean value: 0.442 ± 0.43 °C (Max: 0.7 °C; Min. 0.1 °C).
Root: Mean value: 0.231 ± 0.24 °C (Max: 0.3 °C; Min. 0.0 °C).
We have made the first examination of this work with thermal camera by recording the metal welding processes: the limit of this device lies in the possibility of getting only a jaws surface evaluation. This is the reason for using the four thermocouples system which, even if more difficult and longer to perform, allows checking internal temperature of the structures.
Higher thermal rise was recorded in the pulp chamber; however, for all the twelve samples tested, the maximum temperature rise was lower than 5.5°C, which is considered as critical value for pulp vitality (Oelgiesser et al, 2003- Martins et al, 2006- Sulieman et al, 2006) and, in any case, it was lower than in the other studies regarding brazing process through an electric arc between two electrodes . (Shibuya et al, 2004 and Haney et al, 1996)
In the sulcus, too, the temperature increase was very low and yet lower than 1 degree. In bone and roots, the temperature increase was practically absent and this is particularly interesting because it allows expecting an in vivo clinical use in order to connect titanium bars to fixtures without damaging bone and compromising the Osseo-integration of implants.
Probably, the low number of shots (4 to 5) sufficient to fix the two metallic parts and the low Fluence transferred to the biological structures (Fluence= J/cm2; J=W/s) explain these interesting results for low temperature rise and so the temperature increase values in tissues keep below dangerous limit for integrity.
Laser welding process in metallic structures applied to mandibular molars causes a very low thermal rise in surrounding areas. This technique can be considered as biologically compatible and without the risk of necrosis.

Clinical Cases

1) Extraorals
Case 1
Patient TC, male 8 years old, in treatment in our office with a removable orthodontics appliance of Schwartz, came to us for the periodic check of the appliance and we saw that one of the Adam's hooks was broken (Figure 24).
We welded it without filler metal (Figure 25) and the plastic shield, even if very close to the welded zone, did not result damaged or modified.(Figure 26)
We could re-apply the repaired appliance to the patient only after some minutes.(Figure 27)

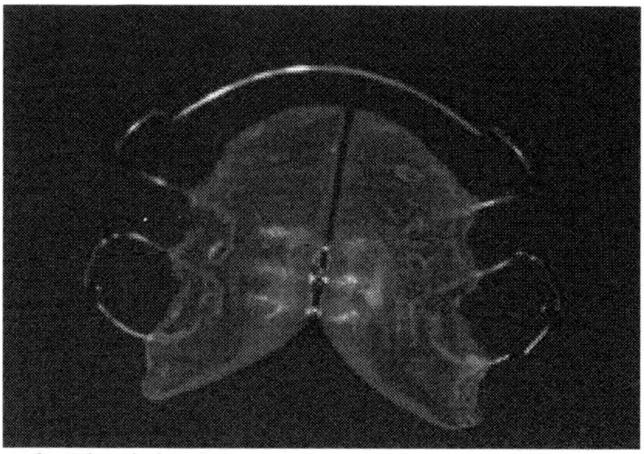

Fig. 24. Appliance with Adam's hook broken.

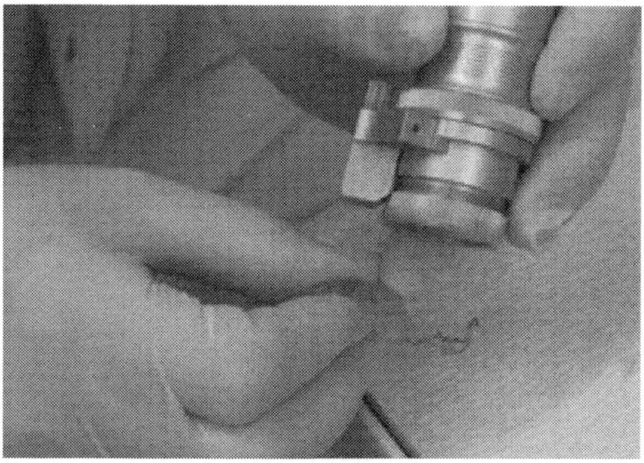

Fig. 25. Laser welding process without filler metal

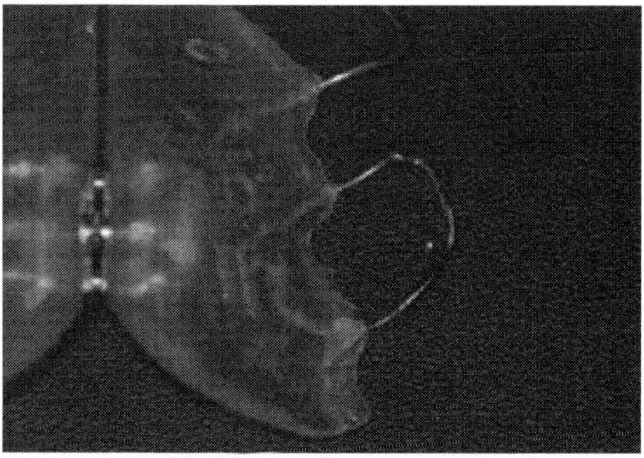

Fig. 26. Hook repaired without damaging acrylic portion close

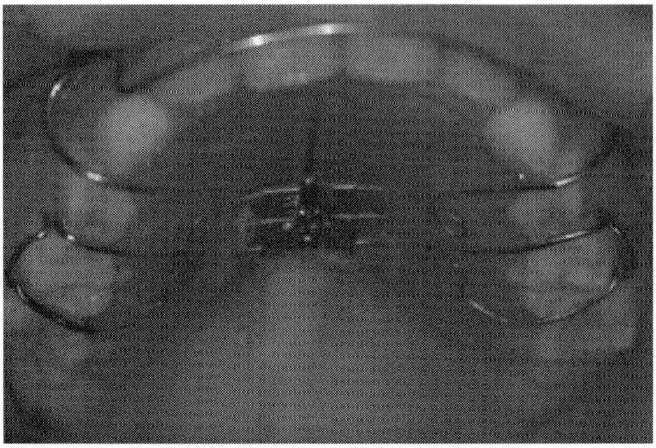

Fig. 27. Appliance replaced into the mouth

Case 2

Patient RK, 49 years old female came to our office with her removable prosthesis broken in the metallic portion. (Figure 28) Due of the impossibility to determine the exact position of the two fragments, we took an impression with the prosthesis inserted into the mouth and we prepared a stone model. (Figure 29)

Fig. 28. Broken removable appliance: Impression

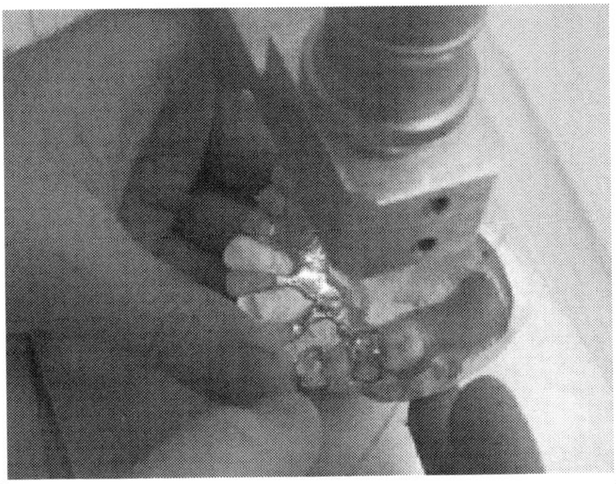

Fig. 29. Laser Welding

Then, we welded the two parts by Fidelis Plus III directly in our office utilizing filler metal without damaging or destroying the acrylic parts. (Figures 30)

After polishing the prosthesis, we applied it into the patient mouth just after half an hour. (Figure 31)

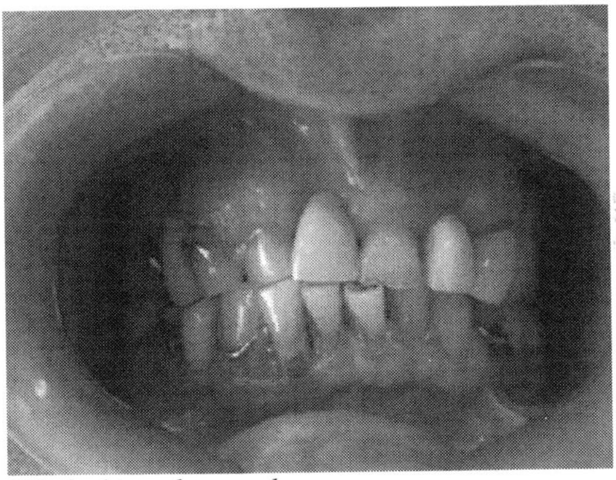

Fig. 30. Prosthesis re-applied into the mouth

Case 3

Patient ML, male, 43 years old, came urgently to our office with acute hypersensitivity, particularly from cold drinks, localized to the left mandibular area. The clinical examination showed that the lower left bridge was damaged in the occlusal surface of the crown of 37 (fig. 31). The problem was that the patient had to go on a three-weeks business trip the day after. So, we decided to remove the bridge, control the vitality of the tooth 37 and being sure it was free of any decay we repaired the crown directly in our office with the laser(fig. 32). Then, the crown surface was polished and the bridge re-cemented onto the teeth (fig. 33).

Fig. 31. The damaged bridge

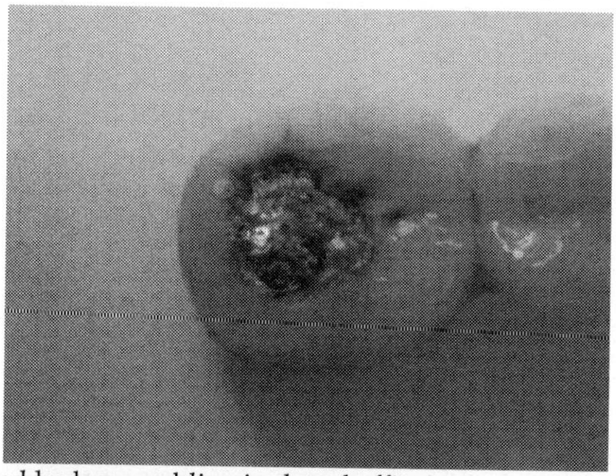

Fig. 32. Bridge repaired by laser welding in dental office

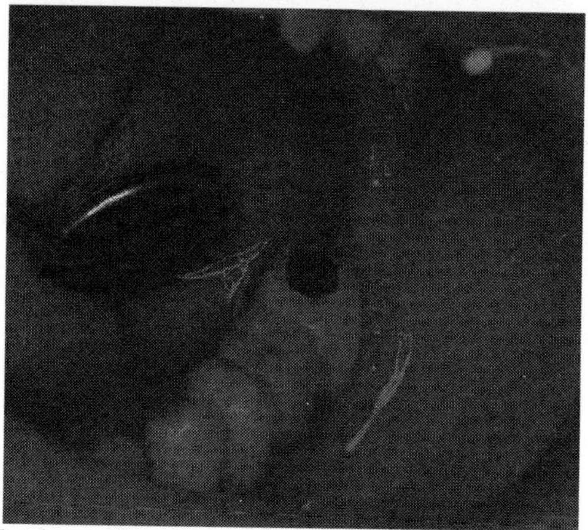

Fig. 33. The repaired bridge replaced into the mouth

2) Intraorals

Case 1:

Patient 59 years old male with fixed prosthetics placed in to the upper arch, with two crowns and five bone implants. (Fig. 34)

After the crowns preparation and the taking of the impressions, dental technician constructed the metallic structure of the bridge in two sections to assure fit.

In order to protect the soft tissues from the ejection of warm metal splinters, we made a sort of mask by silicon normally used to take prosthetic impressions with a little hole corresponding to the contact of the two portions of prosthesis. (Fig. 35)

Then we started to weld the bridge with the apposition of filler metal on the two parts. Filler material used was Bego WiroWeld 2 mm diameter, (Co 65%, Cr 28%, Mo 3.5%, others 3.5%)

After removing the bridge from the mouth, it was sent to the laboratory to complete its realization. (Fig.36)

During and after welding process patient said he did not feel any discomforts.

After three weeks we could seal the bridge and finish the rehabilitation of the patient. (Fig. 37)

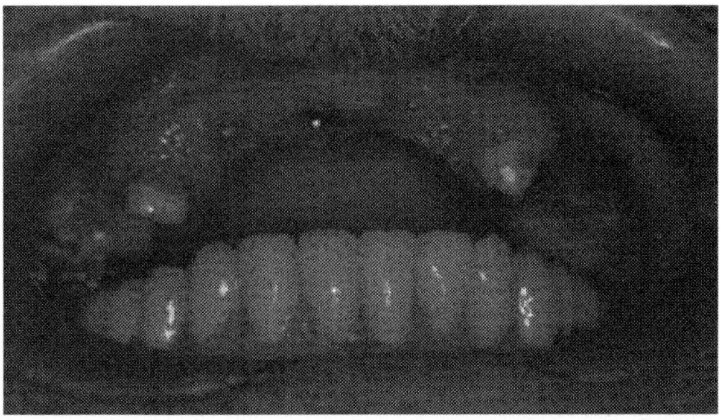

Fig. 34. Patient with preparation of two crowns and five bone implants

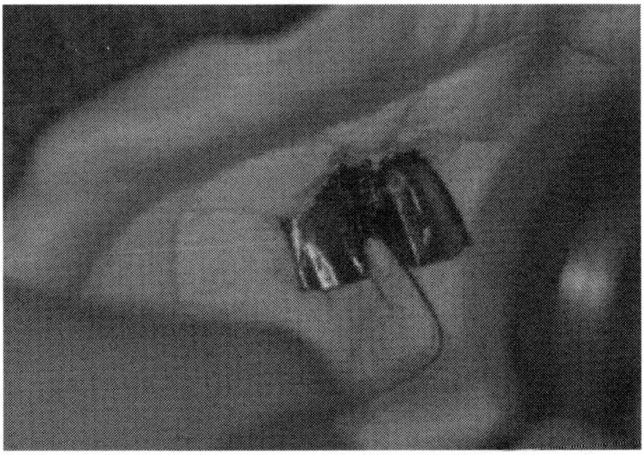

Fig. 35. Intraoral laser welding of the two portions

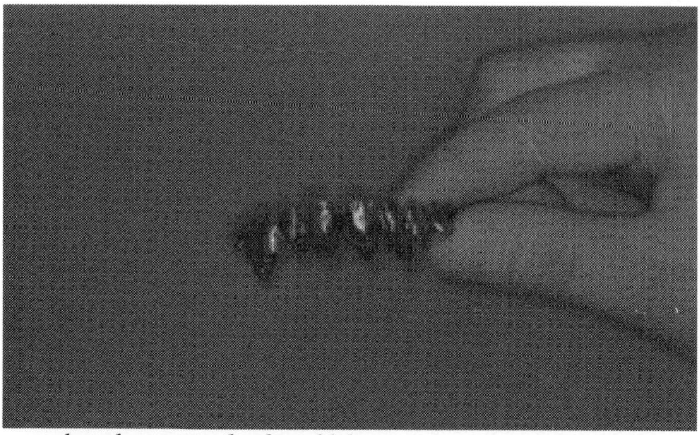

Fig. 36. Bridge ready to be sent to the dental lab

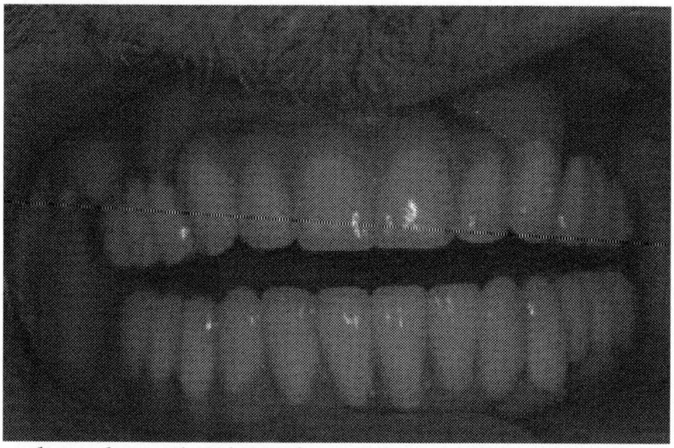

Fig. 37. Completed prosthesis placed into the mouth

Case 2:

Patient SV fourteen years old male come to the office with the lingual wire of the appliance broken. It was an orthodontic appliance called Delaire which consists of two wires, one vestibular and one lingual, connected to two braces on first upper molars. (Fig. 38)

We made the screen in silicon in order to protect soft tissues and we welded the appliance without filler metal.

Laser device, fiber, handpiece and parameters were the same as before, and the entire operation had a duration of four minutes, the welding time was 75 sec. (Fig. 39)

So, after few minutes and without sending the appliance to the dental laboratory and without discomfort for patient, we could repair it. (Fig. 40)

The follow-up, made monthly for six months, demonstrated that the appliance was active and strength-proof.

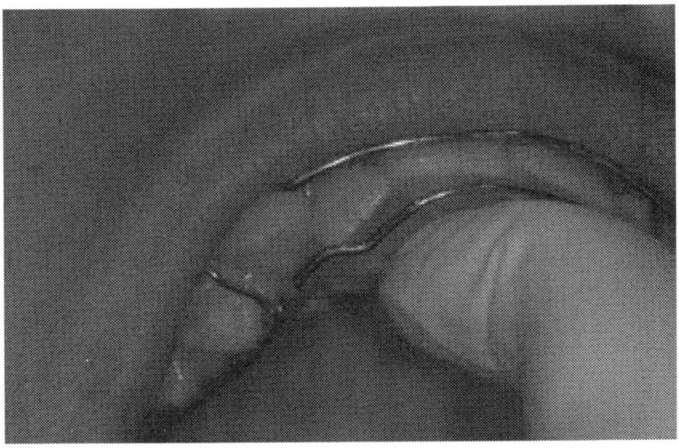
Fig. 38. Broken orthodontics appliance

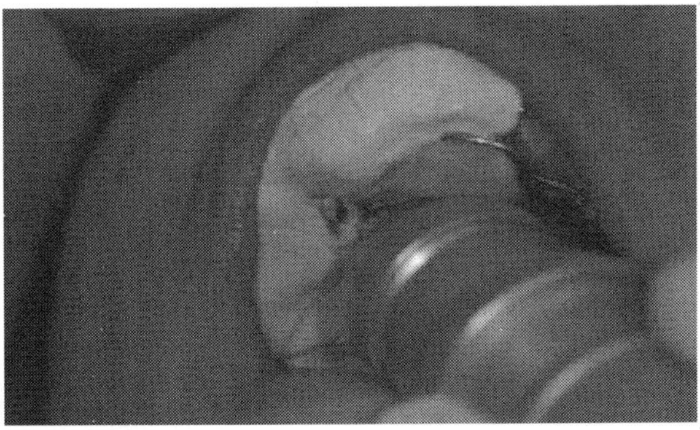
Fig. 39. Intraoral laser welding

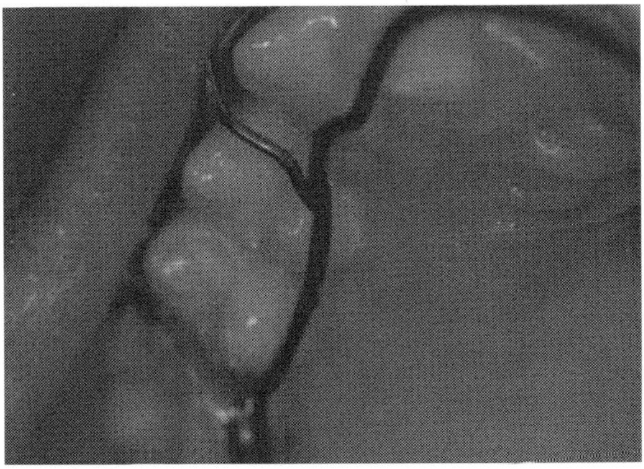
Fig. 40. Appliance repaired

Case 3:

Patient CV 45 male old come to our office to made a prosthetic rehabilitation of lower arch. In upper arch he had a gold-resin fixed prosthetics broken in the middle, between the two central incisors. (Fig. 41)

So, we decided to use our new technique to repair the bridge intraorally. We removed a little portion of resin by the two incisors with a bur and we welded by Fidelis III with filler metal. In this case the protection of soft tissues was done by a plastic cylinder. (Fig. 42) After welding, we put a layer of composite resin to complete, by an aesthetically point of view, the restoration. (Fig. 43)

During welding process, which had a duration of seven minutes, the patient did not feel any discomfort.

Subsequent checks, made after one, two and six months did not evidence any kind of problems.

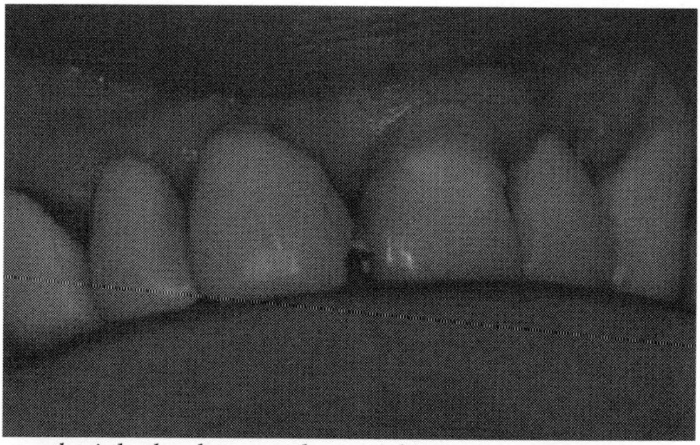

Fig. 41. Fixed prosthesis broken between the central incisors

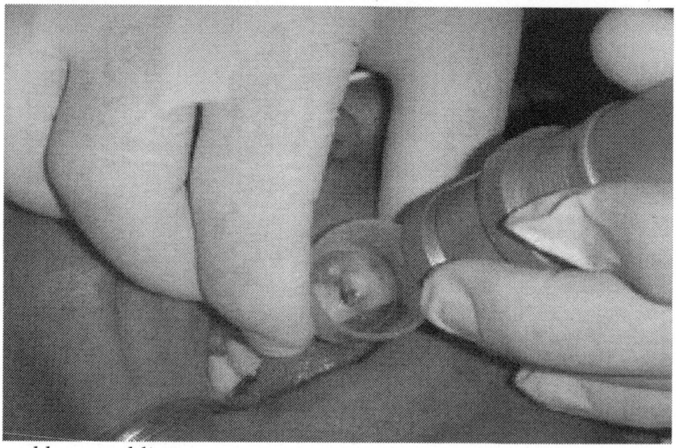

Fig. 42. Intraoral laser welding

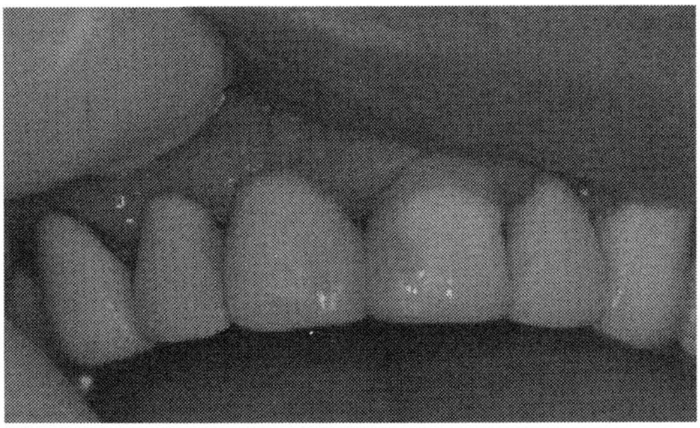

Fig. 43. Prosthesis repaired

Discussion and Conclusion

We have described, in every aspect, the use of Nd:YAG laser to weld appliances, extra- and intra-orally, directly in dental office by dentist himself.

Tests we made seem to demonstrate that Fidelis plus III is able to make a real welding process on different kinds of metal, with and without apport metal, with and without shielding gas and with good strength of the joints, even after months.

This gives the possibility both to repair broken prosthesis without their deplacement from the oral cavity, both the fixation of the position during the process of fabrication, avoiding the necessity of the use of silicon impression and/or resin and so reducing the possibility of inaccuracies due to the transfer of the impression to the laboratory.

During this study we had to solve different problems related to the fact that the appliance used was not projected to weld. So, our aim was to find technical solutions about parameters and device modifications in order to reach a good result, by a point of view of the quality of process welding, and also by the patient safety aspects.

Several problems remain unsolved but, with the help of manufacturer, this may be the aim of future studies.

The first unsolved problem is concerning the possibility to have special chemical glasses, connected with laser, with the possibility to obscure for a very short time during the shot, such in industrial field and in laboratory welding laser.

In fact, the shining of the interaction between the beam and metal is very troublesome for the operator.

The second one is linked to the opportunity to have a contra-angle handpiece in order to weld also in posterior areas of the mouth.

But the most interesting result is that patients, during intraoral welding and in the following days, has not shown trouble, pain and problem to the dental and periodontium structures of the support elements and the vitality tests performed are regular.

This first *in vivo* tests proved that using the laser technique to get the intraoral welding of metal prostheses can be possible with neither particular problems nor risks for the biological structures close to the welding zone.

Further tests will surely be necessary to confirm our work; however, we may say that it opens a new interesting perspective in modern dentistry.

3. References

Baba N, Watanabe I, Liu J, Atsuta M. Mechanical strength of laser-welded cobalt-chromium alloy.J Biomed Mater Res B Appl Biomater. 2004 May 15;69(2):121-4.

Berg E., Wagnere W.C., Davik G., Dootz E.R., J Prosth Dent 74, 250-257 (1995)

Bertrand C., Arts et Technique Dentaires 6, 363-368 (1995)

Bertrand C., Le Petitcorps Y., Albingre L., Dupuis V., Br Dent J 196, 413-418 (2004)

Bozzola J.J., Russell L.D.. Electron microscopy 2nd Edit 1999 Jones and Bartlett Publishers

Dawes C: Laser Welding. Cambridge (England):Abington Publishing Ltd, in association with The Welding Institute; 1992

Duhamel R. F. and Banas C. M., "Laser Welding of Steel and Nickel Alloys," Lasers in Material Processing, American Society of Metals, 1983.

Goldstein J., Newbury D.E., Joy D.C., Lyman C.E., Echlin P., Lifshin E., Sawyer L.C., Michael J.R.: Scanning Electron Microscopy and X-Ray Microanalysis: (Eds.), Kluwer Academic/Plenum Publishers, New York/Boston/Dordrecht/London/Moscow, 2003, pp. 689

Goodhew P.J., Humphreys J, Beanland R., F. J. Humphreys Electron Microscopy and Analysis, 3rd Edition January, 2001 Taylor & Francis

Haney SC, Meiers JC. In vitro study of intraoral welding. Quintessence Int. 1996 Dec;27(12):841-7.

Julian P. Heath Dictionary of Microscopy 2005 John Wiley & Sons

Liu J, Watanabe I, Yoshida K, Atsuta M. Joint strength of laser-welded titanium. Dent Mater. 2002 Mar;18(2):143-8.

Maddox S.J., British Welding Inst. Report, 36-70 (1970)

Martins GR, Cavalcanti BN, Rode SM (2006) Increases in intrapulpal temperature during polymerization of composite resin. J Prosth Dent 96(5):328-31.

Menard K.P. Dynamic Mechanical Analysis: A Practical Introduction, Second Edition. Taylor & Francis, 2008

Oelgiesser D, Blasbalg J, Ben-Amar A (2003) Cavity preparation by Er-YAG laser on pulpal temperature rise. Am J Dent. 16(2):96-8.

Ream SL: Laser materials processing. New York: IFS Publications and Springer-Verlag in association with The Laser Institute of America. 1988, pgg. 69-74

Santos M., Acciari H.A., Vercik L.C.O., Guastaldi A.C., Mater Lett 57, 1888-1893 (2003)

Shibuya I, Nishiyama N, Hayakawa T, Koishi I, Nemoto K. Inraoral laser welding- Part 1- Effect of laser irradiation on temperature rise. J J Dent Mate. 2004; 23 (6): 495-500.

Shinoda T., Matsunaga K., Shinhara M., Welding International 5, 346-351(1991)

Sulieman M, Rees JS, Addy M (2006) Surface and pulp chamber temperature rises during tooth bleaching using a diode laser: a study in vitro. Br Dent J 200(11):631-4.

Taylor JC, Hondrum SO, Prasad A, Brodersen CA. -Effects of joint configuration for the arc welding of cast Ti-6Al-4V alloy rods in argon. J Prosthet Dent. 1998 Mar;79(3):291

Walter M., Reppel P.D., Bonning K., Freesmayer W.B., J Oral Rehabil 26, 91-96 (1999)

Watanabe I, Topham Ds. Laser welding of cast titanium and dental alloys using argon shielding. J Prosthodont. 2006 Mar-Apr;15(2):102-7

Yamagishi T, Ito M, Fujimura Y.. Mechanical properties of laser welds of titanium in dentistry by pulsed Nd:YAG laser apparatus. J Prosthet Dent. 1993 Sep;70(3):264-73

Yamagishi T., Ito M., Fujimura Y., J Prosth Dent 70, 264-273 (1993)

2

Laser nerve welding

Kun Hwang, MD, PhD,[1] and Sun Goo Kim, MD, PhD[2]
[1]*Inha University College of Medicine,* [2]*Gachon University Gil Hospital, Republic of Korea*

1. Introduction

Microsurgial suture (Conventional microsurgical suture repair, CMSR) is the most commonly used method for anastomosis of severed peripheral nerves. Functional recovery after this type of repair is often inadequate, however, even though the peripheral nervous system has a remarkable ability to regenerate adequate sizes and numbers of axons. The most significant problems with microsuture repair are inherent in the technique: (1) surgery traumatizes the nerve by repeated introduction of a needle; (2) suture material at the anastomotic site presents a focus for scar and neuroma formation that may impede the growth of regenerating axons from the proximal segment into the proper distal segment endoneurial tubes and, ultimately, the neuromuscular junction; (3) microsurgical repair inevitably leaves small gaps that allow entry of fibroblasts and other scar tissue forming cells, permit regenerating axons to escape into an improper extraneural space, and promote loss of neurotrophic hormones that may be secreted locally to aid in the conduction of regenerating axons to their proper target; and (4) microsurgery is time-consuming and may be difficult to perform in restricted areas, particularly when supportive epineurial material is scarce. Also, especially in the head and neck region, there are often areas in which poor exposure or difficult surgical access precludes the placement of microsutures in nerve repair (Korff et al., 1992; Eppley et al., 1989; Huang et al., 1992).

A major focus of nerve repair research has been the development of procedures with which to avoid or minimize the use of sutures and prevent fibrous ingrowth at the repair site (Menovsky & Beek, 2001; Maragh et al., 1988). Several sutureless methods have been developed, although none of them has been demonstrated to be consistently superior to sutures.

Most importantly, sutureless methods must fulfill several criteria to have an advantage over suture repair. First, the procedure must result in a sufficient acute tensile strength. Second, it must not compress the nerve and must not involve increased severity of trauma compared with sutures. Third, the early and late tissue reaction of the nerve must be kept to a minimum and the axonal regeneration must not be impaired (Menovsky & Beek, 2001).

The laser was introduced in 1960s and is now widely used in medicine and surgery for the cutting, coagulation, and vaporization of various tissues. At low powers, thermal lasers, such as CO2, argon, and YAG, can be used to 'weld' tissue together by local protein coagulation. This property of the laser has been applied to nerve anastomosis as well as other tubular structures, such as vessel, bowl (Huang et al., 1992; Okada et al.,1987; Neblett et al., 1986). Laser nerve welding (LNW) has great potentials over conventional suture methods, such as a less trauma to tissue, less inflammatory reaction, and a faster surgical procedure (Menovsky et al., 1995). Thus, sutureless repair of peripheral nerves, by using lasers, has stimulated the interest of researchers.

2. General principle

Laser nerve welding had been investigated by several authors, since it offers a significant theoretical improvement to the problems just noted.

The mechanism of laser repair involves protein denaturation and subsequent fusion of the collagenous portion of the proximal and distal segment epineurium by low level thermal coagulation. Laser application has been shown to cause a change in collagen substructure with interdigitation of altered fibrils (Korff et al., 1992). Due to this mechanism, laser nerve welding provides significant theoretical advantages over conventional suturing methods, such as atraumatic nerve handling, avoidance of a foreign body reaction, a water tight seal of the epineurium with prevention of axonal escape and proliferation of connective tissue into the anastomosis, and a reduced operation time. There are several hypotheses to explain the improved results obtained with the laser. First, the irradiation of peripheral nerves with the CO2 laser does not result in severe inflammatory changes, nor does the nerve regeneration capacity seem limited by laser effects. The laser injury to the tissue heals very favorably, without excessive scar tissue or structural morphological alterations. The epineurium heals more favorably due to minimizing the foreign body reaction, and thus the axons are less blocked or misdirected at the repair site; consequently, the neural alignment is improved. Second, because of less instrumental manipulation of the nerve segments and avoidance of an excessive number of sutures, better axonal alignment of the nerve ends is achieved, and the cellular and fibroblastic reaction can be restricted to a minimum. Better alignment may also be achieved because the compressive forces and trauma to the epineurium resulting from sutures are reduced by simply using fewer of them. Intraneural scarring, which is mainly caused by collagen production by fibroblasts and Schwann cells inside the nerve, is less likely to be prevented by laser repair. Third, because of a "sealed" epineurium, the tendency of the regenerating proximal axons to sprout outside the nerve is suppressed, which leads to minimal neuroma formation and less scar tissue than in suture control groups. Also, ingrowth of fibrous scar tissue from outside the nerve is avoided. The epineurial seal is also postulated to provide a more favorable microenvironment for axonal growth by holding neurotrophic factors (Menovsky & Beek, 2001; Menovsky & Beek, 2003).

Laser nerve welding requires precise control of the thermal effects of the laser to permit welding of the epineurium without damage to the delicate underlying axons. This method became technically feasible with the development of the CO2 milliwatt laser, which can operate in power increments of 0.001 W while precise power stability is maintained (Huang et al., 1992). The first attempt at laser neurorrhaphy by Fischer et al. used a CO2 laser at relatively high powers (5 watts in 0.5 second pulses). As a result, dense carbonaceous deposits surrounded by a mild inflammatory reaction were found at the repair site (Fischer et al., 1985). No such carbon deposits were with the use of the CO2 milliwatt laser in the 120 to 150 mW range. This low thermal energy allows welding of superficial tissue without deep thermal penetration (Huang et al., 1992).

Many factors may influence the bonding strength of laser nerve welding, such as the amount of laser exposure, the amount of tissue available for fusion, and the technique that is used. Menovsky performed experimental studies for laser settings and end point determination which produce the greatest bonding strength (Menovsky et al., 1994). In laser welding procedure, the opposite nerve ends were closely approximated, and the epineurium of one of the nerve sections was pulled over the nerve end of the other nerve section and welded around its circumference with 2 to 3 laser pulses; the repair site is welded around its circumference with 5 to 8 laser pulses (Fig. 1) (Menovsky et al., 1994; Hwang et al., 2005).

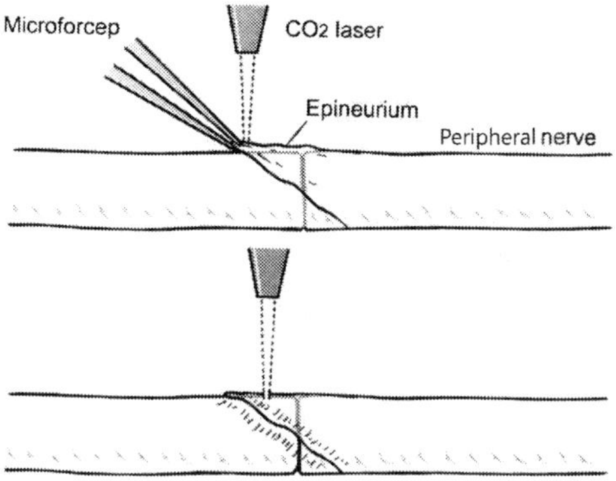

Fig. 1. The procedure of laser nerve welding. The epineurium of one of the nerve sections is pulled over the nerve end of the other nerve section and welded with 2 to 3 laser pulses; the repair site is welded around its circumference with 5 to 8 laser pulses. Reproduced with permission (Hwang et al., 2005).

Several technical points were essential for effective welding with the CO_2 laser. Most importantly, bonding occurred only when the tissues were directly opposed with the epineurium overlying the repair site. Also, dry tissue surfaces were essential to obtain adequate welds. The dryer the tissue surface, the greater the effect of the laser on tissue. Welding of the overlying epineurium has the advantage that the two epineurium of the different nerve stump can be welded tightly without having tension on the repair site. This may be an advantage because tension on the repair site increases the amount of connective tissue that is formed and thus decreases regeneration (Menovsky et al., 1994).

The end point of the welding procedure was defined as the visible fusion of the epineurium, which was observed meticulously with a 40 fold magnification. And the visible effects of laser radiation on the epineurium were scored quantitatively with values varying from 0 to 6, where 0 indicated no visible effect, 1 drying of the epineurium, 2 shrinkage, 3 whitening, 4 caramelization (slightly browning), 5 carbonization, and 6 perforation of the epineurium (Menovsky et al., 1994). In experimental and clinical tissue welding, the endpoint of welding is normally based on visual changes of tissue. Drying and shrinkage, blanching, tanning, and browning of the tissue were indicated as the endpoints for proper welds (Maragh et al., 1988; Menovsky et al., 1994; Poppas et al., 1992). The results of one study show that the strongest welds were produced at powers and exposure times (100mW-1.0s) that gave whitening and a beginning of caramelization (beginning of browning) of the epineurium. The whitening is related to the denaturation of proteins. The use of higher powers and longer pulse durations gave rapid carbonization with weaker bonds. When the epineurium was perforated, the exposed fascicles reacted differently to the laser irradiation. Curling of the fascicles, creating small excavations without bonding of the tissue occurred. The difference in the reaction of epineurium and perineurium to laser radiation can be explained by the difference in composition. The epineurium is mainly composed of densely packed collagen bundles, while the perineurium is formed of several lamellae which consist of closely packed perineurial cells (Menovsky et al., 1994).

Comparable functional and morphometric data with much better results for the laser nerve welding than microsuture have been achieved by only few authors. This is probably due to the difference in the welding technique, taking into consideration the basic requirements of laser welding (Happak et al., 2000). The epineurial edges need to be approximated precisely by two forceps trying to avoid any air gap between the epineurium of the nerve stumps. Furthermore, laser welded nerve coaptations should be performed without tension, as it has been assumed for sutured nerve repair. The bonding effect of welded collagen fibers and other proteins was made responsible for a sufficient tensile strength of coaptations and anastomoses (Schober et al., 1986; Bass et al., 1992). Bass et al. have shown that the welding effect does not depend on the amount of collagen that has been welded, but on different molecular binding mechanisms (Bass et al., 1992).

Also noted were the thermal effects of LNW anastomosis on the underlying fascicular structures. The heat resulted in the deleterious effects of the destruction of myelin and loss of axons immediately adjacent to the anastomotic site. Some authors reported that irradiation of nerves with laser nerve energy does not appear to have any adverse effects on axonal regeneration. Morphometric analysis of myelinated axons around the repair site by Beggs et al. and Maragh et al. revealed no deleterious effects of the laser on the degree of retrograde axonal degeneration or regeneration as shown by the similar numbers and sizes of regenerated fibers present in the distal stump compared with suture repairs (Maragh et al., 1988; Beggs et al., 1986; Eppley et al., 1989).

If a CO2 laser is used for fusing the epineurium, a water tight seal can be established without damage to the underlying neural tissue. For welding of nerves, the CO2 laser should be used because the penetration depth of this laser is small compared to other lasers proposed for LNW and LANR (Nd:YAG laser, and the argon laser) (Schober et al., 1986; Campion et al., 1990).

The major advantages of laser welded nerve coaptation are, under optimal circumstances, the decreased surgical time, the extreme precision, the minor damage to the nerve tissue, and the prevention of foreign-body reactions (Happak et al., 2000; Balies et al., 1989). In contrast, the essential drawback of welding techniques is the initial low tensile strength after coaptation, which leads to a relatively high dehiscence rate, varying between 12 and 41% (Ochi et al., 1995). The clinical application of laser-assisted nerve repair has been limited by the risk of dehiscence in the postoperative period and the inability to achieve consistently successful laser welds (Korff et al., 1992; Maragh et al., 1988; Dubuisson & Kline, 1993). One study showed that the anastomotic strength of the laser weld has been found consistently inferior to that of microsuture repair for at least the first 4 days after surgery. The weak initial weld has caused unacceptable rates of dehiscence (Maragh et al., 1988). As reported by Maragh et al. in the rat sciatic nerve model, there was no significant difference in tensile strength of the laser repaired nerves and the suture repaired nerves at 8 days postoperatively. The critical period is clearly the first week before host-connective tissue elements add the necessary stability. To make laser repair an attractive alternative to suture repair, the tensile strength of the laser repaired nerve must be improved further (Korff et al., 1992).

To overcome this problem, one or more stay sutures for supporting the welds (called laser assisted nerve repair, LANR) or nerve grafts to reduce the tension at the repair site have been used (Fischer et al., 1985; Menovsky et al., 1994; Eppley et al., 1989). However, these adaptations impair the nerve regeneration. Grafting gives the additional complication of two repair sites, and sutures produce a foreign body reaction. Some authors performed a laser assisted nerve repair with the aid of stay sutures which were eventually removed (Fischer et al., 1985; Balies et al., 1989; Ochi et al., 1995; Benke et al., 1989). In 1988, Maragh et al. reported about CO2 laser welded nerve coaptations by using a circumferential epineurial technique. No sutures were used, and the dehiscence rate of the coaptations was 12% (Maragh et al., 1988). Nerve coaptations with the CO2 laser were only achieved without dehiscence when an additional bonding material was used. The coaptations were strengthened with peri-epineurial sheets (Korff et al., 1992; Kim & Kline, 1990), fibrin sheets (Ochi et al., 1995), or fibrin glue (Menovsky & de Vries, 1998).

Although the precise mechanism of nerve welding is not yet understood, tissue welding by the CO2 laser is probably caused by protein denaturation and fusion of collagen, and/or dehydration of the structural proteins (Menovsky et al., 1994; Dew, 1986; Fenner et al., 1992). As these proteins are believed to be the primary component of welding process, topically applied proteins (used as solders) may provide the necessary amount of proteins for welding and result in a greater bonding strength (Menovsky et al., 1994). Another significant development in laser welding is the use of protein solders as an adjunct to the welding process. These solders, which provide extra protein for the fusion process, are melted on the outer surface of the repair site to hold the tissue together, and result in stronger welds and theoretically less thermal damage to the tissue (Menovsky et al., 1994; Menovsky, 2000).

In general, the CO2 laser-repaired nerves appeared to heal with less cellular response and less scar tissue than the sutured nerves. Also, proliferation of the epi- and perineurium was significantly less than that of the CMSR nerves. The alignment of axons and intraneural scars was most favorable in the soldered nerves. The severe inflammatory reaction around the solder observed in the first week was absent, and instead, a well defined epi- and perineurium was present. Consequently, the regeneration in the distal nerve segment was most advanced in the soldered nerves. In this segment there were also fewer extraneural fibers compared to other repair groups (Menovsky & Beek, 2003). The addition of solder seems to result in even better histological results than the CO2 laser alone. In a rat sciatic nerve repair using the CO2 laser and a fibrin film, the repair site revealed a smooth continuity of nerve fibers at 8 weeks (Ochi et al., 1995).

Besides the use of protein solders, extra tissues for improving the bonding strength have been used. Kim and Kline used perineurial and epineurial tissue to serve as a supplement for the welding procedure, resulting in 100% bonding rate (Kim & Kline, 1990). Korff et al. used LNW in combination with subcutaneous tissue wrapped around the nerve (Korff et al., 1992).

On the other hand, Happak et al. suggested that a minimal amount of collagen and other proteins of the epineurium provides sufficient welding strength in case of an adequate laser irradiation being applied without additional bonding materials or stay suture. This has been documented by the lack of dehiscences of the coapted nerves after 6 months of regeneration in their study (Happak et al., 2000). And Hwang et al. performed facial nerve repair by laser nerve welding and microsurgical suture in rat model and they also reported that they did not perform supportive procedures to enhance the laser welding site and there was no dehiscence in any of the 12 rats (Hwang et al., 2005).

3. Experimental study

During the last decade laser has become an increasingly useful surgical tool, and laser welding repair of injured peripheral nerves has been investigated. Since first successful CO2 laser welded coaptations were published by Fischer et al. in 1985 (Fischer et al., 1985), several authors reported that laser nerve welding was at least equal or more successful to microsurgical suture in effectiveness in animal model of facial and sciatic nerve repairs.

3.1 Facial nerve repair and facial – hypoglossal nerve anastomosis

Eppley et al. performed an experimental study to evaluate the effectiveness of laser nerve welding versus conventional suture repair of facial nerve grafts in the rabbit (Eppley et al., 1989). A group of 10 animals underwent bilateral 1 cm facial nerve resections which were primarily repaired with the contralateral resected nerve in which the right side was anastomosed by suture (CMSR, interrupted epinueural 10-0 nylon sutures) and the left side by laser nerve welding (CO2 laser with 150mW, 300u spot size and 0.5s duration assisted by a single suture for traction and rotation). Postoperative assessment was carried out at 1 and 3 months and consisted of electrophysiological recording prior to sacrifice, which was compared to preoperative recordings of the same nerve, and was then followed by histological evaluation of the anastomosis and graft. Electrophysiologic assessment of function revealed no significant difference between the two techniques after 3 months follow up period. It resulted in similar conduction recording to CMSR after 3months and appeared to reduce the amount of axonal intermingling, although quantitative axonal assessment was not performed in this study. Histological differences were apparent between two groups. Unlike sutures, laser nerve welding provides a seal around the epineurium without the potential of the introduction of foreign material into the underlying structures. Theoritically, this seal may prevent axonal escape as well as proliferation of connective tissue into the anastomosis. The author's findings indicate that less entrapment of axons occurred with laser nerve welding, presumably due to the fewer sutures required with this type of repair. However, both types of repair had evidence of axons outside the anastomosis, thus indicating that the laser nerve welding did not provide a watertight epineurial seal in the mW power levels used. Also noted was the thermal effect of laser nerve welding on the underlying fascicular structures. The heat resulted in the deleterious effects of the destruction of myelin and loss of axons immediately adjacent to the anastomotic site. While this appeared to be of little consequence in the 3 month specimens, this tissue destruction could lead to fibrous tissue formation which may be as detrimental to axonal regrowth as that induced by retained suture material. The effects of heat upon neural structures at 1 month were apparent but did not affect the longer term specimens.

The authors concluded that laser nerve welding may offer an alternative to conventional suturing or be useful when suturing is technically difficult due to access (with the intraoral repair of the lingual and mandibular nerves) or when lack of supportive neural structure exists (Eppley et al., 1989).
Hwang et al. performed facial nerve repair by laser nerve welding and microsurgical suture in rat model and evaluated nerve regeneration with an immunochemical detection of the retrograde nerve tracer cholera toxin B (Hwang et al., 2005). In the buccal branches of rat's facial nerves on the both sides were transected, and CO2 laser welding of the epineurium was performed on the right side and microsurgical suture technique was applied on the left side. For the laser nerve welding, the proximal and distal epineurium of the severed nerve ends were pulled together to meet the nerve ends of each other and welded at two directly opposite spots with 2 to 3 pulses of a CO2 laser (Wonderful CO2 Laser, Wonder Laser, Inc., Daejon, Korea) setting at 100mW continuous wave energy, 320 mm spot size, and 1 second duration time and the repair site is welded around its circumference with 5 to 8 laser pulses (Fig.1). In six rats Cholera Toxin B Subunit (CTb) was injected in the epineurium distal to the nerve anastomosis site at postoperative week 4 and 8 (Fig. 2).

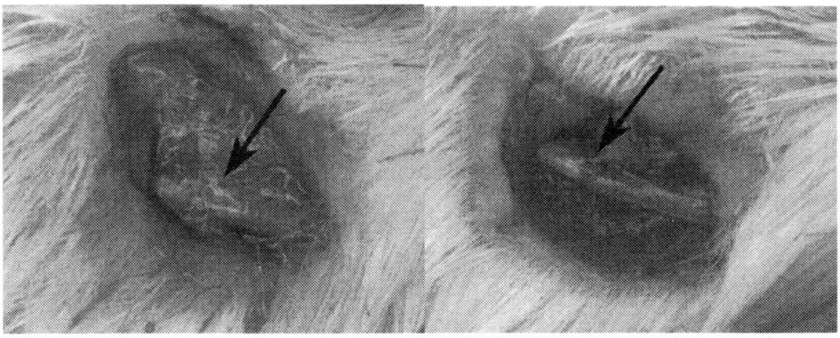

Fig. 2. Four-week postoperative photographs. (Left) Laser-assisted nerve anastomosis. Arrow indicates anastomosis site. (Right) Microsurgical suture anastomosis. Arrow indicates suture site. Reproduced with permission (Hwang et al., 2005).

Neurons of facial nuclei labeled positively by CTb were detected immunohistochemically, and the numbers were counted. CTb-positive neurons were seen significantly more in the group with laser welding than in the group with microsurgical suture in postoperative week, but there was not much difference in postoperative week 8. None of 12 rats showed dehiscence at the nerve anastomosis done by laser welding. This study showed that nerve regeneration is more apparent in the nerve repaired by laser welding than in that repaired by microsurgical suture. And this study showed that laser nerve welding affected regeneration of the repaired nerve equally to or more effectively than microsurgical suturing. Quantitative assessment was carried out with the immunohistochemical detection of the retrograde nerve tracer cholera toxin, which is one of the most widely used probes for studies of neuronal connectivity (Hwang et al., 2005).
The authors did not perform supportive procedures to enhance the laser welding site. There was no dehiscence in any of the 12 rats. This result indicates that the cranial nerves, including the facial nerve and other nerves in the head and neck, are not subjected to

significant stretching or tension as occurs with peripheral nerves in the extremities, such as the sciatic nerve (Huang et al., 1992). Thus, dehiscence is expected low in the head and neck area.

Hypoglossal-facial anastomosis (HFA) is an alternative surgical measure of facial palsy as the facial nerve itself can not be restored. HFA is usually attempted in a delayed repair because it is difficult to suture directly a disrupted facial nerve in most clinical cases (Angelov & Gunkel, 1993; Chen & Hsu, 2000).

Hwang et al. performed another study to compare laser nerve welding to microsurgical suturing of hypoglossal-facial anastomosis (HFA), and a result of immediate to delayed repair, and to evaluate the effect of laser nerve welding on HFA for reanimation of facial palsy in animal model of rats (Hwang et al., 2006). The first group underwent immediate HFA by microsurgical suturing and the second group by CO2 laser welding. The right hypoglossal nerve was transected 1 mm proximal to the bifurcation of the medial and lateral branches. The right facial nerve was transected at near stylomastoid foramen, sparing the posterior auricular branch. Proximal stump of the hypoglossal nerve and distal stump of the facial nerve were approximated by laser nerve welding without additional stay suture in the first group and by microsurgical suturing with three sutures of 9-0 nylon in the second group. The third group underwent delayed HFA by microsurgical suturing, and the fourth group by laser nerve welding. The right facial nerve was transected at near stylomastoid foramen, sparing the posterior auricular branch. The proximal stump was left, but the distal stump was tagged with a 9-0 nylon stitch before closing the wound. Ten days later, the distal stump of the facial nerve was explored. The previous tagged suture was removed, and a 1-mm segment of the stump was cut. Ipsilateral hypoglossal nerve was exposed and transected proximally at the bifurcation of its medial and lateral branch. Intact proximal stump of the hypoglossal nerve was approximated to the predegenerated distal stump of the facial nerve by laser nerve welding in the third group and by microsurgical suturing in the fourth group.

In all rats of the four different treatment groups, cholera toxin B subunit (CTb) was injected in the epineurium distal to the anastomosis site on the postoperative 6th week. Neurons labeled CTb of hypoglossal nuclei were positive immunohistochemically, and the numbers were counted. There was no significant difference between immediate and delayed anastomosis in the laser welding group, but there was significance between immediate and delayed anastomosis in the microsurgical suturing group. No dehiscence in the laser welding site of nerve anastomosis was seen at the time of re-exploration for injection of CTb in all rats. This study showed that the regeneration of anastomosed hypoglossal-facial nerve was affected similarly by laser welding and microsurgical suturing, and more effective, especially in delayed repair.

And tensile strength of the anastomosis site is insecure in the laser welded nerve. The cranial nerves in the head and neck area are not so tense as peripheral nerves in the extremities. A dehiscence rate of the laser welded nerves in the head and neck area is expectedly quite low (Huang et al., 1992). In this study, no dehiscence in the laser welding site of the nerve anastomosis was seen at the time of re-exploration for injection of CTb in all 20 rats. These results were thought to show the possibility of successful application of laser nerve welding in clinical settings, especially of head and neck area (Hwang et al., 2006).

Hwang et al. performed additional study to compare laser nerve welding of hypoglossal-facial nerve to microsurgical suturing and a result of immediate and delayed repair, and to evaluate the effectiveness of laser nerve welding in reanimation of facial paralysis of the rabbit models (Hwang et al., 2008). The first group underwent immediate HFA by microsurgical suturing and the second group by CO2 laser welding. The third group underwent delayed HFA by microsurgical suturing, and the fourth group by laser nerve welding. In laser welding group, one stay-suture of 9-0 nylon was applied temporarily only in mild tensed gap of the severed nerves (Fig. 3 & 4).

In this study, in microsurgical suturing group, the mean number of labeled neurons in immediate HFA was significantly higher than in delayed HFA. The mean number of labeled neurons subjected to laser welding was higher than microsurgical suturing in the delayed HFA. In microsurgical suturing, axons may be caught up and displaced from their connections because of the passage and retention of sutures. And gaps persist in the epineurium between the sutures which can allow for connective tissue invasion or axonal extravasation (Eppley et al., 1989). In the delayed nerve repair, there were degeneration and weakness of epineural tissue and a lot of inflammation and scar tissues around the repair site. This result indicates that the sutures especially could be adverse to the regenerating axon with degeneration in a delayed nerve repair.

In laser welding group, the mean number of labeled neurons was all but same in the immediate and delayed HFAs. Laser welding provides strict thermal effect on the epineurium without adding any damage to the adjacent tissue like underlying axons. Unlike sutures, this provides a seal around the epineurium without any potential for the introduction of foreign material into the underlying fascicular structures and makes no mechanical damage to the degenerated epineural tissue in the delayed repair (Eppley et al., 1989). Therefore, the laser nerve welding may give a less trauma to the regenerated axon and the mean number of labeled neurons in the laser welding group was almost the same in the immediate and delayed HFAs in this study.

There was little difference of nerve regeneration between laser welding and microsurgical suturing in the immediate HFA. In the immediate repair group in which there is little inflammation and scar tissue, the method of neurorrhaphy may not affect the regeneration of anastomosed hypoglossal-facial nerve.

In this study, 1 epineurial stay-suture of 9-0 nylon was applied temporarily only in cases with mild tensed gap of the severed nerves. The stay-suture was used for approximation between both nerve ends and the repair sited was sealed circumferentially by laser nerve welding. No dehiscence was seen on the laser welding site of nerve anastomosis in all the rabbits.

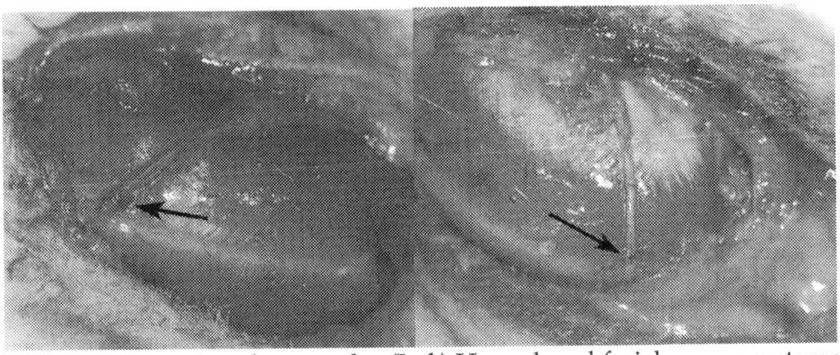

Fig. 3. Immediate postoperative photographs. (Left) Hypoglossal-facial nerve anastomosis (HFA) by laser nerve welding. Arrow indicates anastomosis site. (Right) HFA by microsurgical suturing.

Arrow indicates suture site. Reproduced with permission (Hwang et al., 2008).
This study showed that regeneration of the anastomosed hypoglossal-facial nerve was affected similarly by either laser welding or microsurgical suturing in immediate repair; however, the welding was more effective especially in delayed repair (Hwang et al., 2008).

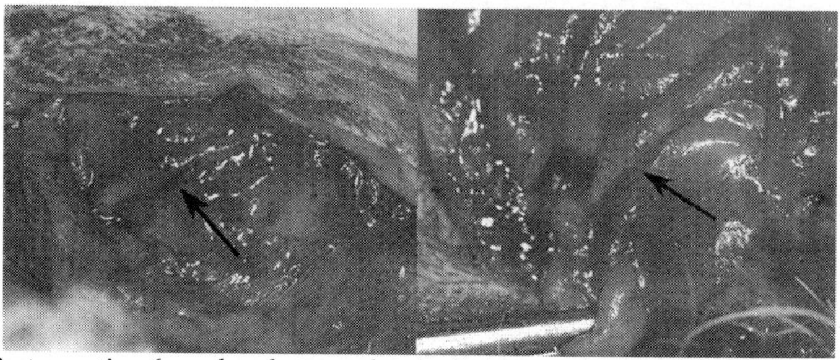

Fig. 4. Postoperative 6 weeks photographs. (Left) Hypo- glossal-facial nerve anastomosis (HFA) by laser nerve welding. Arrow indicates anastomosis site. (Right) HFA by microsurgical suturing. Arrow indicates suture site. Reproduced with permission (Hwang et al., 2008).

3.2 Sciatic nerve repair

Huang et al. performed a comparative study between microsuture and CO2 laser repair of transected sciatic nerves in rats (Huang et al., 1992). The left sciatic nerves of rats were transected and repaired by microsuture (six 10-0 nylon sutures) and laser nerve welding (CO2 laser with 120 to 150 mW power setting, 0.5 mm spot size, 30 to 60 sec duration time, and two epineurial sutures of 10-0 nylon as markers). No attempt was made postoperatively to immobilize the lower extremity of the animals. Nerve regeneration was measured in terms of morphology, electrophysiology, and function. For measurement of motor function of sciatic nerve, 'Sciatic function index's, reflecting footprint length and width, toe spread, and stride length, were measured preoperatively, then weekly for the first 2 months postoperatively, and then every other week until 4 months after nerve repair. At 4 months after repair, the mean nerve conduction velocity of repaired sciatic nerves was measured.
Functional recovery, as determined objectively using measurements of gait footprints, showed no difference between suture and laser repair. EMG and nerve conduction velocity were similar for the two repair methods. On histologic analysis, there was no difference in the size and number of regenerated axons. The laser repaired nerve demonstrated good axonal regeneration and no evidence of tissue charring or carbonaceous deposits. Good axonal regeneration was also found in microsuture repaired nerves. However, foreign body reactions were present surrounding the sutures and appeared to cause distortion of axon fibers and perineurial fibrosis around the suture granulomas. Laser nerve repair using epineurial welding and two anchoring sutures took approximately one-third less time than microsuture repair using six epineurial sutures. However, this study showed a 41% rate of nerve dehiscence in laser repair group, even with placement of two anchoring sutures, probably as a result of inadequate tensile strength of the anastomosis immediately postoperatively. The authors suggested that one possible solution of high dehiscence rate is to be temporarily

immobilize the limb for 7 to 10 days postoperatively to allow adequate healing to occur before tension is allowed on the repaired nerve. However, laser nerve welding was faster and simpler than microsurgical suture repair and required less manipulation of the nerve.

This study showed laser repair of peripheral nerve is possible with results comparable to conventional microsurgical nerve repair and laser nerve repair may be effective alternative to microsurgical suture repair (Huang et al., 1992).

Happak reported an experimental study to obtain functional and morphologic informations by using a nerve coaptation technique by epineurial CO2 laser welding only (Happak et al., 2000). In this study, the sciatic nerves of 24 rats were transected and epineurially coapted with the CO2 laser at 60 mW, 135 um spot size or with microsuture as a control. Walking track analysis was carried out to evaluate the functional recovery, and the nerves were harvested for histology after 6 months of regeneration. None of the 24 nerves showed dehiscence of the coaptations, and all showed good nerve fiber regeneration. Better results were obtained for the functional evaluation of the sciatic function index ($P<0.02$) and the toe spread index ($P < 0.04$) from the laser nerve coaptations. Likewise, the morphologic evaluations of the fiber density ($P<0.04$) and area fraction ($P < 0.002$) were better in the laser group. The better functional and morphologic results of the LNC group might be explained by the avoidance of disturbing effects on nerve tissue and its regeneration due to laser welding. The thermal damage of the underlying nerve fascicles is described, as one concern, connected with the epineurial CO2 laser welding (Trickett et al., 1997; Lauto et al., 1997). In this study, the authors selected the lowest power setting, 60 mW, to prevent the nervous tissue from any possible damage. They have found no hard evidence of such damage, because no carbonization of the epineurium was visible under the microscope during the welding procedure. The CO2 laser parameters chosen for this procedure kept the absorption depth of the light at a minimum, and it seemed to prevent any immediate damage. CO2 laser irradiation of the epineurium does not impair the nervous tissue in contrast to the damage caused by sutures in the control group.

On gross morphology, the authors found neuromatous thickening, which was typical of the coaptation site in the microsuture group. The thickening of the coaptation site is described as the neuromatous regeneration of nerve fibers outside of the epineurium through the gaps between the sutures. Less neuromatous thickening was observed in the laser group, these findings are in accordance with the results of other authors (.Fischer et al., 1985; Ochi et al., 1995; Almquist et al., 1994). In their opinion, this is due to the water tight seal of the epineurium, which prevented the sprouting of nerve fibers outside the epineurial tube. In the laser group, these fibers remain extrafascicular within the epineurium of the distal nerves and, thus, may contribute to the better functional results as a higher number of nerve fibers reach the target organ.

And they conclude that a minimal amount of collagen and other proteins of the epineurium provide sufficient welding strength in case of an adequate laser irradiation being applied. This has been documented by the lack of dehiscences of the coapted nerves after 6 months of regeneration in this study. In this study, in the rat sciatic model, the laser nerve weldings are significantly better in several parameters than the sutured nerve repair (control) and CO2 laser welding may result in a better nerve regeneration with better functional outcome and may be a new approach for clinical trials. The results of this study could be to fulfill the aforementioned advantages by simply welding the epineurial edges of the transected sciatic nerve by using the CO2 laser, excluding additional aids (Happak et al., 2000).

3.3 Complementary methods using stay sutures, protein solders

Korff et al., attempted to improve on the laser welding technique by examining the effect of laser radiation on the compound action potential (CAP) of intact rat sciatic nerves (Korff et al., 1992). In the second phase of the study, a new technique to improve anastomotic strength was developed, S-Q weld, which involved harvesting a sheet of subcutaneous tissue from the experimental animal, wrapping it around the cut nerve ends, and lasering it to the epineurium. The final part of the study examined the long term effectiveness of the technique as compared to microsuture repair in the rat sciatic nerve model.

In S-Q weld procedure, the nerve ends were reapproximated and the subcutaneous tissue was wrapped around the nerve and spot welded to the epineurium using CO2 laser with 1.0 w, 0.05 s, 0.36mm spot size. Multiple pulses were applied to the S-Q sheath around a circumference of 180 degrees to allow adequate stump alignment.

CO2 laser produced almost no decrease in CAP transmission at 0.5 watts. However, this level of irradiation did not provide adequate bonding of subcutaneous tissue to nerve epineurium during S-Q welding. In this study, a higher power of 1.0 w was used, because it produced adequate bonding with minimal anticipated increase in damage to endoneural axons and support structures. The strength of the S-Q weld was considerably greater than that produced by laser welding alone. And the other phase of the study compared regeneration at 2 months in severed rat sciatic nerves repaired by either microsuture or S-Q weld. Analysis of the compound action potential values indicated that the number of regenerating fibers after laser repair was greater than that after suture repair, although a significant difference could not be demonstrated.

The S-Q weld technique uses host tissue, which is allographic, biocompatible, and readily available, to provide an instrument for sealing and binding severed nerve stumps. Using this approach, the laser energy is focused on the subcutaneous wrap at sites removed from the actual nerve juncture, where scarring and neuroma formation should be avoided. The tensile strength of the S-Q weld immediately after repair was far greater than the tensile strength without subcutaneous sheath. The strength of S-Q welded nerves was also inferior to the tensile strength of the acute suture repair, but the improvement over CO2 laser weld alone was substantial and would encourage further refinement of this procedure.

The authors concluded that laser nerve welding using the S-Q weld technique has several theoretical advantages over suture repair. The observed improvement in initial anastomotic strength over laser repair alone warrants the need for further investigation with different laser energies and other improvement in technique (Korff et al., 1992).

Menovsky et al. designed experimental study to investigate the in vitro bonding strength of nerves welded by a CO2 laser at different radiant exposures and exposure times. Laser nerve welding was performed at 15 different laser settings (power outputs of 50, 100 and 150 mW; pulse durations of 0.1, 0.5, 1.0, 2.0, and 3.0s) with a spot size of 320 um. The effect of different solders on the bonding strength was investigated and compared to conventional microsurgical suture repair, laser assisted nerve repair, and fibrin glue repair. As a solder, 5 and 20 % albumin solution, dried albumin solution, egg white, fibrinogen solution, fibrin glue, and red blood cells were used (Menovsky et al., 1994).

The strongest welds (associated with whitening and caramelization of tissue) were produced at 100 mW with pulses of 1.0 s and at 50 mW with pulses of 3 s. The use of a dried albumin solution as a solder at 100 mW with pulses of 1 s increased the bonding strength 9-fold as compared to LNW. However, positioning the nerves between cottons soaked in saline for 20 minutes (rehydration) resulted in a decrease of the bonding strength. Other solders did not increase the bonding strength in comparison to LNW. Understanding of the precise mechanism of the fusion process, however, might lead to an appropriate selection of the concentration and kind of proteins to be used as a solder. Although the bonding strength of LNW in combination with the use of 20% albumin solution and egg white as a solder was lower than in CMSR, improvement over LNW alone was substantial and encourages further in vivo research on the use of solders.

In this study, the bonding strength of LNW performed at optimal laser settings was significantly lower than in CMSR (bonding strength 2.4 +- 0.9 versus 29.6 +- 10.4 g). Comparison between the fibrin glue repair and LNW without solder showed no differences. For LANR using one 10-0 nylon stay suture, the bonding strength was about 20% that of the nerves sutured with four 10-0 nylon stay sutures and was independent of the laser settings used.

Despite the low bonding strength of LNW, it seems likely that the strength of the weld will increase in time in vivo studies (Menovsky et al., 1994; Richmond, 1986). The critical period for dehiscence is the first week postoperatively before the fibroblasts have formed a definite closure of the wound. Maragh et al. reported that LNW (90-95 mW, 200 um spot size, 0.2s exposure time) had a strength of 43.1 g at day 4 postoperatively. At day 8, LNW had a strength comparable to the epineurial suture control group (166.7 g) (Maragh et al., 1988). Thus, for 7 to 10 days postoperatively to allow adequate healing to occur before tension is allowed on the repaired nerve, additional complement such as temporary splint to limb could be one possible solution.

The substantial increase in bonding strength for some solders suggests that it is worthwhile to investigate the dehiscence rate and nerve regeneration of solder enhanced LNW in an in vivo study.

One possible source of complications with the use of solders in general could be the persistence of solder between the nerve ends. If this happens, the solder could block the sprouting axons and could induce scar tissue formation between the nerve ends. Therefore, it is preferable to weld the epineuria first and then to continue the procedure with solder. Also, premature absorption and disintegration of the solder is possible, which may result in early dehiscence of the union.

The authors demonstrated that (1) the operation time of LNW or LNW + solder is short compared to CMSR, (2) the strongest welds are associated with specific changes in tissue appearance, which can be used to determine the end point of the welding, (3) that LNW in combination with 20% albumin solution, dried albumin solution, and egg white as solders gives bonding strengths that may be sufficient for holding the nerve ends together in an in vivo study, and (4) that the strongest welds in LNW and LNW + solder were found at 100mW with pulses of 1.0 s and at 50 mW with pulses of 3.0s (Menovsky et al., 1994).

Menovsky et al. performed a study to evaluate CO2 laser-assisted nerve repair and compare it with nerve repair performed with fibrin glue or absorbable sutures (Menovsky & Beek, 2001). The sciatic nerves of rats were sharply transected and approximated using two 10-0 absorbable sutures and then fused by means of CO2 milliwatt laser welding (power 100 mW, exposure time 1 second per pulse, spot size 320um), with the addition of a protein solder (bovine albumin) to reinforce the repair site. The control groups consisted of rats in which the nerves were approximated with two 10-0 absorbable sutures and subsequently glued using a fibrin sealant (Tissucol), and rats in which the nerves were repaired using conventional microsurgical sutures (four to six 10-0 sutures in the perineurium or epineurium). Evaluation was performed 16 weeks postsurgery and included the toe-spreading test and light microscopy and morphometric assessment. The motor function of the nerves in all groups showed gradual improvement with time. At 16 weeks, the motor function was approximately 60% of the normal function, and there were no significant differences among the groups. On histological studies, all nerves revealed various degrees of axonal regeneration, with myelinated fibers in the distal nerve segments. There were slight differences in favor of the group treated with laser repair. There were no significant differences in the number, density, or diameter of the axons in the proximal or distal nerve segments among the three nerve repair groups, although there was a trend toward more and thicker myelinated axons in the distal segments of the laser-repaired nerves.

Soldering procedures rely on laser energy to produce fixation of the solder to the tissue (Menovsky et al., 1996; Menovsky et al., 1997). The protein behaves the same way in the welding process as does an inorganic solder used to join metal parts with the application of heat. In this way, a sleeve-type joint is formed by the solder, which is much stronger mechanically than a simple edge-to-edge joint. In addition to being mechanically stronger, laser soldering methods may be more technically forgiving than non soldering methods, because the solder may be able to bridge small gaps in coaptation that would otherwise produce a lead point for separation of the weld, and therefore it may reduce the need for stay sutures. Solder may also be beneficial in that it can protect the underlying tissue from the damaging thermal effects seen with non soldering methods.

It was found that CO_2 laser–assisted nerve repair with soldering is at least equal to fibrin glue and suture repair in effectiveness in a rodent model of sciatic nerve repair (Menovsky & Beek, 2001).

The clinical application of laser assisted nerve repair (LANR) is limited by the high dehiscence rate and the inability of achieving consistent successful laser welds (Korff et al., 1992; Huang et al., 1992). So far, two sutures placed equidistantly are thought to be essential to facilitate the initial coaptation and subsequent handling of the nerve during LANR. In this case, an important issue is the choice of suture material which is used in combination with laser repair, and it is important to use sutures which cause the least tissue reaction. Menovsky et al. tried to find an optimal laser assisted technique which would result in the most favorable nerve healing, by choosing several suture materials and adding solder to the repair site. This study was designed to investigate regeneration of peripheral nerves repaired with a CO_2 milliwatt laser in combination with three different suture materials and a bovine albumin protein solder (Menovsky & Beek, 2003).

In the laser repair group, the nerves ends were approximated with two stay sutures, including 10-0 nylon, 10-0 PGA, and 25-um-thick stainless steel. Thereafter, circumferential irradiation of the nerve with a CO_2 milliwatt laser was performed at 100 mW, with pulses of 1.0 s and a spot size of 320 um. A total of 5-8 pulses were needed for each nerve, with a total irradiation time of 5-8 s. In the fourth subgroup of laser repair, the nerves were approximated with two 10-0 nylon stay sutures, and a protein solder consisting of bovine albumin powder dissolved in saline was applied to the repair site. The control group consisted of nerves repaired by conventional microsurgical suture repair (CMSR), using 4-6 10-0 nylon sutures.

Evaluation was performed at 1 and 6 weeks after surgery, and included qualitative and semiquantitative light microscopy. At sacrifice, no dehiscence was observed, and all nerves were in continuity. After 6 weeks, the nylon and stainless steel sutures were visibly detectable; the PGA sutures were not. LANR performed with a protein solder results in good early peripheral nerve regeneration, with an optimal alignment of nerve fibers and minimal connective tissue proliferation at the repair site. All three suture materials produced a foreign body reaction; the least severe was with polyglycolic acid sutures. CMSR resulted in more pronounced foreign-body granulomas at the repair site, with more connective-tissue proliferation and axonal misalignment. Furthermore, axonal regeneration in the distal nerve segment was better in the laser groups. Based on these results, CO2 laser-assisted nerve repair with soldering in combination with absorbable sutures has the potential of allowing healing to occur with the least foreign-body reaction at the repair site.

In this study, the authors concluded that LANR with the addition of a protein solder leads to optimal early histological results. Concerning the choice of suture material, PGA sutures can be used for LANR and have the potential of allowing healing to occur with the least foreign-body reaction at the repair site. In further experiments, the combination of PGA sutures and LANR using a solder may further improve the histological results (Menovsky & Beek, 2003).

4. Conclusion

Laser nerve welding of peripheral nerves may offer several advantages over conventional microsurgical suture repair, such as a less trauma to the tissue, less inflammatory reaction, a water tight seal of the epineurium and a faster surgical procedure (Menovsky et al., 1995).

Nevertheless, the clinical application of laser-assisted nerve repair has been limited by the risk of dehiscence in the postoperative period and the inability to achieve consistently successful laser welds (Korff et al., 1992; Maragh et al., 1988; Dubuisson & Kline, 1993).

In previous studies, the high risk of nerve dehiscence has been overcome by placing one or two stay sutures before laser welding (Fischer et al., 1985; Beggs et al., 1986) or, by the use of protein solders, which are melted onto the outer surface of the repair site, resulting in stronger welds (Menovsky et al., 1994). Besides the use of protein solders, extra tissues for improving the bonding strength have been used, which included perineurial and epineurial tissue as a supplement for the welding procedure (Kim & Kline, 1990), and LNW in combination with subcutaneous tissue wrapped around the nerve (Korff et al., 1992).

As reported in the rat sciatic nerve model, there was no significant difference in tensile strength of the laser repaired nerves and the suture repaired nerves at 8 days postoperatively (Maragh et al., 1988). The critical period is clearly the first week before host-connective tissue elements add the necessary stability. To make laser repair an attractive alternative to suture repair, the tensile strength of the laser repaired nerve must be improved further, especially during this period (Korff et al., 1992). In experimental studies, the bonding strength for LNW with additional aids, as stay suture, protein solder, subcutaneous tissue, was inferior to the tensile strength of the acute suture repair, but the substantial improvement over CO2 laser alone would encourage further refinement of this procedure.

However, some authors suggested that a minimal amount of collagen and other proteins of the epineurium provide sufficient welding strength in case of an adequate laser irradiation being applied without additional bonding materials or stay suture. This has been documented by the lack of dehiscences of the coapted nerves after 6 months of regeneration

in their study (Happak et al., 2000). In some studies of facial nerve repair and facial-hypoglossal nerve anastomosis, no dehiscence in the laser welding site was seen. They did not perform supportive procedures to enhance the laser welding site (Hwang et al., 2005; Hwang et al., 2006). These results indicate that the cranial nerves, including the facial nerve and other nerves in the head and neck, are not subjected to significant stretching or tension as occurs with peripheral nerves in the extremities, such as the sciatic nerve (Huang et al., 1992). Thus, dehiscence is expected low in the head and neck area. And these studies showed that the regeneration of anastomosed nerve by laser nerve welding was affected more effectively especially in delayed repair. These results were thought to show the possibility of successful application of laser nerve welding in clinical settings, especially of head and neck area (Hwang et al., 2006).
Moreover, consistent achievement of successful laser nerve weld can be increased by careful selection of the laser parameters and technique and by the aforementioned use of additional aids.

Acknowledgments
This review is supported by a grant from INHA university.

5. References

Almquist, E.; Nachemson, A.; Auth, D.; Almquist, B. & Hall, S. (1984). Evaluation of the use of the argon laser in repairing rat and primates nerves. *J Hand Surg [Am]*, Vol.9, No.6, 792–799, 0363-5023

Angelov, D. & Gunkel, A. (1993). Recovery of original nerve supply after hypoglossal-facial anastomosis causes permanent motor hyperinnervation of the whisker-pad muscles in the rat. *J Comp Neurol*, Vol.338, No.2, 214-224, 0021-9967

Bailes, J.; Cozzens, J.; Hudson, A.; Kline, D.; Ciric, I.; Gianaris, P.; Bernstein, L. & Hunter, D. (1989). Laser-assisted nerve repair in primates. *J Neurosurg*, Vol.71, No.2, 266–272, 0022-3085

Bass, L.; Moazami, N.; Pocsidio, J.; Oz, M.; LoGerfo, P. & Treat, M. (1992). Changes in type I collagen following laser welding. *Lasers Surg Med*, Vol.12, No.5, 500–505, 0196-8092

Beggs, J.; Fischer, D. & Shetter, A. (1986). Comparative study of rat sciatic nerve microepinuerial anastomosis made with carbon dioxide lasers and suture techniques: Part 2. *Neurosurgery*, Vol.18, No.3, 266-269, 0148-396X

Benke, T.; Clark, J.; Wisoff, P.; Schneider, S.; Balasubramaniam, C.; Hawkins, H.; Laurent, J.; Perling, L. & Shehab ,A. (1989). Comparative study of suture and laser-assisted anastomosis in rat sciatic nerves. *Lasers Surg Med*, Vol. 9, No.6, 602–615, 0196-8092

Campion, E.; Bynum, D. & Powers, S. (1990). Repair of peripheral nerves with the argon laser. *J Bone Joint Surg* , Vol.72, No. 5, 715-723, 0021-9355

Chen, Y. & Hsu, C. (2000). Histologic rearrangement in the facial nerve and central nuclei following immediate and delayed hypoglossal facial nerve anastomosis. *Acta Otolaryngol* , Vol.120, No.4, 551-556, 0001-6489

Dew, D. (1986). Review and status on laser fusion. *Int Soc Opt Eng*, Vol. 712, No.2, 255-257, 0091-3286

Dubuisson, A. & Kline, D. (1993). Is laser repair effective for secondary repair of a focal lesion in continuity? *Microsurgery*, Vol.14, No.6, 398–403, 0738-1085

Eppley, B.; Kalenderian, E.; Winkelmann, T. & Delfino, J. (1989). Facial nerve graft repair: Suture versus laser assisted anastomosis. *Int J Oral Maxillofac Surg*, Vol.18, No.1, 50-54, 0901-5027

Fenner, J.; Martin, W.; Moseley, H. & Wheatley, D. (1992). Shear strength of tissue bonds as a function of bonding temperature: A proposed mechanism for laser assisted tissue welding. *Lasers Med Sci*, Vol.7, No.1, 39-43, 0268-8921

Fischer, D.; Beggs, J.; Kenshalo, D. & Shetter, A. (1985). Comparative study of microepineural anastomosis with the use of CO2 laser and suture technique in rat sciatic nerves. Part 1. *Neurosurgery*, Vol. 17, No.2, 300-307, 0148-396X

Happak, W.; Neumayer, C.; Holak, G.; Kuzbari, R.; Burggasser, G. & Gruber, H. (2000). Morphometric and Functional Results After CO2 Laser Welding of Nerve Coaptations. *Lasers Surg Med*, Vol.27, No.1, 66–72, 0196-8092

Huang, T.; Blanks, R.; Berns, M. & Crumley, R. (1992). Laser VS. suture nerve anastomosis. *OTOLARYNGOL HEAD NECK SURG*, Vol.107, No.1, 14-20, 0194-5998

Hwang, K.; Kim, S.; Kim, D. & Lee, C. (2005). Laser welding of rat's facial nerve. *J Craniofacial Surg*, Vol.16, No.6, 1102-1106, 1049-2275

Hwang, K.; Kim, S. & Kim, D. (2006). Facial-hypoglossal nerve anastomosis using laser nerve welding in the rats. *J Craniofacial Surg*, Vol.17, No.4, 687–691, 1049-2275

Hwang, K.; Kim, S. & Kim, D. (2008). Hypoglossal-Facial Nerve Anastomosis in the Rabbits Using Laser Welding. *Ann Plast Surg*, Vol. 61, No.4, 452–456, 0148-7043

Korff, M.; Bent, S.; Havig, M.; Schwaber, M.; Ossoff, R. & Zealear, D. (1992). An investigation of the potential for laser nerve welding. *Otolaryngol Head Neck Surg*, Vol. 106, No.4, 345-350, 0194-5998

Kim, D. & Kline, D. (1990). Peri-epineurial tissue to supplement laser welding of nerve. *Neurosurgery*, Vol.26, No.2, 211-216, 0148-396X

Lauto, A.; Trickett, R.; Malik, R.; Dawes, J. & Owen, E. (1997). Laser- activated solid protein bands for peripheral nerve repair: an vivo study. *Lasers Surg Med*, Vol.21, No.2, 134-141, 0196-8092

Maragh, H.; Hawn, R. & Gould, J. (1988). Is laser nerve repair comparable to microsuture coaptation? *J Reconstr Microsurg*, Vol. 4, No.3, 189–195, 0743-684X

Menovsky, T.; Beek, J. & van Gemert, M. (1994). CO2 laser nerve welding: optimal laser parameters and the use of solders in vitro. *Microsurgery*, Vol.15, No.1, 44-51, 0738-1085

Menovsky, T.; Beek, J. & Thomsen, S. (1995). Laser(-assisted) nerve repair. A review. *Neurosurg Rev*, Vol.18, No.4, 225–235, 0344-5607

Menovsky, T.; Beek, J. & van Gemert, M. (1996). Laser tissue welding of dura mater and peripheral nerves: a scanning electron microscopy study. *Lasers Surg Med*, Vol.19, No.2, 152–158, 0196-8092

Menovsky, T.; van der Bergh Weerman, M. & Beek, J. (1997). Transmission electron microscopy study of acute laser welds in dura mater and peripheral nerves. *Lasers Med Sci*, Vol.12, No.2, 131–136, 0268-8921

Menovsky, T. & de Vries, J.(1998). Use of fibrin glue to protect tissue during CO2 laser surgery. *Laryngoscope*, Vol.108, No.9, 1390–1393, 0023-852X

Menovsky, T. (2000). CO2 and Nd:YAG laser-assisted nerve repair: a study of bonding strength and thermal damage. *Acta Chir Plast* , Vol.42, No.1, 16-22, 0001-5423

Menovsky, T. & Beek, J. (2001). Laser, fibrin glue, or suture repair of peripheral nerves: a comparative functional, histological, and morphometric study in the rat sciatic nerve. *J Neurosurg*,Vol. 95, No.4, 694-699, 0022-3085

Menovsky, T. & Beek, J. (2003). Carbon dioxide laser-assisted nerve repair: effect of solder and suture material on nerve regeneration in rat sciatic nerve. *MICROSURGERY*, Vol.23, No.2, 109–116, 0738-1085

Neblett, C.; Morris, J. & Thomsen, S. (1986). Laser-assisted microsurgical anastomosis. *Neurosurgery*, Vol.19, No.6, 914-934, 0148-396X

Okada, M.; Shimizu, K.; Ikuta, H.; Horii, H. & Nakamura, K. (1987). An alternative method of vascular anastomosis by laser: experimental and clinical study. *Lasers Surg Med*, Vol.7, No.3, 240-248, 0196-8092

Ochi, M.; Osedo, M. & Ikuta, Y. (1995). Superior nerve anastomosis using a low-output CO2 laser on fibrin membrane. *Lasers Surg Med*, Vol.17, No.1, 64–73, 0196-8092

Poppas, D.; Rooke, C. & Schlossberg, S. (1992). Optimal parameters for CO2 laser reconstruction of urethral tissue using a protein solder. *J Urol*, Vol. 148, No.1, 220-224, 0022-5347

Richmond, I. (1986). The use of lasers in nerve repair, In: *Advanced Intraoperative Technologies in Neurosurgery*, Fasano VA (Ed.), 175-183, Springer Verlag, 0387818804, New York

Schober, R.; Ulrich, F.; Sander, T.; Du¨ rselen, H. & Hessel, S. (1986). Laser induced alteration of collagen substructure allows microsurgical tissue welding. *Science*, Vol.232, No.4756, 1421–1422, 0036-8075

Trickett, I.; Dawes, J.; Knowles, D.; Lanzetta, M. & Owen, E. (1997). In vitro laser nerve repair: protein solder strip irradiation or irradiation alone? *Int Surg*, Vol.82, No.1, 38-41, 0020-8868

3

Low speed laser welding of aluminium alloys using single-mode fiber lasers

Jay F. Tu and Alexander G. Paleocrassas
North Carolina State University
USA

1. Introduction

Laser welding of aluminium alloys is an important industrial technology and yet many challenges still lie ahead. Laser welding studies were reported almost within two years since the first laser was invented in 1960. However, practical metal seam welding was not feasible until the early 1970s when multi-kilowatt, continuous wave CO_2 lasers were developed to allow for deep penetration keyhole welding (Duley, 1999). Unfortunately, the application for deep penetration welding of aluminium was limited due to its very high reflectivity at the relatively long wavelength (10.6 micron) of CO_2 lasers. Flash-pumped Nd:YAG lasers with a 1.06 micron wavelength were not suitable due to their low power and extremely poor efficiency at the time. Since the 1980s, high power seam welding of carbon steels using multi-kilowatt CO_2 lasers has become a regular industrial practice, in particular in the automotive industry. In the mid 1990s, diode pumped Nd:YAG lasers were developed that offered kilowatt power and high efficiency. As a result, aluminium laser welding became more feasible because the beam absorption of aluminium alloys at 1.06 micron is three times as much as it is at 10.6 micron. Nevertheless, the poor beam quality and high cost of diode-pumped Nd:YAG lasers still hinders their acceptance in industry. In the early 2000s, with the arrival of single-mode and multi-mode high power fiber lasers at a 1.075 micron wavelength, along with excellent beam quality and low maintenance cost, the expectation was that the advantage of laser welding aluminium components could be better realized.

The requirement of very high laser power for aluminium welding is not only due to its high reflectivity and high heat conductivity. Aluminium has been known to be one of the most challenging metals to weld successfully (Mandal, 2002). Other factors affecting the weld quality of aluminium alloys include different kinds of porosity formation, hot tearing, solidification cracking, oxide inclusions and loss of alloying elements. It has been found that weld porosities can be significantly suppressed at high welding speeds. In order to maintain a stable keyhole at high speeds, very high laser power is needed. What has been less explored is the reason why the welding process becomes less stable and prone to defects as the speed is reduced. There are many applications where high speed welding is not suitable. With the expansion of modern miniaturized consumer products, the weld path can be short and with intricate shapes. High welding speed may not be effective due to the short paths and constant accelerations and decelerations required to follow the path precisely. One such application is the fusion of fatigue cracks in aluminium parts, where a crack path is irregular. It also has been shown that welding at a lower processing speed can reduce the tendency of transverse solidification cracking. Finally, with the availability of better laser sources such as high power fiber lasers, it is important to expand laser welding of aluminium to wider processing conditions for various applications.

This book chapter will discuss latest research results in extending laser welding of aluminium in the low speed range by investigating the welding instability phenomena. The following topics will be discussed:

- Aluminium alloys and welding defects
- Brief review of high speed laser welding of aluminium
- High power fiber lasers and optical setups
- Process modelling of laser welding of aluminium
- Experimental process characterization of low speed welding
- The instability and defects at low speed welding
- Applications of fatigue crack repair in aluminium

First, the properties of aluminium alloys and the cause of welding defects are discussed. Prior to discussing low speed welding, a brief review of high speed laser welding of aluminium is provided. The characteristics of high power fiber lasers and their optical setups for welding applications, such as focusing lens, assist gas, alignments, and damage prevention due to beam reflection by aluminium are then presented. The chapter then proceeds to present recent research results in low speed laser welding of aluminium, which includes theoretical process modelling, experimental process characterization, in-process monitoring of several critical signals, such as plasma radiation and beam reflection, as well as the causes and consequences of process instability at low speeds. The transition of process stability from medium welding to a low speed threshold and its mechanisms are explored. Finally, an application of low speed laser welding of aluminium for fatigue crack repair is given. A discussion on different applications and future development conclude the chapter.

2. Background and Reviews

2.1 Aluminium Alloys

Aluminium alloys can be separated into two major categories: Non heat-treatable and heat-treatable. The initial strength of non heat-treatable alloys depends primarily upon the hardening effect of alloying elements such as silicon, iron, manganese and magnesium. The non heat-treatable alloys are mainly found in 1xxx, 3xxx, 4xxx and 5xxx series. Additional strength is usually achieved by solid-solution strengthening or strain hardening.

The initial strength of heat-treatable alloys depends upon the alloy composition, just like the non heat-treatable alloys. In order to improve their mechanical properties they need to undergo solution heat treating and quenching followed by either natural or artificial aging (precipitation hardening). This treatment involves maintaining the work piece at an elevated temperature, followed by controlled cooling in order to achieve maximum hardening. The heat-treatable alloys are found primarily in the 2xxx, 6xxx and 7xxx alloy series (ibid).

The 7xxx series alloys contain zinc in amounts between 4 and 8 % and magnesium in amounts between 1 and 3 %. Both have high solid solubility in aluminium. The addition of magnesium produces a marked increase in precipitation hardening characteristics. Copper additions between 1 and 2 % increase the strength by solid solution hardening, and form the basis of high strength aircraft alloys. The addition of chromium, typically up to 0.3 %, improves stress corrosion cracking resistance. The 7xxx series alloys are predominantly used in aerospace applications, 7075-T6 being the principal high strength aircraft alloy (Ion, 2000). This chapter will focus on fiber laser welding of 7075-T6 because of its predominant use in aircraft components.

2.2 Laser Welding Defects in Aluminium

Laser welding is one of the most promising metal joining methods because it can provide high productivity, high weld quality, high welding speed, high weld aspect ratio, low heat input, low distortion, manufacturing flexibility and ease of automation (Duley, 1999, Mandal, 2002). According to a study on phase transformations in weldments (Cieslak, 1992), solidification rates of 10^2 to 10^3 °C/sec encountered in conventional arc welding processes, are much lower compared to high-energy density laser processes which reach 10^5 to 10^6 °C/sec as a result of high heat input experienced at high travel speeds. Under these processing conditions the weld metal microstructure bears no resemblance to that expected as the result of arc welding. Consequently, the weldment is mostly comprised by fine-grained microstructures.

There are four major types of weld defects in laser welding of aluminium: a) porosity, b) cracking, c) inclusions and d) loss of alloying elements (Matsunawa, 1994, Cao, *et al.*, 2003).

Hydrogen porosity: Hydrogen is very soluble in aluminium and its alloys. Most gas porosities precipitated in aluminium alloys are attributed to hydrogen. The solubility of hydrogen in liquid aluminium is an exponential function of temperature, which is why its porosity is a much bigger problem in laser welding (than in conventional welding) due to increased temperatures. Also, the high cooling rate is very unfavorable because it does not allow for diffusion (i.e. "floatation") of the trapped hydrogen. Normal hydrogen levels in molten aluminium vary from approximately 0.10 to 0.40 mL/100g. It is worth noting here that in order for an aircraft part to pass aerospace quality inspections, the gas contents have to be less than 0.06 mL/100g. A critical lower welding speed possibly exists at which formation and growth of hydrogen porosity can be prevented. Also, another way to reduce hydrogen porosity is to increase power density, because it keeps the keyhole stable and increases solidification time, allowing the hydrogen to escape (Cao, *et al.*, 2003).

According to a study conducted on porosity formation (Kutsuna and Yan, 1998), the rate of hydrogen porosity shows a tendency to rise considerably as the magnesium content increases. This happens because magnesium in aluminium alloys raises the hydrogen solubility in the molten pool and hence the segregation of magnesium enhances the segregation of hydrogen during solidification.

Porosity caused by collapse of unstable keyholes: Even with proper material surface preparation, laser parameters, shielding gas and material compositions, aluminium alloys are susceptible to random porosities after laser welding (Weeter, 1998). Keyhole instability and the coupling of the laser beam into the metal are suspected to cause these random events. These porosities have irregular or turbular form and are large enough to be visible with x-ray analysis (Dausinger, *et al.*, 1997). They are usually located in the keyhole path, whereas hydrogen pores are more or less equally distributed with slight enrichment at the melting line. The number of cavities is strongly influenced by processing parameters such as the

power, focusing and wavelength. Most likely, keyhole stability is increased with the shorter wavelength lasers (Nd:YAG) because the beams are not as drastically affected by the weld plume as the ones with longer wavelength lasers (CO_2). In the latter case, the weld plume periodically blocks the beam from impinging on the metal and thus causes an instability in the keyhole. The shorter wavelength laser beams can pass through the plume and can provide a more consistent heat input into the metal (Weeter, 1998). It has also been observed that the highest level of porosity is concentrated in the regions where an unstable keyhole is formed. They are mainly composed of metal vapor but will condense at room temperature. The way to reduce this type of porosity is to keep the keyhole as stable as possible; this can be achieved by welding at high speeds and the addition of filler wire. Also, the use of high-power continuous wave (CW) can improve the stability of keyholes (Cao, *et al.*, 2003b).
Results based on a study that was conducted in vacuum and under low pressure welding with a tornado nozzle were reported to reduce or suppress porosity. It was also reported that the forward welding with about 15 to 20 degrees of beam inclination was able to reduce porosity (Katayama, *et al.*, 2003). However, full penetration and pulse-modulated welding approaches were not completely effective.

Cracking: Aluminium alloys exhibit a strong propensity for weldment crack formation because of their large solidification temperature range, high coefficient of thermal expansion, and large solidification shrinkage. The restrained contraction of a weld during cooling, sets up tensile stresses in the joint which may cause cracking. There are two types of hot cracking: a) cracking that occurs in the weld fusion zone during solidification of the weld metal which is known as solidification cracking, and b) cracking that takes place in the primary melting zone due to tearing of the liquate, called liquation cracking (Zhao, *et al.*, 1999). These cracks are detrimental to the integrity of the weld since they form areas of high stress concentration and will significantly reduce the strength of the weld, probably leading to catastrophic failure.

Oxide Inclusions: Oxides are one of the main types of inclusions in aluminium alloys. During keyhole laser welding, the inherently unstable keyhole flow may entrap shielding gas or even air because of imperfect gas shielding (Matsunawa, *et al.*, 1998). Additionally, the shielding gas cannot be truly pure; therefore, some oxide particles may be present in the keyhole vapor. The surface of liquid metal in weld pools (strictly speaking, the surface here should be referred to as the interface between the liquid metal in the weld pool and metal vapor or shielding gas) may also be partly oxidized to form oxide films because of the entrapment of air or shielding gas into the pools. Depending on the magnesium contents in aluminium alloys, oxides such as Al_2O_3, Al_2MgO_4, MgO, or their combination may occur. When aluminium alloys contain magnesium, because it is surface active in liquid aluminium, the oxidizing tendency of the molten aluminium increases rapidly with magnesium contents. When the aluminium alloys contain a trace of magnesium, a mixed oxide ($MgAl_2O_3$), spinel, is formed. When the magnesium content of the alloy exceeds approximately 2%, the liquid oxidizes rapidly to form MgO. The oxides entrained into welded metal because of surface turbulent flow in welding are referred to as young oxides. Oxides in base metal, originally from primary processing of aluminium alloys are termed old oxides.

Loss of Alloying Elements: The high power density used for laser welding may cause selective vaporization of some alloying elements with a low fusion point such as lithium, magnesium, and zinc because of their higher equilibrium vapor pressure than aluminium. Selective vaporization of alloying elements can take place in both keyhole and conduction mode laser welding. The vaporization mechanism is divided into three stages. The first stage involves transport of vaporization elements from the bulk to the surface of the molten weld pool. Then the vaporization of elements occurs at the liquid/vapor interface, and finally the vaporized species are transported into the surrounding gas phases (Cao, *et al.*, 2003ab). This will also cause a void on the top of the weld called underfill. It was found that the intrinsic vaporization of alloying elements at the weld pool surface, controls the overall vaporization (Zhao, *et al.*, 1999).

The loss of alloying elements can be minimized by controlling the beam power density distribution during continuous wave (CW) laser welding, which can influence the temperature of the molten metal in the welding pool (Cao, *et al.*, 2003ab). Another way of reducing this loss is through the use of filler metal, which is used as an auxiliary source of material to fill the gap. It also provides a means of controlling the metallurgy of the weld bead and ensures weld quality (Ion, *et al.*, 2001), by helping replenish the loss of alloying elements and also prevent solidification cracking. The use of filler metal in laser welding is justified only if the joint gap and sheet metal thickness is larger than the beam and loss of alloying elements is significant (Molian, 2004). Some skepticism remains about the process, mainly because it is considered to be complex and requires high precision. The interactions between the large number of process variables involved are also not fully understood (Ion, *et al.*, 2001).

2.3 High Speed Laser Welding of Aluminium

Dausinger, et al., (1996) reported that with a 2.2 kW Nd:YAG laser, weld depths of up to 3.7 mm in AA 6082 have been obtained at approximately 16.7 mm/s, at a power density of 3 MW/cm^2. Also, Yoshikawa, et al. (1995), report that successful butt welds of 3 mm thick 5 and 6 series aluminium alloys can be obtained. They also used high duty cycle power modulation (pulsing) in order to prevent cracks. In a different study, a 3 kW CO_2 laser has been able to achieve approximately 2.5 mm weld depth in aluminium alloy 7075-T6 at about 25 mm/s (Katayama and Mizutani, 2002). Also, a 4.5 kW CO_2 produced penetration depths of 3.5 mm in aluminium alloy series 5000 (non heat-treatable) and 6000 (heat-treatable), at a speed of approximately 33 mm/s; in comparison, a 4 kW Nd:YAG produced weld depths of 4 mm at the same speed (Cao, et al., 2003a). In addition, Ramasamy and Albright (2000) showed that when welding with a pulsed 2 kW Nd:YAG, or a 3 kW continuous wave Nd:YAG, or a 3-5 kW CO_2 laser, vaporization of magnesium and/or silicon can occur from aluminium alloy 6111-T4 and also the metal hardness was reduced. This means that when operating at very high power densities, loss of alloying elements is a significant problem.

Some other more recent studies are also worth being reviewed. Oi et al. (2006) used a slab CO_2 laser for bead-on-plate and filler wire welding of 2.4 mm thick AA 7075-T6. Mechanical properties of these welds were examined and, after heat treatment, achieved tensile strengths between 75 and 84 percent (depending on the filler metal added) of the base metal. Another study on the mechanical properties of laser welded heat-treatable aluminium was conducted by Xu et al. (2008). They used a CO_2 laser to weld AA 2519-T87 and examined the weld's microstructure and tensile strength. Results showed that the grains in the welds were very fine and the tensile strength, after heat treatment, reached up to 75 percent of the base metal as compared to 61 percent for MIG welding.

2.4 Single-Mode High Power Fiber Lasers

The power source used in this study is a 300 W, Single-Mode, Ytterbium Fiber Optic Laser. The power unit consists of six 50 W modules which are combined to produce an output power of 300 W. Each module contains a fiber optic bundle through which the laser light resonates. The fiber is doped with a rare earth called ytterbium, and acts as the laser gain medium. Each module contains several diode lasers which serve as the pump (IPG, 2003).

	Fiber Laser	Nd:YAG	CO_2	Disc
Wall Plug Efficiency	30%	~5%	~10%	15%
Output Powers	to 50 kW	to 6kW	to 20 kW	to 4 kW
BPP (4/5kW)	< 2.5	25	6	8
Diode Lifetimes	100,000	10,000	N.A.	10,000
Cooling	Air/Water	Dionized	Water	Water
Floor Space (4/5kW)	< 1 m^2	6 m^2	3 m^2	> 4 m^2
Operating Cost/hour	$21.31	$38.33	$24.27	$35.43
Maintenance	Not Required	Often	Required	Often

Table 1. Characteristics comparisons between major high power industrial lasers (Industrial Laser Solutions, 2005).

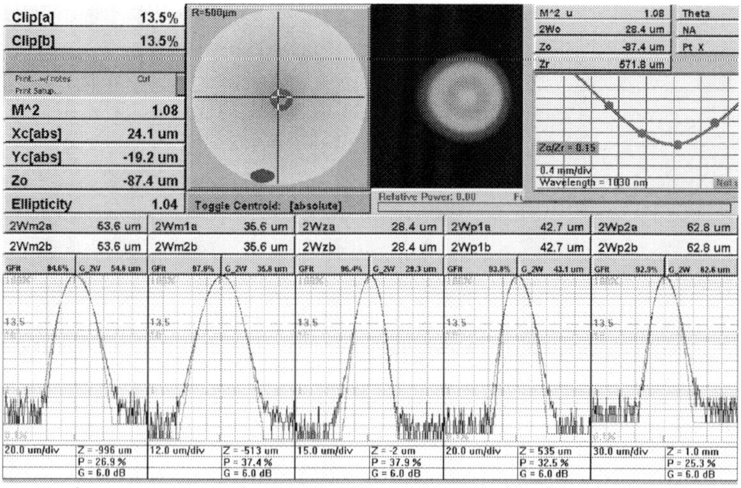

Fig. 1. Beam mode measurement of a 300 W, CW, single-mode fiber laser, provided by IPG, Inc.

A comparison of a high power single-mode fiber laser and other major industrial lasers is listed in Table (1). It is clear that high power fiber lasers have large advantages in wall plug efficiency, maximum output powers, beam quality, reliability and operating cost. Among these advantages, the beam quality, as measured by Beam Parameter Product (BPP) (beam waist radius times divergence angle) and the M^2 value are most impressive. Their M^2 value of approximately 1.08 (Figure 1) is very close to a Gaussian (normal) power distribution. This allows the beam to be focused to a very small spot to achieve very high power densities. In addition to the excellent beam quality, this laser has a wavelength of approximately 1.075 μm (near infrared spectrum), which is relatively short compared to CO_2 lasers. This allows for increased absorption, useful for welding highly reflective materials like aluminium. The beam diameter exiting the collimator is approximately 7 mm, and can be focused down to about 10 μm. At 300 W, a maximum power density of about 382 MW/cm^2 can be achieved.

2.5 Review of Laser Welding with Single-Mode Fiber Lasers

Limited research has been conducted on laser welding using fiber lasers. Prof. Miyamoto was one of the first to realize the advantages of the fiber laser and propose that it be used in laser welding (Miyamoto, et al., 2003). His experiments were performed on stainless steel foil with a limited output power (~50 W). Ever since, there have been others who have recognized the value of fiber lasers in laser material processing. Allen at al. (2006) used a high power fiber laser as part of a broader study in welding of 7000 series aluminium alloy of thicknesses between 6 and 12 mm. The processing parameters of power and welding speed were not mentioned, however, for proprietary information purposes. Another more recent study (Brown, 2008) focused on keyhole welding on several different metals, including AA 1100, using a moderate power fiber laser (600 W). Uniform high aspect ratio welds were observed, which were in reasonable agreement with the two-dimensional Rosenthal model for a moving-line heat source that was used for comparison. Also, Katayama et al. (2008) used a high power fiber laser to investigate the various welding conditions on penetration and defect formation, on several aluminium alloys and in particular AA5083. Power densities ranged from 40 kW/cm^2 to 90 MW/cm^2. At 64 MW/cm^2 and 10 m/min (166.7 mm/s) 10 mm thick plates were penetrated fully. Porosity was generated at certain processing conditions, reasons for which were given by interpreting the keyhole and molten metal behaviors, observed using a high speed camera and micro-focused X-ray transmission. It was found that nitrogen gas was more effective than argon, in minimizing or even preventing porosities.

Other research using fiber lasers includes a study on micromachining using a 100 W, single mode fiber laser (Naeem and Lewis, 2006). This research group has focused their study on micro joining and micro cutting various metals using both continuous wave and pulsed modes. Similarly, Wagner (2006) studied high speed micro welding of thin sheets of various metals including aluminium, assessing the potentials for low distortion at high speeds. The processing speeds employed reached 100 m/min (1667 mm/s).

2.6 Experimental Setup for Fiber Laser Welding

An optical isolator was attached to the collimator and is used to divert any reflected light away from the collimator in order to avoid damage to the fiber due the high reflectivity of aluminium. Consequently, the beam diameter and beam quality were changed slightly. The beam diameter increased from 5 mm in diameter to 7 mm, while the M^2 value goes from 1.08 to 1.15.

A 3x beam expander is used in combination with the 100.1 mm triplet lens to obtain a minimum focus spot size of 12.0 μm. Equation 1 shows how to calculate the minimum spot size.

$$\text{Spot size} = \frac{\text{Lens Focal Length}}{\text{Collimator Optics Focal Length} * \text{ Beam Expantion Factor}} * \text{Fiber Diameter}$$
$$= \frac{100.1 \text{ mm}}{25 \text{ mm} * 3} * 9\ \mu\text{m} = 12.01\ \mu\text{m} \tag{1}$$

The laser beam is centered with respect to the beam expander and the laser head. The laser head contains the focusing triplet and can be adjusted using the outer ring. At the bottom of the cutting head there is a chamber that allows for shielding to flow out through the welding nozzle. This chamber is sealed by a special cover glass and a rubber gasket.
The determination of the laser beam's focusing position was done by using a laser drilling technique. One of the fiber laser's particular characteristics is that when a laser pulse is released, there is an approximately 1,500 W power spike that is output for about 1 μs before it drops to the steady-state power value of 300 W. By pulsing the laser for a very short time, approximately 3 μs, we can take advantage of this power spike and create a very high power density at the focus plane. This enables us to perform laser ablation to form a crater into a stainless steel plate. The focusing technique utilizes this process, by creating spot welds or holes at different z-positions, every 10 μm. A picture is taken of each group of welds/holes and using special calibrated software, the radii are measured and plotted versus the z-focus position.

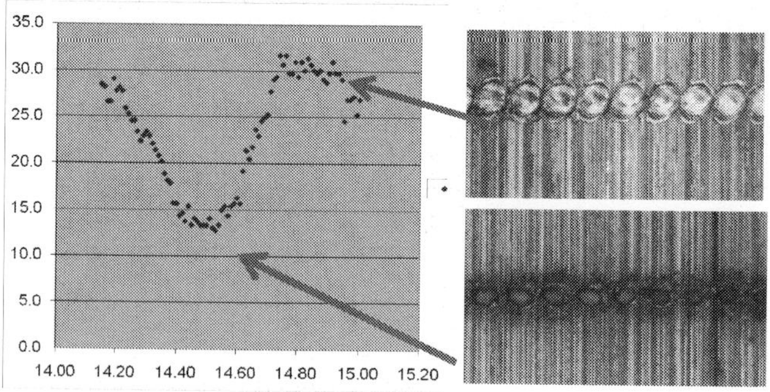

Fig. 2. Spot welds or holes at different focusing locations

Since the laser spot size is very different at different focusing positions, the pulsing will either create very small holes, approximately on the order of the focused beam spot size, or larger spot welds. When all the radii are plotted, the minimum of the resulting curve shows the approximate location of the focusing plane. This is a relatively quick and effective way to find the location of the focus. This technique can be further expanded to obtain the beam profile along its propagation axis (Harp et al, 2008).

3. Low Speed Laser Welding of Aluminium

3.1 Modeling of an Idealized Welding Process

A 2-D heat conduction model for laser welding is reported in Lankalapalli, Tu, and Gartner (1996). This model makes several assumptions which significantly reduce its complexity. The general idea of the model is to calculate the heat conduction over an infinitesimally thin layer of thickness (depth) dz at a specific distance from the top of the surface (Figure 3).

One of the assumptions made, is that the walls of the keyhole within this layer are perpendicular to the surface and that heat conducted in the z-direction is much less than the heat conducted in the radial direction. Therefore, a conical keyhole can be divided into an infinite number of such infinitesimally thin layers and the depth can be approximated by cylindrical heat sources of varying radii, moving together at a constant speed in each of these thin layers. Another assumption made is that there is a quasi-steady state environment in which a cylindrical surface of radius a, at uniform temperature T_V, is moving with a constant speed, v, along the x direction, in an infinite medium initially at constant temperature, T_0. Finally, assuming that the thermal properties of the medium are constant and that the axis of the cylindrical surface passes through the origin of the coordinate system, the governing differential equations and boundary conditions for the temperature distribution can be written as:

$$\frac{\partial^2 T}{\partial x^2} + \frac{\partial^2 T}{\partial y^2} + \frac{v}{\alpha}\frac{\partial T}{\partial x} = 0 \tag{2}$$

$$T = T_V \text{ at } x^2 + y^2 = a^2 \tag{3}$$

$$T(x, y) \rightarrow T_0 \text{ as } x \rightarrow \pm\infty \text{ and y} \rightarrow \pm\infty \tag{4}$$

where x and y are the surface coordinates, z is the depth coordinate, a is the keyhole radius, v is the welding speed, α is the thermal diffusivity, T_0 is the initial temperature and T_V is the vaporization temperature of the material (Carslaw and Jaeger, 1962).

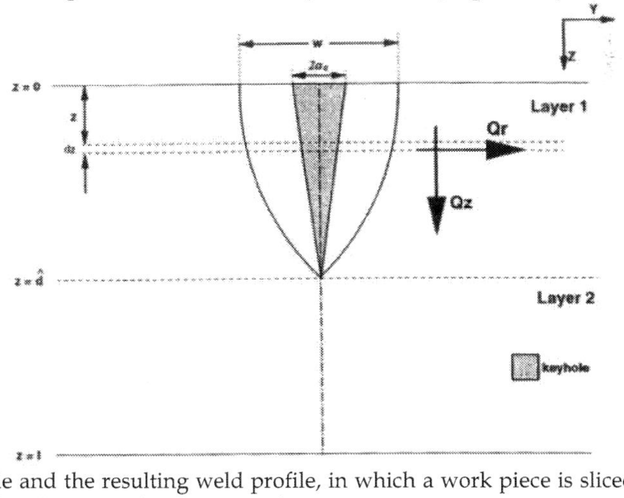

Fig. 3. Keyhole and the resulting weld profile, in which a work piece is sliced to many thin layers (Lankalapalli, Tu, and Gartner, 1996)

After several derivations, the following equation which estimates penetration was found as (Lankalapalli, Tu, and Gartner, 1996)

$$d = \frac{P_i}{k(T_V - T_0)} \frac{1}{\sum_{i=1}^{6} \frac{c_i}{i} (Pe_0)^{i-1}} \tag{5}$$

where k is the thermal conductivity of the material and c_i are coefficients to a polynomial fit to the equation that was evaluated numerically for 100 different values of Pe in the operating range of 0 - 0.025:

$$g(Pe) = \int_0^{2\pi} G(\theta, Pe) d\theta = C_1 + C_2 Pe + C_3 Pe^2 + C_4 Pe^3 + C_5 Pe^4 + C_6 Pe^5 \tag{6}$$

where

$$G(\theta, Pe) = \quad Pe * e^{(\;Pe\cos\theta)} * \quad \sum_{n=0}^{\infty} \varepsilon_n I_n(Pe) cos(n\theta) \quad \frac{n}{Pe} \quad \frac{K_{n+1}(Pe)}{K_n(Pe)} \quad cos\,\theta$$

$$= \frac{}{r^*} \frac{T_V \quad T}{T_V \quad T_0} \Big|_{r^*=1} \tag{7}$$

$$\frac{T_V - T}{T_V - T_0} = 1 - e^{(-Pe*r^* \cos\theta)} * \sum_{n=0}^{\infty} \varepsilon_n \frac{I_n(Pe)}{K_n(Pe)} K_n(Pe * r^*) \cos(n\theta) \tag{8}$$

is the closed-form solution in polar coordinates (r,θ) of the aforementioned governing differential equation with the specified boundary conditions for the temperature distribution, where Pe = v*a / (2α) is the Péclet number, r^* = r/a is the normalized radial coordinate, $\varepsilon_n = 1$ for n = 0 and 2 for $n \geq 1$, I_n is a modified Bessel function of the first kind, of order n and K_n is a modified Bessel function of the second kind of order n. Note that the above model is not material specific. With proper material parameters and process parameters incorporated, this model allows for very rapid simulation of the temperature field at the top surface (Equation 8) and for an estimation of penetration depth (Equation 5). This model has been validated over a wide range of speed and laser power, different materials, and different lasers (Lankalapalli, Tu, and Gartner, 1996; Paleocrassas and Tu, 2007), as shown in the next section.

3.2 Model Validation through High Speed Welding of SUS 304

Several SUS 304 specimens, 300 microns thick, were welded at relatively high speeds (200—1000 mm/s) (Miyamoto, *et al.*, 2003). In order to determine the operating Péclet number, apart from the welding speed and the thermal diffusivity, the keyhole radius is also required. Determining the keyhole radius is not trivial. There exists a method (Lankalapalli, Tu, and Gartner, 1996) to estimate the Péclet number from the weld width. The idea is that a contour plot of isotherms can be generated for specific Péclet numbers for the top surface using Equation 8 , and by measuring the width of the curve corresponding to the melting temperature range, the normalized weld width (w/a) and Péclet number can be correlated.

The normalized weld width is obtained by taking twice the maximum y value (due to symmetry) of the melting temperature isotherm curve. Therefore, an equation can be calculated numerically which can be used to determine the Péclet number at the surface of the specimen, for a corresponding weld width.

Figure 4 shows the model prediction compared to the experimental results from Miyamoto et al (2003). The model predicts a satisfactory trend of penetration change versus the Péclet number for different laser powers. However, the data of a specific laser power usually match the predictions of a lower laser power. For example, the data of 170 W laser power match the predictions of 130 W laser power, data of 130 W match better with prediction at 90 W, etc. Based on this observation, it can be stated that approximately 70-80% of laser power is absorbed. This absorption is relatively low, likely due to the very high welding speed.

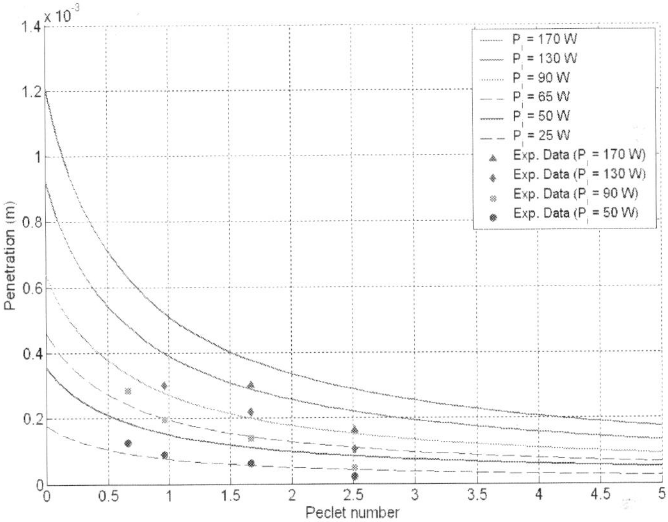

Fig. 4. Theoretical estimate vs. experimentally measured penetration depth in SUS 304.

3.3 Model validation using low speed welding of AA 7075-T6

Figure 5 presents the data/simulation comparison based on the same model for low speed welding of AA7075-T6. As in Figure 4, the model predicts a satisfactory trend of penetration versus the Péclet number. The laser beam absorption is about 90% for welding speeds from 2 mm/s to 10 mm/s. Note that Figures 4 and 5 cover a wide range of Péclet numbers (from 0.5 to 2.5 in Figure 4 and 0.001 to 0.08 in Figure 5). The absorption in Figure 5 is higher than those in Figure 4, likely due to slower welding speed and deeper penetration even though stainless steel is used in Figure 4, while aluminium is used in Figure 5. Once keyhole is formed, the laser beam is absorbed efficiently. For those conditions in Figure 4 with very high welding speeds, the keyhole is shallower and likely tilted to reflect beam power (Fabbro and Chouf, 2000).

However, the point corresponding to 1 mm/s shows a significant decrease in penetration, with its absorbed power being only 68% of the input power. Also, by observing the cross-sections of the welds at three different processing speeds, it can be seen that the 1 mm/s weld is significantly different from the other two. The 1 mm/s weld shows a significant decrease in aspect ratio. In some cross-sections, large blowholes and porosities were present. The other two welds show more of a conical shaped cross-section, a higher aspect ratio and the absence of any major defects.

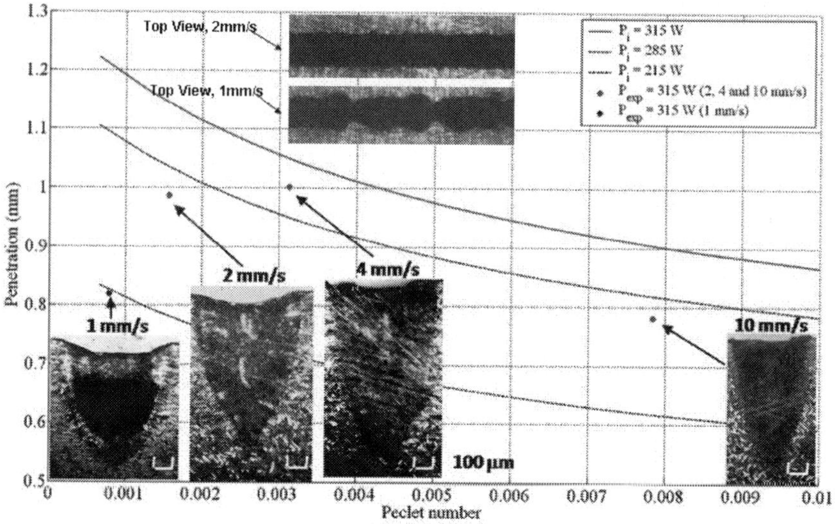

Fig. 5. Model validation for low speed welding of AA 7075-T6

This observation leads to the suspicion that at extremely low speeds the process breaks down and the laser energy is not coupled as efficiently. If this is the case, the model no longer applies to speeds below 2 mm/s.

3.4 Effect of Focusing Positions on Low Speed Welding

Figure 6 shows the change in weld penetration as the focusing position changes (positive indicating the beam is focused into the workpiece).

The general trend is that the best focus position corresponds with the maximum weld depth. This goes along with the recommendation for most welding processes, which is that the focus should be positioned at the desired weld depth (Steen, 2003). Another observation that can be made is that, as the beam is focused past the maximum depth location, the penetration drops at a much higher rate, with the exception of the 10 mm/s condition. This is an indication that the slower speeds are much more sensitive to focusing changes, which means that higher focusing is required to produce adequate and repeatable weld penetrations.

Specifically, for the 10 mm/s processing speed, the maximum weld penetration is approximately both ~ 0.8 mm and this occurs when the focus is approximately 0.9 mm into the workpiece. For the 2 and 4 mm/s speeds the weld penetration is deepest (~ 1 mm) when the beam is focused approximately 1 mm into the workpiece. The difference in weld penetration (~ 0.8 mm) between these two speeds is not much, with the 4 mm/s weld being slightly deeper. However, the 1 mm/s welds show a significant drop in penetration.

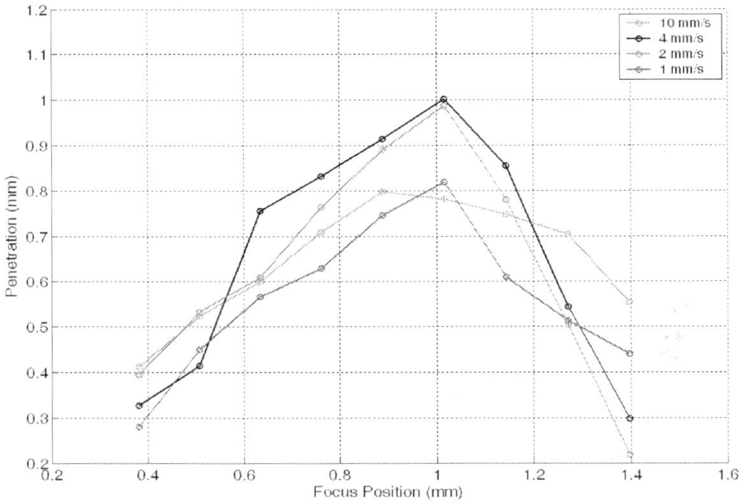

Fig. 6. Effects of focusing position on penetration for different welding speeds.

3.5 Energy-Based Process Characterization

Paleocrassas and Tu (2007) proposed metrics to characterize welding process efficiency. One such metric was defined as keyhole fluence per weld length (KF) which has since been slightly modified and is redefined as follows:

$$\mathrm{KF} = \frac{P_i}{A_b} \cdot \frac{l_w}{v} \cdot \frac{1}{l_w} = \frac{P_i}{A_b \cdot v} \tag{9}$$

where P_i is total incident power, A_b is the outer surface area of the immersed laser beam (as calculated from the beam profile approximation, also shown in Figure 8.7), l_w is the length of the weld and v is the processing velocity. This metric represents the total irradiated energy density per weld length.

As mentioned before, due to different types of power losses during welding, the total irradiated energy density per weld length (KF) from the laser is not going to be completely absorbed by the material. Therefore it is of interest to determine the "weld efficiency" by looking at the total energy used to create the weld and how "well" it is used; for example, the same amount of absorbed weld energy could translate into a shallow and wide weld, or a deep and narrow weld.

Specific weld energy per weld length was also defined by Equation 10 to define how well the amount of energy that used to created the weld was used. In this paper, this metric is denoted as Effective Weld Energy (EWE):

$$EWE = \frac{m_{weld} \cdot \zeta}{\pi \cdot r_{profile}^2} = \frac{\rho \cdot V_{weld} \cdot \zeta}{\pi \cdot r_{profile}^2} = \frac{\rho \cdot A_{weld} \cdot \zeta}{\pi \cdot r_{profile}^2} \tag{10}$$

where m_{weld}, V_{weld}, A_{weld}, and $r_{profile}$ are the mass, volume and radius of the top profile (or half of the weld width) of the weld (Figure 7), respectively, ρ is the density and ζ is the specific energy of AA 7075-T6, which is determined by

$$\zeta = C_p \cdot \Delta T + \text{Latent Heat of Fusion} \tag{11}$$

where C_p is the specific heat capacity and ΔT is the temperature change between ambient temperature and the melting point.

Figure 8 was generated by applying the above energy based process characterization to the experimental data. The EWE of each data point is plotted with respect to the input KF. There are four sets of data and each set is connected by a different colored line, corresponding to a different processing speed. The 1, 2, 4 and 10 mm/s data are shown in red, green, black and cyan, respectively. Each data point in each set corresponds to a weld created with a different focusing position. The number next to each data point represents how deep the beam is focused into the workpiece, in thousands of an inch. The first observation that can be made is that for each processing speed, the point that has the highest EWE is the one where the beam was focused at approximately 1 mm (.040 in.) into the workpiece, which indicates that it is the focusing condition that produces the best energy coupling. This is the case because the majority of the vapor pressure used to maintain a certain depth is created at the bottom of the keyhole.

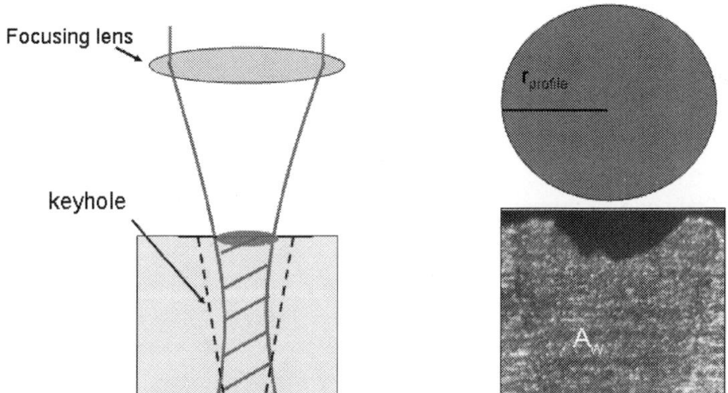

Fig. 7. Schematics showing the submerged beam surface area (A_b), the weld's cross-sectional area (A_w) and the profile radius ($r_{profile}$)

Therefore, placing the focus at the desired weld depth ensures that the maximum power density will be at the bottom of the keyhole, creating the majority of the metal vapor. However, focusing too deep can have an adverse effect on EWE because a minimum power density at the surface is required to create and maintain vaporization of the metal. This explains why the EWE decreases when the laser beam is focused too deep.

By examining the processing speed trend, it was observed that as the speed decreases, the EWE increases, until the speed drops below 2 mm/s. It can therefore be seen that the process is not only dependent on the amount of KF, but also in the manner it is deposited into the workpiece. This leads to the examination of the efficiency of the process.

Global Efficiency: One of the primary concerns in any process involving energy exchange is how efficient it is; in this case, that is, how much of the irradiated energy density per weld length was translated into a desirable, high aspect ratio weld. This is where we can define the "global efficiency" of the process. It is simply the ratio between EWE and KF, as stated in Equation 12.

$$\text{Global Efficiency} = \frac{EWE}{KF} \tag{12}$$

With this metric, we can determine the efficiency at each speed and at each focusing position. If we look at the actual percentages, we will see that the highest efficiency does not exceed 3 percent of the total KF. This might seem extremely low at first, but it is important to remember that this number corresponds only to the energy used to create the weld itself.

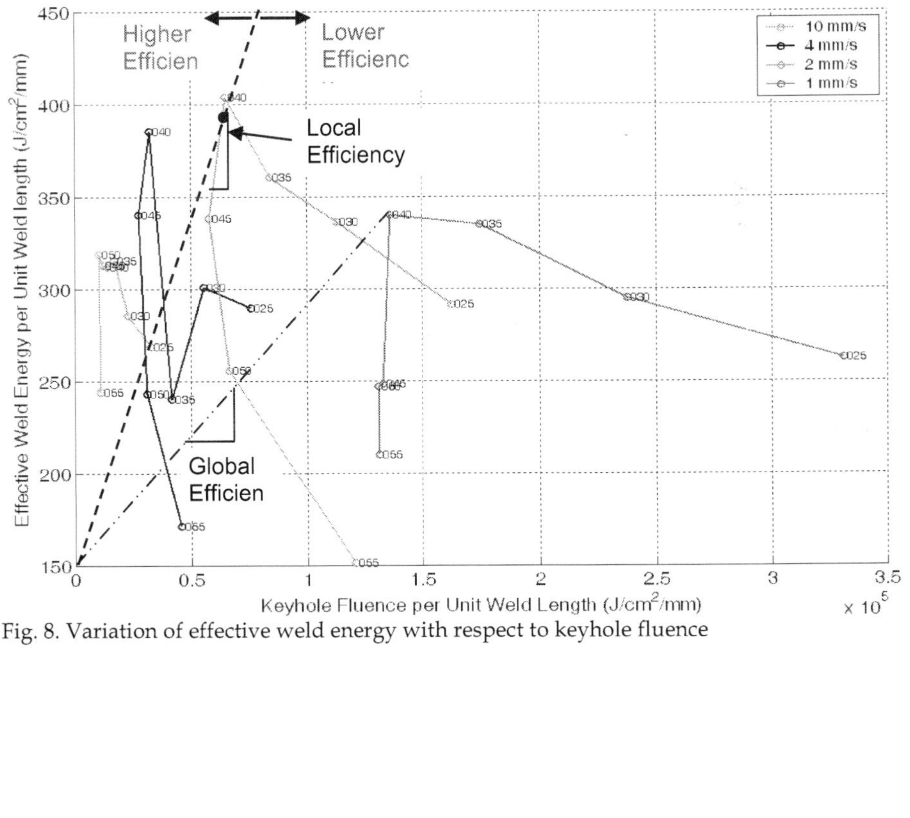

Fig. 8. Variation of effective weld energy with respect to keyhole fluence

During the process, a substantial portion of the absorbed power is conducted away. Therefore the relative change in global efficiency is of more interest than the actual number itself. Looking at the four speeds we observe that the global efficiency decreases slightly from the 10 mm/s data to the 4 mm/s and then slightly lower to 2 mm/s, but drops significantly at the 1 mm/s data. This is another indication that even though there is an increase in KF, the energy is not used as effectively to create a deep and narrow weld. This phenomenon, i.e. the process breakdown of laser welding at extremely low speeds, requires further investigation to explain the reasons behind this drastic change.
Local Efficiency: Another metric we can define to measure the efficiency between different focusing points for a specific processing speed is the "local efficiency." The slopes of these lines can be defined as a "local efficiency" which signifies how efficient the process is, as the focusing changes and the KF increases. In other words it is the ratio of the change in EWE and the change in KF for a particular processing speed (Equation 13).

$$\text{Local Efficiency} = \frac{dEWE}{dKF} \tag{13}$$

It is apparent that the local efficiency is only positive between the weld with the best focusing position and the one that is focused slightly deeper. Again, this is evidence that increasing the KF is not enough to create a good weld, it has to be deposited correctly. It seems that this happens because, when the beam is focused too deep, the incident power density at the surface of the workpiece is not sufficient to create enough vaporization to sustain a keyhole. The process, therefore, switches to conduction welding mode, where a weld is created solely from melting, resulting in a shallow and wide weld. Conversely, if the focus is too close to the surface of the workpiece, the process is again inefficient because the power density at the bottom of the keyhole is too low and cannot sustain the vaporization required for a deeper keyhole.

4. Inherent Process Instability

Table (2) lists the EWEs, global efficiencies, and power efficiencies for the data shown in Figure (8). They clearly showed that the welding became less efficient (from 2.25 % to 0.25 % as the speed drops from 10 mm/s to 1 mm/s) and the quality and aspect ratio of the weld started deteriorating after the processing speed was decreased below 2 mm/s. Large porosities were observed in these very low speed welds in aluminium. This phenomenon is denoted as inherent process instability. In the following sections, the potential contributors of this instability are examined.

Speed (mm/s)	EWE (J/mm^3)	Power Efficiency	"Global" Efficiency
10	312.0	~ 90 %	~ 2.25 %
4	385.0	~ 90 %	~ 1.19 %
2	403.4	~ 90 %	~ 0.62 %
1	340.0	~ 68 %	~ 0.25 %

Table 2. Decrease in EWE and efficiency when speed drops to 1 mm/s.

4.1 Laser Power Distribution

Assuming the laser welding process reaches a quasi-steady state condition, the power distribution, rather than energy distribution, is used to break down the laser power into several components:

$$P_{in} = P_{weld} + P_{cond} + P_{evap} + P_{ref} + (1-\alpha)\cdot P_{vap/plasma} + P_{scat} \quad (14)$$

where P_{in} is the input power from the laser radiation, P_{weld} is the power used to form the weld similar to the definition of EWE, P_{cond} is the power absorbed by the workpiece and then conducted away into the bulk material, P_{evap} is the power absorbed to produce vapor/plasma, P_{ref} is the power reflected away by the workpiece, $P_{vap/plasma}$ is the power absorbed by the vapor/plasma plume hovering above the workpiece, P_{scat} is the power which is scattered away by the vapor/plasma, and α is the fraction of the power absorbed by the vapor/plasma that is re-radiated on the workpiece and absorbed by the workpiece. All six terms on the right hand side of Equation 14 are unknown. No attempt is made to solve this equation or to measure each of these unknowns precisely. Dividing both sides by P_{in}, Equation 14 becomes

$$1 = \frac{P_{weld}}{P_{in}} + \frac{P_{cond}}{P_{in}} + \frac{P_{evap}}{P_{in}} + \frac{P_{ref}}{P_{in}} + \frac{(1-\alpha)\ P_{vap/plasma}}{P_{in}} + \frac{P_{scat}}{P_{in}} \quad (15)$$

Each term on the right-hand side of Equation 15 represents the respective percentage of the laser input power. Based on Table (2), it has been confirmed that P_{weld}/P_{in} drops significantly, which should result in changes in some of the rest of the five terms. Therefore, instead of determining the precise value of each term in Equation 15, the attempt is made to determine how the rest of the five terms change as the welding speed drops from 10 mm/s to 1 mm/s.

4.2 Laser Beam Reflectivity Measurements

Among those losses in Equation 14, we first investigate the reflective loss, P_{ref}, to determine if it is a major factor to cause process instability.

Figure (9) shows the reflected laser beam measured by a photodiode at different welding speeds. For each test, the laser beam is first irradiated at the target, remaining stationary for 5 seconds, before actual welding started at speeds from 10 mm/s to 1 mm/s. In every plot, during this 5 second duration in the beginning of the process, there is a large, sudden increase in intensity which gradually dies off to almost a zero state. The substantial reflected laser radiation in the beginning is due to the beam being reflected by the flat surface of the workpiece. As a keyhole forms, the laser beam penetrates deeper into the workpiece and eventually is absorbed by multiple reflections by the keyhole wall. As a deep keyhole acts like a black body, trapping nearly 100 percent of the laser beam, no reflected laser radiation is detected after about 2 seconds, when the keyhole becomes deep enough. After 5 seconds have passed, the workpiece is then translated at the specified welding speed.

When the workpiece begins to move, the reflected signal appears, again, as a series of high frequency spikes, but with a low average intensity, between 0.25 and 0.4 (a.u.). This is pretty common for processing speeds 10, 4 and 2 mm/s. The spikes in the signals can be attributed to the fact that when the laser beam moves over a solid front, the reflectivity suddenly increases; as soon as that happens, the keyhole is created and it absorbs the beam completely, which causes a decrease in measured intensity. This pattern repeats at a frequency of approximately 30 Hz.

The 1 mm/s processing speed shows a significantly different type of reflected signal, where the spikes are not as strong, but the average intensity of its signal was significantly larger than the other processing conditions' signals, namely approximately 1.4 (a.u.).

On the other hand, when the processing speed is 1 mm/s, the beam does not necessarily move over a solid front, since the molten pool is large. This causes a reduction in the strength of the spikes, because the molten metal has a higher absorptivity than the solid. The increase in average intensity can be explained by the fact that the keyhole is much shallower and wider. Therefore, much more of the laser beam is reflected back, due to the inability of the shallow keyhole to "trap" the laser beam entirely through multiple reflections.

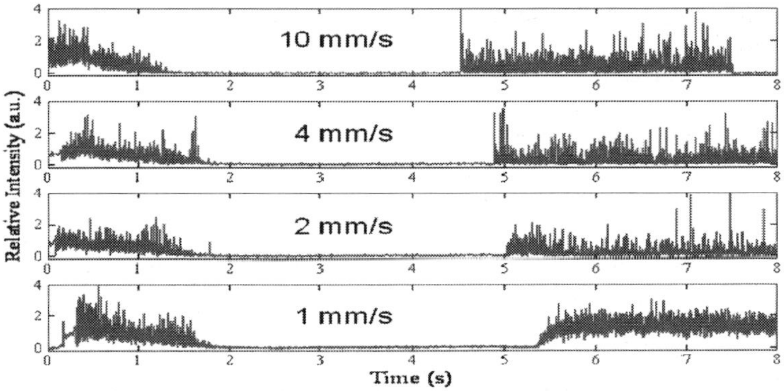

Fig. 9. Measured reflected laser radiations at an angle above the workpiece at different welding speeds.

From Figure (9) it is concluded that the power loss due to reflectivity increases significantly when the speed is lowered down to 1 mm/s. Therefore, the term, P_{ref} / P_{in}, in Equation 15 increases. However, this does not necessarily prove that this increase in reflectivity is the cause of the reduction of P_{weld} / P_{in} in Equation 15.

4.3 Vapor/Plasma Characterization

We conducted a spectroscopic analysis to identify the vapor/plasma effect on the distribution of laser power. The key question is to find out if the vapor/plasma plume, which hovers over the keyhole, is optically "thick" enough to absorb or scatter the laser beam, resulting in reduced laser radiation to reach the work piece. In order to have sufficient signal to noise ratio, we conducted spot welding with a peak power of 1,500 W, which is five times higher than the normal CW welding at 300 W. Even with this much higher peak power, only three Cr I lines (Figure 10) could be detected in the spectroscopic experiment using the Ocean Optics HR4000 spectrometer. Even so, it was deemed worthwhile to come up with an electron temperature estimation using the Boltzmann plot method. Based on the slope of the fitted line, the resulting temperature estimation is approximately 1,200 degrees Kelvin. This value is well below the vaporization temperature of aluminium (~ 3,275 Kelvin), which indicates that the calculation is not valid. This is probably because the upper energy levels of the measured chromium lines are very close to each other, thus introducing significant errors. Nevertheless, based on this calculation and the fact that the rest of the Cr I lines with higher upper energy levels could not be detected, it is likely that the temperature of the vapor/plasma is fairly low.

To confirm the above finding, we reviewed available literature on the vapor/plasma temperature for Nd:YAG (similar wavelength as fiber laser) laser welding of aluminium alloys. Kim and Matsunawa (1996) used a pulse shapeable YAG laser with irradiations of up to 1 MW/cm^2 on 5000 series aluminium alloys and determined that the vapor/plasma plume was very weakly ionized, with approximate temperatures around 3280 K (barely above the vaporization temperature of aluminium) and electron densities of approximately $1.85 \cdot 10^{13}$ cm^{-3}. Kim et al. (2004) did a similar study and found similar results for even higher irradiations (~ 32 MW/ cm^2), namely the vapor/plasma temperature was very close to the boiling point of aluminium.

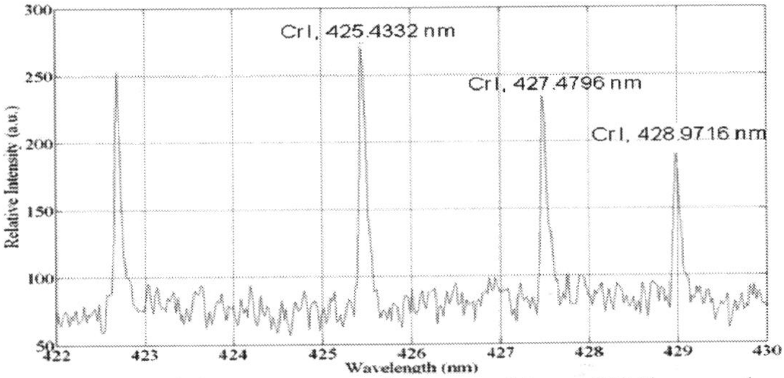

Fig. 10. Aluminium vapor/plasma spectrum for 1 ms pulsing at 100 % power (peak power ~ 1,500 W).

Another group (Lenk et al., 1996) experimented with a Q-switched Nd:YAG laser, operating at power densities of 300 MW/ cm^2, determined electron temperatures of approximately 14,000 K and electron densities of $3 \cdot 10^{16}$ cm^{-3}, and concluded that they "are not high enough for significant absorption by inverse bremsstrahlung." There have been several other studies (Barthélemy et al., 2005, Lu et al., 1999, Knudtson et al., 1987) that have found electron temperatures ranging from 5,000-15,000 K and electron densities up to the order of 10^{18} cm^{-3}, all reaching the same conclusion, namely that IB absorption is not significant. Therefore, it is reasonable to say that the vapor/plasma can be considered to be optically thin. This leads to the conclusion that the last two terms of Equation 15 are of low values and, therefore, their changes, if any, should not be a major factor in the reduction of P_{weld} / P_{in}. This conclusion is different from CO_2 laser welding in which vapor/plasma can grow larger and hotter, and becomes optically thick for the CO_2 laser beam (Tu et al., 2002 and 2003).

The remaining two terms in Equation 15, P_{cond} / P_{in} and P_{evap} / P_{in}, cannot be identified separately. Their changes are considered in the next section.

4.4 The Probable Cause of Process Instability

One probable cause of the process instability which can contribute to increase in P_{cond} / P_{in} and P_{evap} / P_{in}, resulting in the increase in P_{ref} / P_{in}, is if the laser beam mainly irradiates at the molten pool at very low speeds. The molten pool absorbs a large portion of the beam energy near the surface, subsequently transferring the energy into the bulk material via convection, conduction, and evaporation, increasing both P_{cond} / P_{in} and P_{evap} / P_{in} in Equation 15. Thus, this energy is wasted, as it is not used to create keyhole. As a result, the keyhole may become unstable, leading to un-quasi-static behaviors of the welding process (as seen in the top view of the weld in Figure (1)). Consequently, the welding process becomes inefficient and the welds become shallow, uneven, and wide (Figure (5) and Table (2)). As the weld becomes shallow, the laser beam is more easily reflected, resulting in the increase of P_{ref} / P_{in} in Equation 15. As the speed is further reduced, the excessive energy absorbed by the molten pool can also lead to boiling, resulting in large porosities. Readers are referred to Paleocrassas and Tu (2010) for additional tests to investigate this probable cause of instability.

4.5 Significance of the 1 mm/s Threshold

The above analysis helps explain the phenomenon that occurs when the laser welding processing speed drops below a certain low speed threshold. However, a valid question still remains: "why does the laser welding process break down at the particular speed range of about 1 mm/s?"

When the laser beam first irradiates the solid aluminium, it starts to melt and propagate outward with a certain speed. If the welding speed is less than this melting front speed, the laser would fall behind and irradiate on the molten pool. This speed will decrease non-linearly as the surface area surrounding the molten pool also increases non-linearly and therefore the change in molten pool volume.

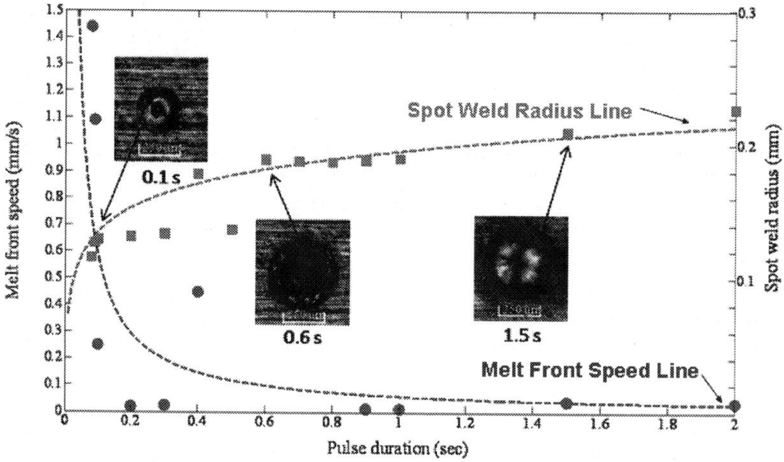

Fig. 11. Change in average melt front speed, as pulse duration is increased

An experiment was conducted to determine the melting front speed. Stationary laser pulses at 100% power, were shot at a AA7075-T6 target with pulse durations ranging from 80 ms to 2000 ms. Figure (11) shows the resulting plot of the melt front speed along with the original data (spot weld radii) for each pulse duration condition.
These melt front speeds are estimated using Equation 16:

$$V_{\text{melt front, } i} = \frac{\left(r_i - r_{i-1}\right)}{\left(t_{p,i} - t_{p,i-1}\right)} \tag{16}$$

where r_i is the radius of the spot weld corresponding to the pulse duration $t_{p,i}$. The first velocity point was calculated by dividing the first spot weld by the corresponding pulse duration. The first spot weld was observed after a laser pulse of 80 ms was irradiated. This was the minimum pulse duration required to produce a spot weld at 300 W (100 %) input power. The speed at which the spot weld formed and propagated was just over 1.4 mm/s.
The experimental results show that from 80 - 90 ms, the melt front speed is estimated to be 1.1 mm/s and for higher pulse durations it drops significantly. The melt front speed at the early stages of molten pool propagation is clearly higher than the 1 mm/s processing speed, which indicates that it initially advances faster than the processing speed. The melt front speed will gradually slow down to a speed below 0.1 mm/s which means that the laser beam will eventually surpass the molten front and irradiate on the solid.
This observation further confirms the fact that at speeds of 1 mm/s or lower, the laser will irradiate directly over the molten pool for a certain period of time, as opposed to the faster processing speeds, which will typically stay slightly ahead of the molten front the whole time. By irradiating over the molten pool the energy will be absorbed more efficiently, further increasing its size and reducing the amount of energy used to maintain keyhole welding.

5. Applications to Aluminium Fatigue Crack Repair

In this section, the low speed laser welding of aluminium is applied to fuse fatigue cracks. There is a serious concern in the aviation industry because airplanes' lifting surfaces undergo millions of cyclic loads throughout their lifetime. After a certain amount of cycles, cracks start to form in the high stress concentration areas. Initially cracks propagate in a stable and predictable manner. After the crack exceeds a certain critical length, it will start growing much faster, in an unstable manner, eventually leading to brittle fracture and catastrophic failure (Sanford, 2003).
Currently, cracks are monitored between flights until they exceed a certain length well below the critical length, after which the cracked part is replaced. This method is very costly due to the loss of flight operation time of the aircraft, as well as the part replacement labor costs.
Reinforcing cracked aluminium structures with composite patches has been recognized as an efficient and economical method to extend the service life of cracked aluminium components (Baker and Jones, 1988; Sun et al, 1996; Daghyani et al, 2003). To further enhance the effectiveness of composite patches, it is envisioned that the crack can be first fused by laser welding to remove the high stress concentration at the crack front before

applying the composite patch (Sun, 2008). The stress intensity factor could be reduced significantly if the fusion is sound.
One challenge is that cracks never propagate in straight lines. This means that the welding speed needs to be reduced and changed in order to trace the crack. Attempts to operate high speed welding would require changing directions abruptly, which will require high accelerations and decelerations. This "jerky" motion in laser welding could lead to inconsistencies in weld width and penetration, thereby compromising the integrity of the weld.
Also, as most of the laser welding experience focused on thick-sheet partial penetration welding, the experience cannot be directly transferred to thin-sheet full penetration welding. In addition, the crack may be skewed across the cross-section of the plate, making it different from welding prepared butt joints.
In this section, the feasibility of the envisioned fusion repair is investigated.

5.1 Cracked Sample Preparation

Fatigue cracks were generated in 2"x 10" thin aluminium sheets (AA7075-T6) with 800 μm thickness using an MTS tensile testing machine. A notch was first machined on one side and then cyclic loads were applied to produce hairline cracks. Depending on the loading, it usually took about 2-3 hours to generate one sample with a 1-1.5" long hairline crack. Care had to taken so that the sample will not crack through and break. Due to the high cost, a total of 20 samples were generated for this study. These hairline cracks were not in straight lines and many of them are skewed across the cross-section as described above.

5.2 Focusing and Workpiece Flatness

Focusing Position: The accepted practice for focusing in thick-sheet partial penetration welding is to focus the beam into the workpiece, without exceeding the maximum penetration that can be achieved for the corresponding power. This is because power density is highest at the focusing plane of the laser beam and by focusing it deep into the material, we can ensure that it will help keep drilling into the molten pool through evaporation of the metal. In the meantime, the power density at the surface of the workpiece should be maintained above the threshold required to melt the solid.
For thin-sheet full penetration conditions, focusing becomes more complicated. If the focus is placed at the bottom surface of the workpiece, violent evaporation may occur, due to the absence of surrounding material to conduct the excess heat away. This may result in a severe disruption of the molten pool, where molten metal gets ejected out of both sides of the workpiece, leading to crude laser cutting rather than laser welding. A similar result will occur if the focus is placed at the top surface of the workpiece. In this case, the best way to focus the laser beam is to create a power density, at the top of the surface, that is just high enough to sustain melting. This will shift the focus to some distance below the workpiece, thereby reducing the power density at the lower surface enough to prevent this violent ejection of the molten pool as well as drop-out.
Flatness Requirement: Because thin sheet aluminium welding is highly sensitive to focusing changes, it is essential to have tight control over the plate's flatness and its position with respect to the laser head. A dial gage was used to check the flatness and the height of the workpiece and a fixture was designed to prevent warping due to thermal distortion during welding.

For thick-sheet, partial penetration welding, changes in focus of about 25-50 μm do not yield significantly different penetration results because the majority of the energy is conducted away and therefore a slight change in power density will not translate into a big percentage of penetration loss.
However, when welding thin-sheets, the slightest increase or decrease in power density could mean the difference between a very violent welding process containing many defects or a weld with insufficient penetration.

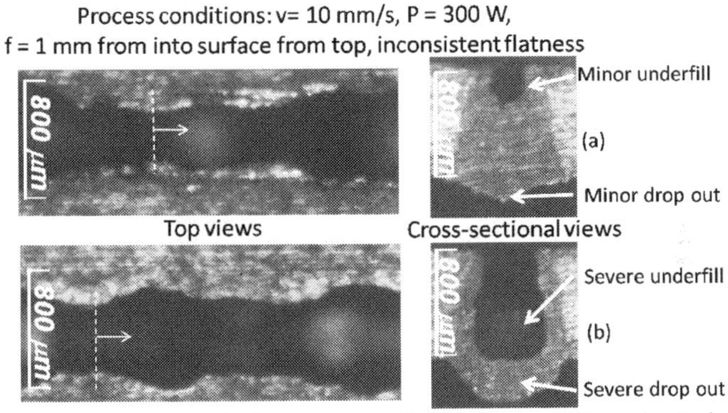

Fig. 12. Top and cross-sectional views of insufficient flatness, thin-sheet, full penetration welds

For example, Figure (12) shows two thin-sheet welds, 1.5 mm apart, on a workpiece with insufficient fixturing. This caused poor flatness, which was about 25 μm for a span of about 50 mm. The resulting welds did not have the desired weld characteristics from beginning to end; namely, consistent width, smooth top and bottom beads and minimal underfill. The welds would start out having the desired consistency and shape and after 20 or 30 millimeters would transition into a violent, unstable process. The welds in Figure (12) show some undesirable characteristics. For both welds, the widths are inconsistent. In Figure (12a) the weld shows a cross-section with minor underfill and drop-out, whereas in Figure (12b) the cross-section reveals significant drop-out and underfill.
Because of this high focusing sensitivity, it is very important that extra care is taken to ensure the flatness of the workpiece is within a tolerance of 10 μm. Also, the relative height between each workpiece needs to be checked to ensure that the focus will lie in the same location. Variations in thickness between workpieces could also interfere with process repeatability if they are larger than the aforementioned tolerance. Due to this high focusing sensitivity, for practical implementation of fusion repair, auto-focusing technology must be used because the structure component may not be flat.

5.3 Crack Tracing

As cracks are not in straight lines, the laser beam must trace the crack precisely. In this study, an off-line method is used for crack tracing. First, the crack sample to be repaired is mounted onto the fixture and a guide beam is used to determine the position coordinates of many points on the crack. A line is then fitted by connecting these points. Usually about 20 to 30 points are identified to trace a crack of 1-1.5" length. This fitted line is then uploaded to the controller of the x-y table for position and welding speed control. Linear motors are used as the driving motors for the x-y table. Figure (13) shows the variation of speed of the x-y table as the crack is being traced.

Once they are finished tracing the crack, the linear motors return to their original positions. We can compare the smoothness of the speed during crack tracing with that of the return, which follows a straight line. The average speed during the crack tracing is 9.34 mm/s and the standard deviation is approximately 2.4 % of the average value. In comparison, during the straight line return the average speed is 9.76 mm/s and the standard deviation is less than 1 % from the average value. Therefore, we can be assured that at 10 mm/s the processing speed stays relatively consistent, without having large deviations during the changes in direction.

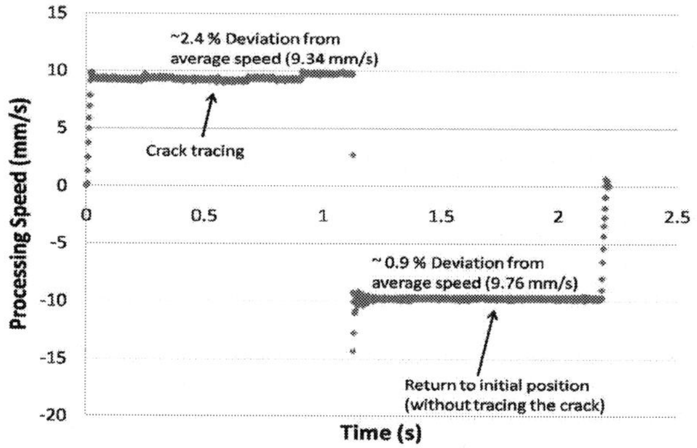

Fig. 13. Variation of the linear motor speed as it traces the crack; comparison with the return of the linear motor to its original position by following a straight line.

5.4 Process Parameter Modification for Thin-Sheet Welding

The first step in transitioning from thick-sheet, partial penetration, bead-on-plate laser welding to thin-sheet, full penetration crack repair, is to find out what parameters combination will yield the best weld results. Several bead-on-plate experiments were performed with the topics discussed in Section 5.2 in mind. The final process parameters are summarized in Table 3.

Initially, the partial penetration welding process parameters were selected as a starting point. Since the transferred welding conditions yielded results resembling laser cutting, it made the most sense to first increase the distance between the nozzle and the workpiece. Doing so, would decrease the pressure exerted on the molten pool by the nitrogen gas. This changed the process from resembling laser cutting to laser welding. By raising the nozzle (3 mm from the surface of the workpiece), the pressure applied by the nitrogen gas to the molten pool was reduced significantly, while still maintaining its shielding from the atmosphere.

Also, the inconsistent results (shown in Figure 2) confirmed that thin-sheet, full penetration welding is very sensitive to small changes in flatness and workpiece height. Therefore, the fixturing method had to be adjusted to improve the flatness tolerance. Since workpiece warping was also a concern, the workpiece was "sandwiched" between two thicker aluminium plates, containing a rectangular slot in the area where the workpiece was to be welded. Not only did this prevent warping, but it also significantly improved the flatness of the workpiece.

	Thick-Sheet, Partial Penetration	**Thin-Sheet, Full Penetration**
Output mode, Power (Watts)	CW, 300	CW, 300
Focus position from top surface (mm)	1	1.2
Speed (mm/s)	4	10
Nozzle position from top surface (mm)	1	3
Flatness tolerance per 50 mm span (μm)	> 25	> 10

Table 3. Process parameter comparison between thick-sheet, partial penetration and

thin-sheet, full penetration laser welding conditions

By ensuring the flatness to be within 10 μm and adjusting the focusing based on the relative height of the workpiece, the resulting welds turned out to be much more consistent from beginning to end and at the ideal focusing position (1.2 mm from the top surface of the workpiece). The ideal focusing position, as opposed to thick-sheet welding, resulted in deeper than the desired weld penetration (800 μm), which would be at the bottom surface of the workpiece. When focused at the bottom surface, the power densities were too high at both, the top and bottom surfaces, causing a violent process and severe underfill.

The speed also had to be changed, because the excess energy led to a larger molten pool, causing larger drop-out. At 10 mm/s, full penetration was achieved with minimal drop-out.

5.5 Crack Repair by Fusion

A single-pass laser welded crack is shown in Figure (14). The weld was created with minimal underfill and drop-out and with a consistent width and defect free weld bead. Double pass crack repairs, offset by 600 μm (center to center), were used to repair significantly skewed cracks that required larger weld width.

5.6 Mechanical Testing for Determining Repaired Crack Strength

Tensile tests were conducted for single and double pass welds, as well as unwelded (baseline) AA 7075-T6 sheets. Both the ductility and the ultimate strength were recorded. The results are shown in Figure (15). Four samples for each condition were tested due to limited numbers of crack samples and due to the fact that the results were highly repeatable (see below).

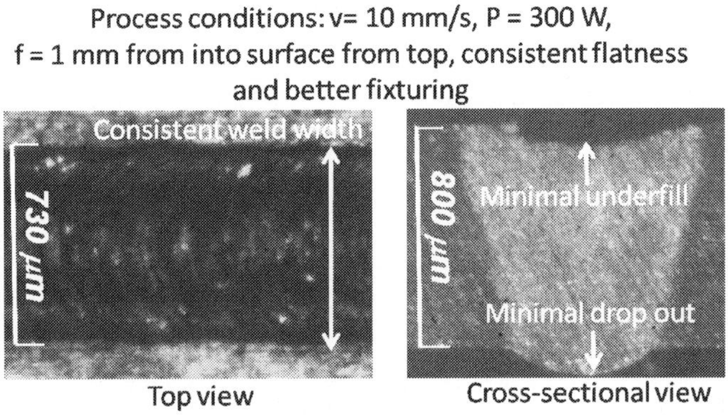

Fig. 14. Single-pass laser welded fatigue crack

The average UTS of the base AA 7075-T6 is 579 MPa, which is identical to the documented value (Sanford, 2003). As shown in figure 7, the lowest strength is 571 MPa and highest is 584 MPa, equivalent to –1.4 % to +0.9% in variation from the average value. The average UTS of the single-pass weld is 430 MPa, which is 74 percent of the base alloy's strength. The lowest is 411 MPa and the highest is 443 MPa, equivalent to –4.4% to +3.0% variation from the average. The average UTS of the double-pass weld is 395 MPa, which is 68 percent of the base alloy's strength. The lowest is 379 MPa and the highest is 422 MPa, equivalent to -4.1% to +6.8 % variation from the average value. These values are extremely encouraging because the UTS for AA 7075-O, which is the same alloy without heat treatment, is only 220 MPa (Sanford, 2003). The single pass weld is 95% higher than that of the untreated alloy, while the double pass weld is 72% higher. These high strength values indicate that the laser did not completely destroy the heat treatment temper. The strength is probably retained due to the use of the fiber laser which, with its highly concentrated energy deposition, produces narrow welds to allow for fast cooling by the bulk material.

As the results of the UTS strength results were highly repeatable, no more samples were tested. For the tensile tests to determine UTS, four samples with highly repeatable results are justified. However, it should be noted that if fatigue life were to be tested, more samples would be needed as fatigue life tests usually exhibits wider statistical distribution. In this study, no attempt was made to test the fatigue life as crack fusion alone should not be considered a viable repair technique unless it is used together with composite patches. It should be clear to see the benefit of the because without it crack fusion, the composite patch is bonded to a part with zero UTS at the crack region and with a high stress intensity factor at the crack front. On the other hand, with crack fusion, the UTS is 172% - 195% that of the untreated alloy and there is no crack front with a high stress concentration.

In addition to tensile tests, the ductility was also measured. The average elongation for the base alloy was measured to be 4.9 percent, while the single pass weld was just under 1 percent and the double bead-on-plate was approximately 0.5 percent. The significant drop in ductility was probably due to the rapid cooling of the weld as described earlier. For the double-pass weld, the second pass re-melted some weld of the first pass, which probably caused the ductility to drop even further. This effect has been observed before (Verkat et al.,

1997) in laser welding and can be improved slightly with the appropriate addition of filler wire (Yoon and Wallach, 2008).

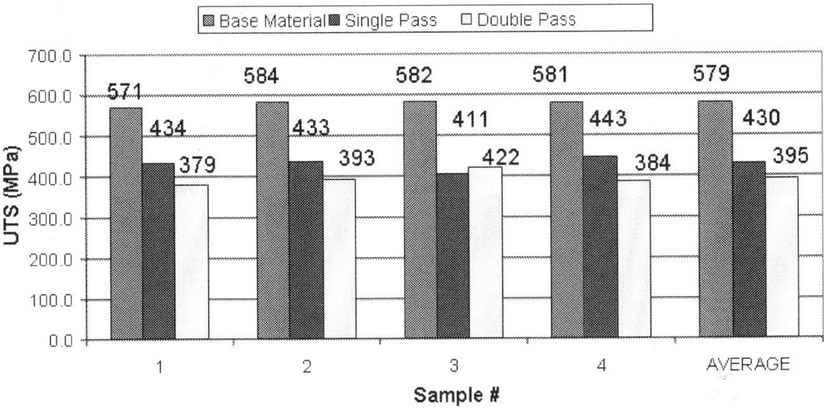

Fig. 15. Base AA 7075-T6 compared to single and double pass bead-on-plate welds

6. Conclusion

In this chapter, topics related to extending fiber laser welding of aluminium in the low speed range were discussed. General topics, such as the properties of aluminium and welding defects, review of high speed laser welding of aluminium, and fiber laser characteristics and optical setups for safety, were first reviewed. Recent research results on the modelling and validation of laser welding of aluminium, experimental characterization of low speed welding processes, and the instability phenomenon and its probable causes were then presented. Finally, an application of low speed fiber laser welding of aluminium for repairing fatigue cracks was discussed.

The difficulty in extending fiber laser welding of aluminium to low speeds is that there exists a low speed threshold (1 mm/s), below which the process becomes unstable. This threshold appears to be related to the molting front propagation speed, which was found to be approximately 1.1 - 1.4 mm/s. Therefore, when the welding speed is less than 1 mm/s, the laser irradiates over the molten pool longer, resulting in less efficient energy coupling for forming high aspect ratio welds. One possible solution to avoid this problem is to employ pulsed welding to reduce the laser power when it is irradiating over the molten pool and to resume the laser power when it moves over a solid surface. More research is needed to devise suitable pulsing schemes based on this instability observation.

Additionally, more studies are needed to understand the overheating of the molten pool and the keyhole stability, as well as the keyhole's interaction with the melting front. It is also important to study how the laser energy density and the currently observed low speed threshold, 1 mm/s, are related. It would be useful to devise new experiments to quantify each term in Equation 14 to further understand the overheating of molten pool.

Finally, the most challenging factor for successful crack fusion is related to the plate flatness which critically affects the relative position between the laser focusing point and the plate location. It has been found that the plate flatness needs to be maintained to be within +- 10 μm for the welding process to be consistent. In practical implementation, an auto-focus system must be developed to maintain proper focusing so that crack fusion can be achieved even for curved plates. Another important technical difficulty is related to crack tracing. An automatic crack tracing system needs to be developed for practical implementation.

7. References

Allen, C. M., Verhaeghe, G., Hilton, P. A., Heason, C.P., Prangnell, P.B. (2006) Laser and hybrid laser-MIG welding of 6.35 and 12.7mm thick aluminium aerospace alloy. *Materials Science Forum*, 519-521, (2), pp.1139-1144.

Baker, A.A. and Jones, R. (1988) *Bonded repair of aircarft structures*, Martinus Nijhoff Publishers.

Barthélemy, O., Margot, J., Chaker, M., Sabsabi, M, Vidal, F., Johnston, T.W., Laville, S., Le Drogoff, B. (2005) Influence of the laser parameters on the space and time characteristics of an aluminium laser-induced plasma. *Spectrochimica Acta Part B*, 60, pp. 905-914.

Brown, R. T. (2008) Keyhole welding studies with a moderate-power, high brightness fiber laser. *Journal of Laser Applications*, 20 (4), pp. 201-208.

Cao, X., Wallace, W., Immarigeon, J.-P., and Poon, C. (2003a) Research in laser welding of wrought aluminium alloys. I. laser welding processes. *Materials and Manufacturing Processes*, 18(1), pp. 1-22.

Cao, X., Wallace, W., Immarigeon, J.-P., and Poon, C. (2003b) Research in laser welding of wrought aluminium alloys. II. metallurgical microstructures, defects, and mechanical properties. *Materials and Manufacturing Processes*, 18 (1), pp. 23-49.

Carslaw, H. S., Jaeger, J. C. (1962) *Conduction of Heat in Solids*, 2nd edition, Oxford: Clarendon, pp. 390.

Cieslak, M. J. (1992) Phase transformations in weldments: new materials and new perspectives. *3rd Int. Conf. on Trends in Welding Research*, Gatlingburg, TN, pp. 229.

Dausinger, F., Rapp, J., Beck, M., Faisst, Hack, R., Hugel, H. (1996) Welding of aluminium: a challenging opportunity for laser technology. *Journal of Laser Applications*, 8, pp. 285-290.

Dausinger, F., Rapp, J., Hohenberger, B., Hugel, H. (1997) Laser beam welding of aluminium alloys: state of the art and recent developments. *Proc. Int. Body Engineering Conf. IBEC '97: Advanced Technologies & Processes*, 33, pp. 38-46.

Daghyani, H.R., Sayadi, A., and Hosseini Toudeshky, H. (2003) Fatigue crack propagation of aluminium panels repaired with adhesively bonded composite laminates. *Proceedings of the institution of Mechanical Engineers, Part L: Journal of Materials: Design and Applications*, pp. 291-293.

Duley, W. W. (1999) *Laser Welding*, 1st edition, John Wiley & Sons, Inc., pp. 4-65.

Freudenstein, S., Cooper, J. (1979) Stark broadening of Fe I 5383 Å. *Astron. Astrophys.*, 71. pp. 283-288.

Fabbro, R. and Chouf, K. (2000) Dynamical description of the keyhole in deep penetration laser welding. *Journal of Laser Applications*, 12 (4), pp. 142-148.

Industrial Laser Solutions (2005)Jan.

Ion, J. C. (2000) Laser beam welding of wrought aluminium alloys. *Sci. Technol. Weld. Joining*, 5 (5), pp. 265-276.

IPG, Inc. (2003) 300W single-model fiber laser operation manual.

Katayama, S., Mizutani, M. (2002) Laser weldability of aluminium alloys. *Trans. JWRI*, 31 (2), pp. 147-155.

Katayama, S., Mizutani, M., Matsunawa, A. (2003) Development of porosity prevention procedures during laser welding. *Proc. of SPIE*, 4831, pp. 281-288.

Katayama, S., Nagayama, H., Mizutani, M., Kawahito, Y. (2008) Fiber laser welding of aluminium alloy. *Keikinzoku Yosetsu/J. of Light Metal Welding and Construction*, 46 (10), pp. 34-43.

Kim, J.D. and Matsunawa, A. (1996) Plasma analysis in laser welding of aluminium alloys. *International Institute of Welding*, pp. 1-9.

Kim, J.D., Oh, J.S., Lee, M.H., Kim, Y.S. (2004) Spectroscopic analysis of plasma induced in laser welding of aluminium alloys. *Material Science Forum*, 449-452, pp. 429-432.

Knudtson, J.T., Green, W.B., Sutton, D.G. (1987) The UV-visible spectroscopy of laser produced aluminium plasmas. *J. Appl. Phys.*, 61 (10), pp. 4471-4780.

Kutsuna, M., Yan, Q. U. (1998) Study on porosity formation in laser welds of aluminium alloys (Report 2). mechanism of porosity formation by hydrogen and magnesium. *J. Light Met. Weld. Constr.*, 36 (11), pp. 1-17.

Lankalapalli, K. N., Tu, J. F., Gartner, M. (1996) A model for estimating penetration depth of laser welding processes. *J. Phys. D, Appl. Phys.*, 29, pp. 1831-1841.

Lenk, A., Witke, T., Granse, G. (1996) Density and electron temperature of laser induced plasma - a comparison of different investigation methods. *Applied Surface Science*, 96-98, pp. 195-198.

Lu, Y.F., Tao, Z.B., Hong, M.H. (1999) Characteristics of excimer laser induced plasma from an aluminium Target by spectroscopic study. *Jpn. J. Appl. Phys*, 38, pp. 2958-2963.

Mandal, N. R., 2002, Aluminium Welding, 1st edition, Narosa Publishing House, pp. 1-19 Martukanitz, R. P., Smith, D. J. (1995) Laser beam welding of aluminium alloys. *Proc 6th Int. Conf. on Aluminium Weldments, AWS*, pp. 309-323.

Matsunawa, A. (1994) Defects formation mechanisms in laser welding and their suppression methods. *Proc. of ICALEO*, pp. 203-219.

Matsunawa, A., Katayama, S., Fujita, Y. (1998) Laser welding of aluminium alloys— defects formation mechanisms and their suppression methods. *Proc. 7th Int. conf./INALCO '98: Joints in Aluminium*, Cambridge, pp. 65-76.

Miyamoto, I., Park, S.-J., Ooie, T. (2003) Ultrafine-keyhole welding process using single-mode fiber laser. *Proc. of ICALEO*: 203-212.

Molian, A. (2004) Private conversation.

Naeem, M and Lewis, S. (2006) Micro joining and cutting with a single mode fiber laser. *Proc. of PICALO*, pp. 400-405.

Oi, J.F., Tian, S., Chen, H., Xiao, R.S., Zuo, T.C. (2006) Slab CO_2 laser welding of 7075-T6 high strength aluminium alloy. *Zhongguo Jiguang/Chinese J. of Lasers*, 33 (SUPPL). pp. 439-444.

Paleocrassas, A.G. and Tu, J.F. (2007) Low-speed laser welding of aluminium alloy 7075-T6 using a 300-W, single-mode, ytterbium fiber laser. *Welding Journal*, 86 (6), pp. 179.s-186.s.

Paleocrassas, A.G. and Tu, J.F. (2010) Inherent instability investigation for low speed laser welding of aluminium using a single-mode fiber laser. *J. Material Processing Technology*, doi:10.1016/j.jamatprotec2010.04.002.

Poueyo-Verwaerde, A., de Frutos, A.M., Orza, J.M. (1993) Experimental study of laser induced plasma in welding conditions with continuous CO_2 laser. *J. Appl. Phys.* 74 (9), pp. 5773-5780.

Ramasamy, S., Albright, C. E. (2000) CO_2 and Nd:YAG laser beam welding of 6111-T4 aluminium alloy for automotive applications. *J. of Laser Appl.*, 12 (3), pp. 101-115.

Salminen, A. S., Kujanpaa, V. P., Moisio, T. J. I. (1994) Effect of use of filler wire on requirements of laser welded butt joints. *Proc. of ICALEO*: 193-202.

Sanford, R.J. (2003) *Principles of Fracture Mechanics*, 1st edition, Prentic Hall, pp. 386-387.

Steen, W. M. (2003) *Laser Material Processing*, 3rd edition, Springer-Verlag London Limited, pp. 61-106.

Sun, C.T., Klug, J., and Arendt, C. (1996) Analysis of cracked aluminium plates repaired with bonded composite patches. *AIAA Journal*, 54, pp. 369-374.

Sun, C.T., School of AAE, Purdue University, 2008, private conversation.

Tu, J.F., Inoue, T., Miyamoto, I. (2003) Quantitative characterization of keyhole absorption mechanisms in 20 kW-class CO_2 laser welding process. *J. Phys. D: Appl. Phys.* 36, pp. 192-203.

Tu, J.F., Miyamoto, I., Inoue, T. (2002) Characterizing keyhole plasma light emission and plasma plume scattering for monitoring 20 kW class CO_2 laser welding processes. *J. Laser Applications*, 14 (3), pp. 146-153.

Venkat, S., Albright, C.E., Ramasamy, S., Hurley (1997) CO_2 laser beam welding of aluminium 5754-O and 6111-T4 alloys. *Welding Journal*, 76(7), pp. 275.s-282.s.

Wagner, F. (2006) Laser beam welding with single mode fibre lasers. *Proc. Of PICALO*, pp. 339-343.

Weeter, L. (1998) Technological advances in aluminium laser welding, *Pract. Weld. Today*, 2 (1), pp.56-58.

Xu, L., Tian, Z., Peng, Y., Xiao, R., Yang, W. (2008) Microstructure and mechanical properties of high strength aluminium alloy laser welds. *Zhongguo Jiguang/Chinese J. of Lasers*, 35 (3), pp. 456-461.

Yoon, J.W., Wallach, E.R. (2008) CW CO_2 laser welding of Al-Mg alloys with filler wires. *Material Science Forum*, 580-582, pp. 539-542.

Yoshikawa, M., Kurosawa, T., Nakata, K., Kimura, S., Aoki, S. (1995) YAG laser welding of aluminium alloys. *Journal of Light Metal Welding & Construction*, 32 (9), pp. 15-23.

Zhao, H., White, D. R., DebRoy, T. (1999) Current issues and problems in laser welding of automotive aluminium alloys. *Int. Mater. Rev.*, 44(6), pp. 238-266.

4

Laser welding of aluminium-steel clad materials for naval applications

Roberto Spina and Luigi Tricarico
Dept. of Mechanical & Management Engineering - Politecnico di Bari
Italy

1. Introduction

Several electronic, naval, aeronautic and automotive components are made by different materials joined together in order to improve mechanical and functional properties. Functionalities provided by clad metals can be grouped into structural, thermal expansion management, thermo-mechanical control, electrical, magnetic, corrosion resistant, joining and cosmetic applications to cite as few (*Chen et al., 2005*). The demand for dissimilar material joints continuously grows because one material can provide only a small spectrum of chemical, physical and mechanical characteristics required for the investigated application respect to the bi- or multi-layer material joints. Moreover considerable weight savings can be achieved by using lightweight materials clad to strength ones directly. For these reasons, researchers and manufacturers continuously evaluate the application of traditional and/or advanced joining processes to clad dissimilar materials and obtain transition joints optimally. Focusing the attention on steel/aluminium joints and shipbuilding industry, the development of lightweight and fast-speed vessels requires a great number of aluminium/steel structural transition joints (STJs) in order to connect aluminium superstructures to the steel hull (*Chao et al., 1997*). Using this solution, the total weight of a ship is reduced due to the lighter aluminium superstructure. However, problems in service may occurred by relations at the atomic level between iron and aluminium and differences existing in physical and chemical properties of the base metals. One of the most undesired effect derives from the large electrochemical difference of 1.22 volts between iron and aluminium that causes a high susceptibility to both inter-crystalline and galvanic corrosion along the STJ interface.

Fusion welding processes, initially used to produce the aluminium/steel STJs with desired physical and mechanical features, are narrowly applied because the subsidiary precipitates and brittle Al/Fe inter-metallic phases, created during fusion and solidification and located along the interface, are severely exposed to corrosion, troubling joint cohesion (*Durgutlu et al., 2005*). The high heat input affects the different thermal properties of the two materials—thermal expansion, heat capacity and thermal conductivity—and may lead to very complex stress fields. Moreover, the heat input causes the lattice transformation and the formation of inter-metallic phases. In iron (cubic body-centred up to 911 °C) and aluminium (cubic face-centred) joints, the inter-metallic phases present a high hardness and low ductility. The

welding procedures of STJs must be carefully controlled in order to avoid disbonding during construction and/or failure during service. The thickness of the Al/Fe inter-metallic layer between parent materials plays an important role in obtaining joints with optimum performances. Thus, the thickness minimisation of Al/Fe inter-metallic phases represents one of the most important problems to solve. This is why all the heat-intensive processes used up until now have been designed to keep the formation of inter-metallic phases within tight limits or even to prevent them from occurring in the first place (*Bruckner, 2003; Chen & Kovacevic, 2004*).
Solid-state processes seems to be more likely for producing STJs because thin inter-metallic thicknesses are achieved. Processes normally employed are roll bonding, pressure welding, friction welding, ultrasonic welding, diffusion bonding and explosive welding (*Deqing et al., 2007*). Explosion-welding is a fast and efficient process to bond two or more different metals with satisfactory corrosive properties. The energy of an explosive detonation is used to create a metallurgical weld between dissimilar materials. In preparation, the cladding plate is placed over the backer plate with a small gap between the two, ground and fixtured parallel at a precise spacing. A measured quantity of a specifically formulated explosive is spread on top of the cladding plate. On detonation, the cladding plate collides progressively with the backer plate at a high velocity. This collision removes the contaminating surface films like oxides and absorbed gases in the form of a fine jet, bringing together two virgin metal surfaces to form a metallurgical bond by electron sharing. The detonation front then uniformly travels across the surface until the end of the plates (*Durgutlu et al., 2005; Bankers & Nobili, 2002*). The combination of surface cleaning and extreme pressure produces a continuous metallurgical weld (*Young & Banker, 2004*). Although the explosion generates intense heat, there is no sufficient time for the heat to conduct into the metals, avoiding bulk heating (*ASM Handbook Vol.6*). Furthermore, there are no changes in the metallurgical characteristics or specification compliance of the component metals.
The objective of the present research is the evaluation of the process feasibility of applying laser welding to explosion-bonded STJs for the final ship assembly. This paper reports results achieved for as-simulated laser welded conditions by imposing severe thermal cycles to specimen obtained from structural transition joints with time periods longer than those normally recorded during laser welding. Metallurgical and mechanical characterisation of heat treated specimen are performed to evaluate the influence of the heat treatments on final joint properties. The analysis was then extended to the bead on plate and double side/double square fillet T-joints.

2. Problem Position

From the chemical point of view, iron reacts with aluminium forming several Fe_xAl_y inter-metallic compounds, as the Fe-Al phase diagram shows (Figure 1).
Only small amounts of iron can be dissolved in aluminium and only small amounts of aluminium can be dissolved in iron. The $FeAl_2$, Fe_2Al_5, Fe_2Al_7 and $FeAl_3$ are Al-rich inter-metallic compounds while FeAl and Fe_3Al are Fe-rich inter-metallic compounds (Table 1). The presence of Al-rich inter-metallic phases must be accurately control to reduce their influence on joint performances, respect to the Fe-rich phases with higher toughness values. In fact the complex lattice structures and too high micro-hardness values (up to 800 HV or more) of Al-rich inter-metallic compounds can cause a high interface fragility.

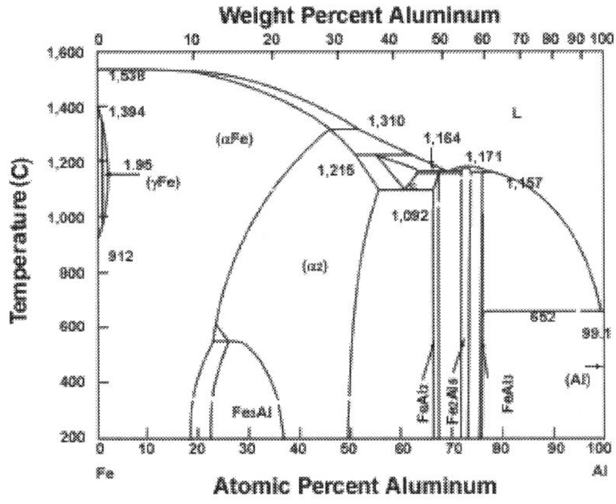

Fig. 1. Fe-Al phase diagram at equilibrium.

Phase	Al Content (atomic %)	Structure	Micro-hardness (HV)	Density (g/cm³)
Fe_3Al	25	Ordered BCC	250-350	6.67
FeAl	50	Ordered BCC	400-520	5.37
Fe_2Al_7	63	Complex BCC	650-680	NA
$FeAl_2$	66-67	Complex rhombohedral	1,000-1,050	4.36
Fe_2Al_5	69.7-73.2	BCC orthorhombic	1,000-1,100	4.11
$FeAl_3$	74-76	Highly complex monoclinic BCC	820-980	3.95

Table 1. Inter-metallic compounds (*Bruckner, 2003*).

The inter-metallic phases occurs at temperatures below the melting point of aluminium not only during explosion welding but also during fusion welding necessary to connect STJs to the steel hull and aluminium superstructure. The formation rate of the inter-metallic phases is diffusion-driven, thus dependent from time and temperature variables. For this reason, the evaluation of joint characteristics before and after fusion welding is necessary. The mechanical and metallurgical properties of the bond zone are determined by means of tests made in the following conditions (*American Bureau of Shipping, 2000*):

- As-clad condition: No preliminary treatment is given to the specimens to represent the as-clad product.
- As-simulated welded condition: A preliminary heat treatment is performed to the specimens in order to represent the product after welding.

The simulated welded specimens are heat-treated at 315±14°C (600F±25 °F) for 15 minutes, as suggested by American Bureau of Shipping. This temperature-time limit is settled-on by considering that a STJ exposed to a higher time or higher temperature than this limit can present a lower performance life than any as-clad explosion-welded STJs. However, two main considerations have to be made on this temperature-time limit such as: (i) the interaction between the temperature and time variables is not accurately evaluated and (ii) the welded condition normally is refereed to TIG or MIG welding processes, both characterised by high heat input profiles. The main hypothesis to verify is whether a very short time at a high temperature may sufficient to compromise and, in the worst condition, destroy bond properties of explosion-welding STJs, making the application of laser welding unfeasible. All above considerations shift the manufacturing problem form suppliers to shipbuilders. In fact, the interest of an STJ is its direct application instead of the way it is produced.

3. Specimen preparation of heat treatment

A tri-metallic transition joint was chosen for this study due to its industrial importance for the fast vessel construction. The rough material was the Triclad® STJ, a trade name of Merrem & la Porte for aluminium/steel transition joints, produced with open-air explosion welding. In particular, the selected rough material consisted of an ASTM A516 steel backer plate clad to an AA5083 flyer plate, with commercial purity aluminium (AA1050) interlayer plate placed between the former two. The presence of the AA1050 interlayer was necessary to improve STJ diffusion resistance with both iron and aluminium (*Bankers & Nobili, 2002*). The investigated STJ, realised by the supplier in compliance with specification ASTM B898 (*Chen et al., 2005*), was analysed with ultrasonic inspection from the manufacturer to confirm the whole weld interface integrity.
STJ specimens for metallographic and micro-hardness evaluation of about 28·13·3 mm^3 (Figure 2) were sectioned by using an abrasive wheel cut-off machine in transverse direction to the length of the rough plate, taking care of minimising the mechanical and thermal distortions of the Al/Fe interface. The specimen surfaces were smoothly ground to give a uniform finish and cleaned before putting them in the heat treatment oven. Each specimen was heated at specific temperature and time in compliance with the Central Composite Design (CCD) experimental plan and cooled outside the heat oven to the room temperature. The CCD design is a factorial or fractional factorial design (with centre points) in which "star" points are added to estimate curvature (*Montgomery, 2000*). The main CCD factors were the temperature and time, ranging between 100 and 500°C and 5 and 25 minutes respectively. The centre point of the CCD, replicated five times, was fixed at 300°C for 15 minutes, according to the limit of the as-simulated welded condition. The entire plan, shown in Table 2, also included the as-clad condition (ID 12) and near-melted condition of aluminium alloys (ID 13). The temperature of the heat treatment oven was rapidly reached by applying a high heat power and then maintaining this temperature for time sufficient to guarantee stationary conditions. The specimen was then inserted into the oven. This process was repeated for all specimens.

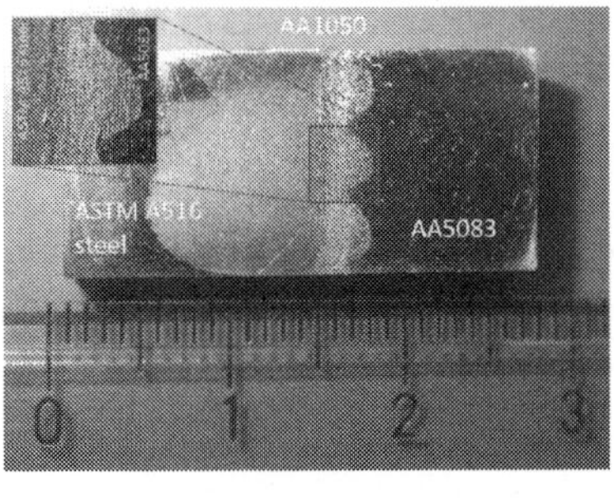

Fig. 2. Triclad® STJ.

Specimen ID	Temperature (°C)	Time (minutes)
1	100	5
2	100	25
3	500	5
4	500	25
5-6-7-8-9	300	15
10	300	0.86
11	300	29.14
12	17.16	15
13	582.4	15

Table 2. Central Composite DOE.

The heat-treated specimens were then prepared by grinding with 200 to 1000-grit silicon carbide papers, followed by mechanical polishing from 6-µm to 1-µm diamond abrasive on short nap clothes. Etching was then performed on the steel side of specimens with Nital solution (2 mL HNO_3 and 98 mL of C_2H_5OH) in distilled water for 15 seconds in order to highlight grain structures as well as inter-metallic phases. Keller's reagent (5 mL HNO_3 and 190 mL of H_2O) was applied for 15 seconds to aluminium side to point macro-structures.

4. Metallographic examination of heat treated specimens

The visual inspection of the STJ specimens by using the metallographic microscope was very useful to investigate modifications of Al/Fe interface due to heat treatments. The as-clad specimen was initially analysed and different areas were detected, as Figure 3 shows. Ripples with different morphological characteristics were located at the interface. These ripples, formed from the rapid quenching of melt regions caused by explosion, consisted of a mixture of different inter-metallic phases, as the grey scale variation suggests (Figure 3-A/B). Areas surrounding these ripples, and sometimes located inside them, exhibited the typical dendrite morphology of a slow cooling process after melting . Small-sized clusters of inter-metallic compounds, formed in not equilibrium cooling conditions, were also observed along the Al/Fe interface, pointing out the interface discontinuity. The cluster thickness ranged between 50 and 160 µm. Along the Al/Fe interface, the inter-metallic phases were detected as a discontinuous narrow band, less than 5 µm wide (Figure 3-C/D). This band was thick in areas submitted to high thermal gradient while it was very thin or absent in areas subjected to very low thermal gradient. The very brittle inter-metallic phases identified in this band at room temperature in the as-clad STJ were the $FeAl_3$ and Fe_2Al_5 on the aluminium side and steel side respectively, as confirmed by quantitative analysis (x-ray diffraction) performed with SEM (Figure 4). Further metallographic features were noted for the STJ base materials. The micro-structure of the ASTM A516 steel consisted of ferrite (lighter constituent) with pearlite (darker constituent), as Figure 3-E shows. Small-sized elongated grains, characteristic of the cold-working conditions, were observed near the interface while medium-sized regular ones were identified in areas immediately after the Al/Fe interface until to the specimen boundaries. As concern the AA1050 side, the micro-structure consisted of insoluble $FeAl_3$ particles (dark constituent) dispersed in the aluminium matrix (lighter constituent), as Figure 3-F shows. The morphology of these particles seemed to be not influenced by explosion welding.

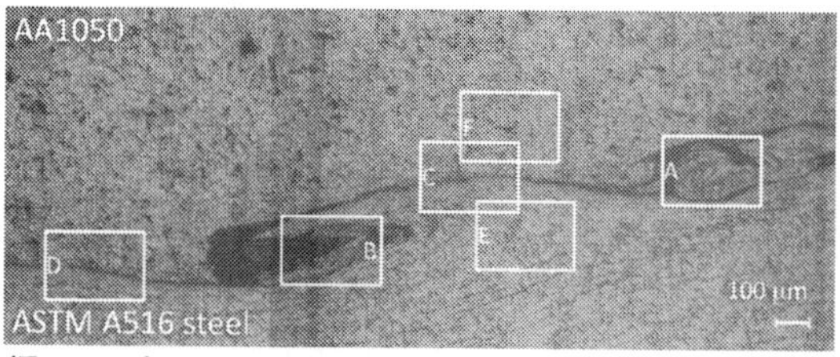

Fig. 3. Details of Al/Fe interface (as-clad condition).

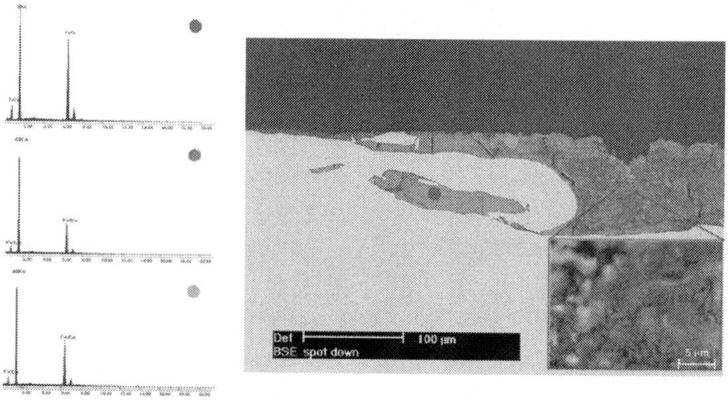

Fig. 4. SEM Observation.

The heat-treated specimens were accurately examined to measure changes in Al/Fe interface. A well-designed measurement process, divided into calibration, acquisition and computation steps, was applied to quantify the extension of the inter-metallic phases along the Al/Fe interface. The micro-structural measurements involved the use an optical microscope connected to a digital camera and a computerised image tool. At the end of the acquisition process, the entire Al/Fe interface of the specimen was captured by shooting multiple images at different locations, performing the brightness/contrast adjustment, joining them in a single frame and finally over-laying a 100 µm grid (Figure 5).

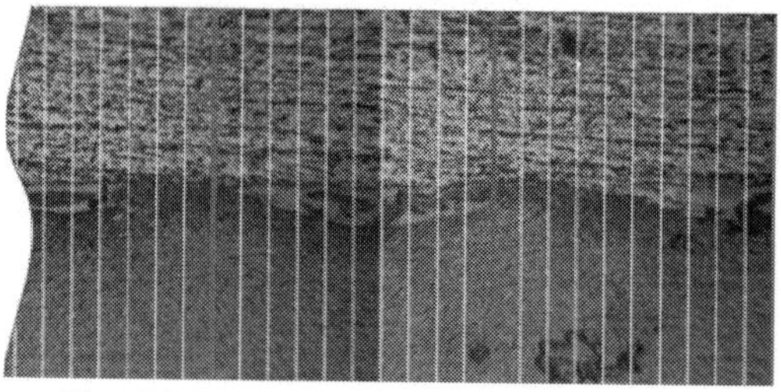

Fig. 5. Al/Fe interface with grid.

In the measurement step, the presence of inter-metallic phases was evaluated for each sector of 100 μm length. These phases, darker than aluminium and lighter than ferrite, were searched at interface. In case of a not very clear distinction between light and dark zones, the inter-metallic phases were considered as not present. The above procedure was repeated for all specimens and the results reported in Table 3 in terms of real inter-metallic extension and percentage respect to the entire specimen length of 13 mm.

Specimen ID	Temperature (°C)	Time (minutes)	Fe_xAl_y length	
			(mm)	(%)
1	100	5	5.4	41.54
2	100	25	6.1	46.92
3	500	5	12.4	95.38
4	500	25	12.9	99.23
5	300	15	5.7	43.85
6	300	15	6.7	51.54
7	300	15	5.4	41.54
8	300	15	7.1	54.62
9	300	15	6.0	46.15
10	300	0.86	5.2	40.00
11	300	29.14	6.6	50.77
12	17.16	15	5.9	45.38
13	582.4	15	11.5	88.46

Table 3. Inter-metallic extensions.

The expected outcome was the increase of the inter-metallic layer length with the increase of both temperature and time. The analysis of variance (ANOVA) of the Fe_xAl_y length response variable pointed-out the temperature as the main factor influencing the extension growth of the inter-metallic phases along the Al/Fe interface while time was negligible (Table 4).

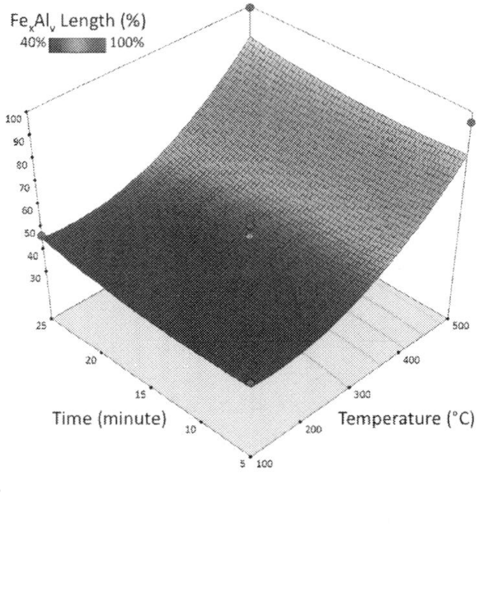

Fig. 6. Fe_xAl_y extension.

Source	SS	DF	MS	F	Prob>F
Model	80.62	5	80.62	8.69	0.0065
Temperature	58.97	1	58.97	31.79	0.0008
Time	0.95	1	0.95	0.51	0.4976
Temperature·Time	10^{-2}	1	10^{-2}	$5.4\cdot10^{-3}$	0.9435
Temperature^2	20.49	1	20.49	11.05	0.0127
Time^2	1.07	1	1.07	0.57	0.4732
Residual	12.99	7	1.86		
Total	***93.60***	***12***			

Equation of the Response Surface

Fe_xAl_y Length=+6.221-1.179·10^{-2}·Temperature -7.550·10^{-2} ·Time-2.5000·10^{-5} ·Temperature·Time+4.291·10^{-5}

Table 4. ANOVA – Inter-metallic extensions.

The good agreement between experimental and predicted results was confirmed by the high value of R^2 and adjusted R^2 of the quadratic model, respectively more than 86% and 76% (Figure 6). The experimental data highlighted that the specimens exposed to a thermal stress for temperatures below 300 °C did not exhibited a significant growth of the inter-metallic band extension respect to the as-clad specimen. The extension of inter-metallic phases covered less than 50% of the entire specimen Al/Fe interface length and the interface consisted of unconnected ripples and sacks. The situation drastically changed in case of specimens submitted to temperature equal or greater than 500 °C. The Al/Fe extension was more than 88% of the entire, sometimes covering the entire interface. The aspect of the interface was constitute by ripples and sacks connected by a continuous thin band of inter-metallic phases (Figure 7).

Fig. 7. Al/Fe interface (specimen ID 13).

Based on above observations, the analysis was extended to measure the inter-metallic layer thickness of the specimens subjected to the temperature of 500°C (ID 3 and 4). The measurement process previous applied for Fe_xAl_y extension was adapted by mounting 25x objective and 20x eyepieces lenses into the microscope, acquiring multiple images of the Al/Fe interface. Five shots per specimen were randomly selected and two grids of 10 µm spacing were over-laid for computing inter-metallic thickness of each shot, for a total number of measurements per specimen equal to 50. The results achieved pointed out the importance of the time variable on inter-metallic thickening. In fact, the average Fe_xAl_y layer thickness of specimen ID 3 (Temperature=500°C and time=5 min) was equal to 5.58 µm, quite an half of the average thickness of specimen ID 4, equal to 11.23 µm, subjected to the same temperature for a total time of 25 minutes. The observations of Fe_xAl_y growth and thickening were very important to justify the application of laser welding. In fact during laser welding, the high thermal input is localised in very narrow zone and the high travel speeds of the laser beam minimise the heat conductivity into the surrounding metal. The Fe_xAl_y layer should become larger and thicker in area just below laser beam while it should remain quite the same dimensions in areas away from laser beam.

5. Micro-hardness evaluation of heat treated specimens

Micro-hardness test were performed to measure the strength of Al/Fe interface in a non-destructive way as well as its variation under heat thermal loads. A Vickers micro-tester (HX-1000TM of REMET) mounting a 40x objective and 15x eyepieces lenses was used, setting the indentation load to 200 g for 15 s, in compliance with ASTM standard E92. Analysis points were sufficiently spaced to eliminate the effect of neighbouring indentations while evaluating existing materials and phases. Five indentations were taken at specific locations at and far from Al/Fe interface (Figure 8) for three times and results then averaged.

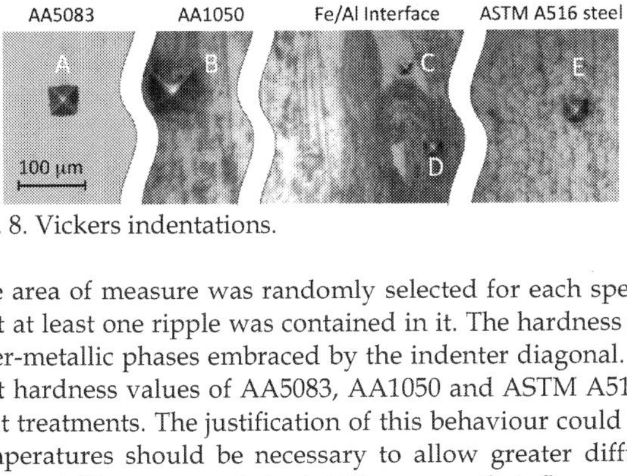

A=+ 3.2 mm
B =+ 1.0 mm
C =+ 0.0 mm
D =- 0.1 mm
E = - 1.5 mm

Fig. 8. Vickers indentations.

The area of measure was randomly selected for each specimen by respecting the condition that at least one ripple was contained in it. The hardness result was the average value of all inter-metallic phases embraced by the indenter diagonal. The results of Table 5 pointed-out that hardness values of AA5083, AA1050 and ASTM A516 steel were slightly affected from heat treatments. The justification of this behaviour could be that longer time at investigated temperatures should be necessary to allow greater diffusion fluxes. The hardness of the inter-metallic compounds (point C) was greatly influenced by temperature and time factors.

Specimen ID	Temperature (°C)	Time (minutes)	Indentation ID (HV)				
			A	B	C	D	E
1	100	5	105	46	583	266	195
2	100	25	107	41	534	245	178
3	500	5	81	31	430	242	182
4	500	25	83	28	483	239	166
5	300	15	91	44	554	235	186
6	300	15	81	45	593	238	180
7	300	15	97	44	618	247	183
8	300	15	89	42	582	278	175
9	300	15	81	39	582	294	183
10	300	0.86	107	43	543	246	186
11	300	29.14	85	40	527	245	174
12	17.16	15	100	41	545	246	179
13	582.4	15	72	30	476	192	158

Table 5. Vickers results.

The analysis of variance (ANOVA) of the inter-metallic hardness response variable pointed-out that increases in time and temperature lower hardness (Figure 9). In particular, temperature square, followed in order of importance from temperature and time square, were the main significant terms of the quadratic model (Table 6). This model presented R^2 and adjusted R^2 values equal to 88% and 79%. The reduction in hardness could be caused by the inter-metallic transformation from hard to soft phases. The fragility of the inter-metallic layers was also witness from the cracks present inside inter-metallic ripples. The investigated specimen, exposed to 300°C for 15 minutes in compliance with as-simulated condition, was characterised by the presence of cracks localised caused only by mechanical polishing.

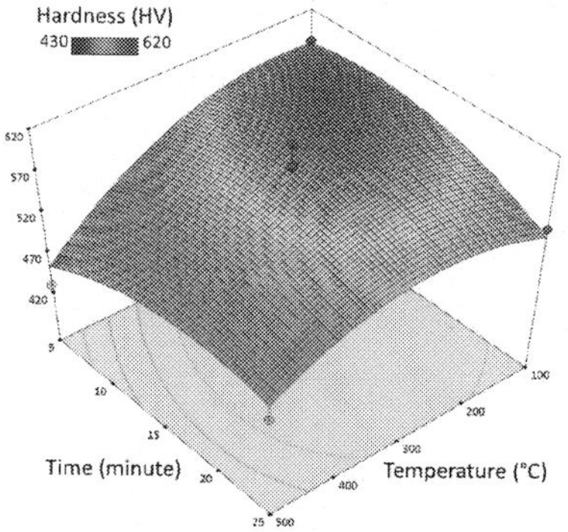

Fig. 9. Inter-metallic hardness.

Source	SS	DF	MS	F	Prob>F
Model	27854.0	5	5570.81	11.75	0.0027
Temperature	8916.54	1	8916.54	18.81	0.0034
Time	34.79	1	34.79	0.07	0.7942
Temperature·Time	4726.56	1	4762.56	9.97	0.0160
Temperature^2	10780.7	1	10780.7	22.75	0.0020
Time^2	5084.84	1	5084.84	10.73	0.0136
Residual	3317.41	7	473.92		
Total	***31171.4***	***12***			

Equation of the Response Surface

$$\text{HV of Point C} = +560.674 + 1.658\cdot10^{-1}\cdot\text{Temperature} + 3.165\cdot\text{Time} - 1.719\cdot10^{-2}\cdot\text{Temperature}\cdot\text{Time} - 9.842\cdot10^{-4}\cdot\text{Temperature}^2 - 0.270\cdot10^{-1}\cdot\text{Time}^2$$

Table 6. ANOVA – Vickers Results of Point C.

6. Mechanical strength of heat treated specimens

STJ specimens for mechanical evaluation were fabricated with machining processes from Al/Fe STJ (Figure 10) in compliance with ASTM A264–03 standard by removing the majority of the weakest clad materials (AA5083 and AA1050) and leaving a small nub with length w and thickness α. The ratio between w and α was 1½ with the maximum allowed dimension of w equal to 3.18 mm (⅛"). Other important dimensional features were the specimen width equal to 25.40 mm (1"), length equal to 63.50 mm (2½"), distance between side surfaces equal to 19.05 mm (¾") and minimum steel thickness t greater than 2w. The heat treatments of these specimens were performed by following the same temperature-time parameters of those defined for the CCD experimental plan of the metallographic evaluation.

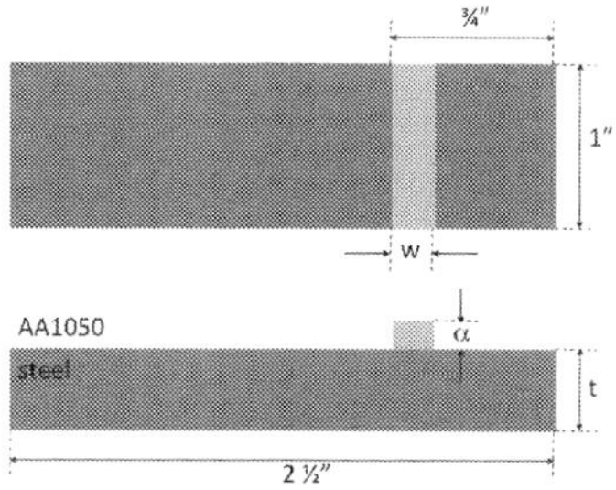

Fig. 10. Shear test - Specimen.

Shear test was very important to assess transition joint technical conformity by measuring the maximum shear strength of the bond interface and comparing it to the maximum shear strength of the weakest material of the STJ. The shear strength of the test piece was defined as the peak shearing load divided by the sheared area. Interfacial bonding of the clad specimen was evaluated by shear stress measured with a universal test machine INSTRON 4485 with a 200 KN loading cell and a home designed testing equipment in which the test piece and the jig were fitted, assuring the easy specimen blocking and high positioning accuracy as well as flexibility of use for other required tests. The shear test were performed with the mobile crosshead moving at 3.0 mm/min. The evolution of stress-displacement curve initially presented a rapid increase of the stress value, its stabilisation and finally its rapid reduction. The rupture was localised in the AA1050 and not at interface AA1050/steel, justifying the trend of this loading curve (specimen BM in Figure 11). The same figures reports the loading curves of specimen ID 8, heated at 300°C for 15 minutes, and specimen ID 4, heated at 500°C for 25 minutes.

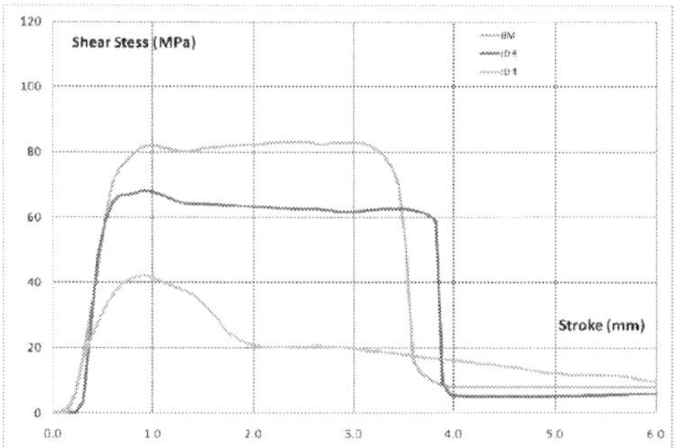

Fig. 11. Loading curve of heat treated specimens.

As figure shows, the shear strength decreased with the increase of both processing temperature and time, causing the rapid lowering of the maximum shear stress values. This trend was also confirmed by analysing the other tests.
Table 7 reports the final results of all tests in term of maximum values of the shear stress. All stress values recorded during tests were decidedly higher than 50-60 MPa prescribed from Lloyd's Register of Shipping, revealing the good fabrication quality of the observed STJ. The worst results were achieved when specimens were subjected to high temperature for a long time (Specimens ID 4 & 13).

Specimen ID	Temperature (°C)	Time (minutes)	Shear stress	
			(MPa)	Lloyd's
1	100	5	69.57	Positive
2	100	25	70.37	Positive
3	500	5	75.95	Positive
4	500	25	44.50	Negative
5	300	15	80.64	Positive
6	300	15	72.09	Positive
7	300	15	70.52	Positive
8	300	15	69.73	Positive
9	300	15	71.63	Positive
10	300	0.86	73.65	Positive
11	300	29.14	69.79	Positive
12	17.16	15	69.57	Positive
13	582.4	15	5.56	Negative

Table 7. Shear stress values.

The analysis of variance (ANOVA) of the shear stress response variable pointed-out that increases in time and temperature lowering shear strength (Figure 12). In particular, temperature, followed in order of importance from temperature square and time, were the main significant terms of the quadratic model (Table 8). This model presented R^2 and adjusted R^2 values equal to 78% and 63%.

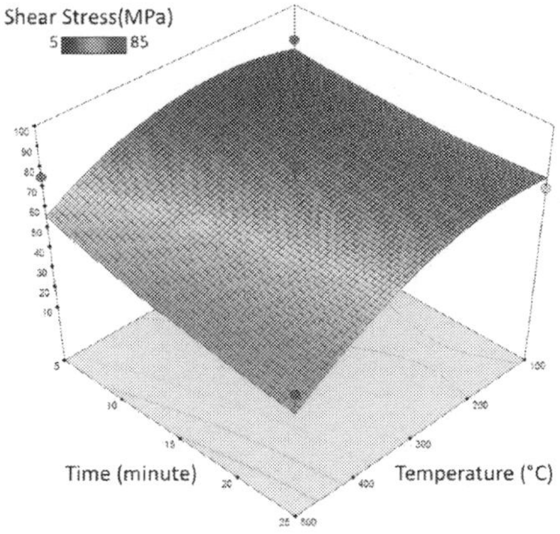

Fig. 12. Shear stress.

Source	SS	DF	MS	F	Prob>F
Model	3626.63	5	725.33	5.04	0.0281
Temperature	1512.87	1	1512.87	10.52	0.0142
Time	163.00	1	163.00	1.13	0.3224
Temperature·Time	260.02	1	260.02	1.81	0.2207
$Temperature^2$	1578.79	1	1578.79	10.98	0.0129
$Time^2$	28.21	1	28.21	0.20	0.6712
Residual	1006.53	7	143.79		
Total	***4633.16***	***12***			

Equation of the Response Surface

Shear Stress=+52.815+2.177·10^{-1} ·Temperature+1.537·10^{-1} ·Time-4.031·10^{-3} ·Temperature·Time

Table 8. ANOVA – Shear stress values.

Some interesting considerations were performed by comparing results of shear stress (Figure 12), inter-metallic extension (Figure 6) and micro-hardness (Figure 9). The behaviour of the shear stress response surface was opposite to that of the inter-metallic extensions. In particular the highest and lowest values of the shear stress was achieved for the lowest and highest temperatures, respectively. On the contrary the highest and lowest values of the inter-metallic extension were achieved for the highest and lowest temperatures. For this reason it was possible to suppose that shear strength decreased with the increase of the inter-metallic length. In contrast, the relation between shear stress and inter-metallic hardness was not so clearly evident. In fact the highest and lowest values of the shear stress were achieved for the lowest and highest temperature-time couple. However in the other areas this trend was not quite evident and further investigation will be necessary. Based on the above results, it was possible to suppose that the main cause of the lowering of the shear strength with the temperature increase was the inter-metallic layer growth, even if the inter-metallic phases become softer.

The strength of the welding zone is evaluated by Ram Tensile Test. During the test, the force is applied to the direction perpendicular to the planes delimiting different material interface. A specimen with a particular parallelepiped/cylindrical shape is used in order to concentrate stresses in the section just above the transition area that corresponds to the bond interface, as Figure 13 shows. The MIL-J-24445A standard, specifies maximum dimensions of the bi-material specimen but its principles can be easily extended to tri-material joints. In this last case, the strength of the steel/AA1050 bond interface is estimated rather than that of the AA5083/AA1050 one. The specimen, punch and matrix prescribed by the above cited standard are reported in Figure 14, together with the designed testing equipment. The matrix is fixed while the punch moves in contact with the bond interface, compressing it until rupture. The MIL-J-24445A standard does not give values or suggestions linked to testing parameters (e.g. punch speed) and/or specimen clamping. The specimen is simply rested on the matrix top surface, allowing its rotation around the cylindrical axis. In the hypothesis that the pressure between punch and specimen during contact is uniform, the applied load can be considered as axial-symmetric. The thickness D of the bond zone is always equal to 1.72±0.1 mm, independently from the values of steel and aluminium thicknesses. The thickness D is achieved by simply varying the hole depth. The testing equipment, designed and realised in compliance with the MIL-J-24445A standard, allowed the easy set-up on an universal tensile testing machine. The designed equipment consisted of a screwed punch and a two elements matrix with a changeable cylindrical insert. The punch is mounted on the top mobile crosshead of the INSTRON 4485 machine while the matrix is fixed to the machine basement with bolts and nuts. The choice of the changeable insert was justified by the need to substitute the insert in case of high wear conditions as well as to select the proper tolerance gap with the specimen. The correct alignment between the punch and specimen axes to avoid bending effects was realised by means of a guided groove in the low part of the matrix. The pressure of the punch caused tensile loads on specimen bond interface. The maximum tensile stress value recorded during test was used to assess the joint strength for several temperature-time cycles. The minimum value of the tensile stress to consider the joint as good should be greater than 75 N/mm^2.

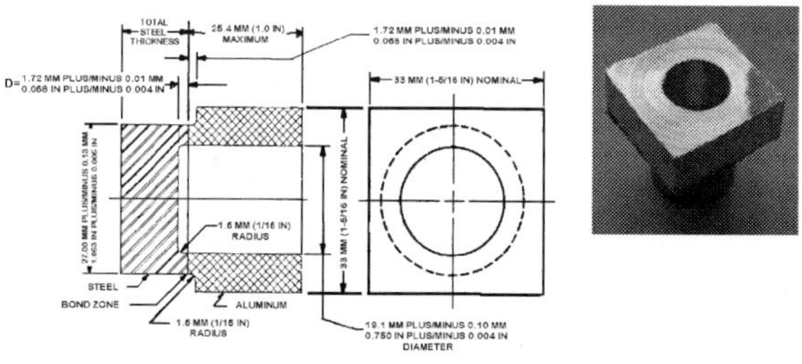

Fig. 13. Ram test specimen.

Fig. 14. Testing equipment

The designed experimental plan for mechanical testing was extended respect to the plan used for micro-hardness evaluation to preserve continuity with previous investigation, evaluate more critical conditions associated to higher time and/or temperatures and improve knowledge about mechanical strength limits of tri-material transitions joints. Metallographic examinations revealed that the heat-treated joints properties remained quite equal to those of as-clad specimen for temperatures lower than 300 °C and processing times lower than 15 minutes. The new experimental plan is reported in Table 9, in terms of temperature-time couples. The punch speed was set to 0.05 mm/s for all tests.
Some important considerations can be carried-out by analysing these results. The maximum strength of the as-clad specimen as well as specimens processed at 300 °C were greater than

minimum value of 120 MPa certified by the material supplier. In addition, the specimen resistance was mainly supported by the small thickness of AA1050, considering that the maximum strength of this commercial alloy was about 180-190 MPa at high cold work conditions. These values suggested that the thermal cycles may reduced the cold work conditions, with consequent reduction of the maximum strength of the tri-material joint. The specimen behaviour changed when processing temperature was greater than 300°C. Based on the above results, temperature was the main factor causing the lowering of the specimen mechanical strength.

Specimen ID	Temperature (°C)	Time (minutes)	Tensile	
			Force (KN)	Stress (MPa)
HTRT1	25	5	29.4	233.7
HTRT2	300	15	30.1	239.3
HTRT3	500	25	4.3	34.2
HTRT4	300	25	25.2	200.3
HTRT5	400	20	16.4	130.4
HTRT6	500	15	4.8	38.2
HTRT7	500	5	17.8	141.5
HTRT8	25	25	29.9	237.7

Table 9. Ram tensile test values.

The same considerations arose from the analysis of the load-displacement data. HTRT1 and HTRT2 specimens were characterised by the same curves, with equivalent values of the maximum load and displacement, as Figure 15 reports. This behaviour pointed-out that mechanical properties were not greatly altered from processing temperature and time, even if HTRT2 specimen stayed in the oven at 300°C and HTRT1 specimen was thermally untreated. The maximum supported load started to drop when specimen were treated at temperatures higher than 300 °C. The HTRT5 specimen, processed at 400 °C for 20 minutes, and HTRT7 specimen, processed at 500 °C for 5 minutes, were accomplished by the reduction of the maximum load value with the increase of the maximum displacement (Figure 16). In fact, the load-displacement curves of the two specimens were perfectly equivalent from the mechanical point of view. The maximum load drastically drop for higher temperatures, as HTRT3 specimen, processed at 500 °C for 25 minutes, witnessed.

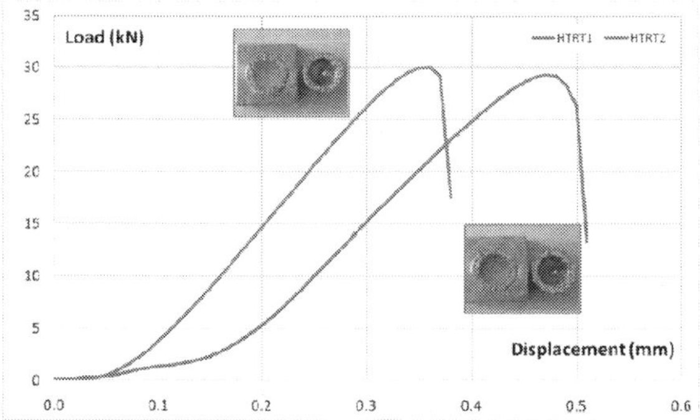

Fig. 15. Load/displacement curves of HTRT1 and HTRT2 specimens.

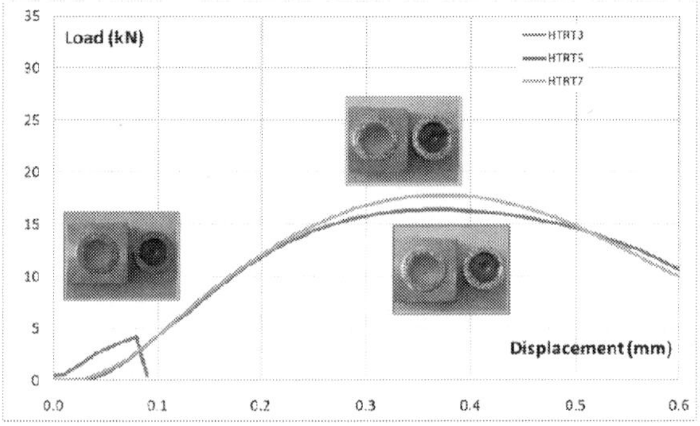

Fig. 16. Load/displacement curves of HTRT3, HTRT5 and HTRT7 specimens.

The reduction of the mechanical strength with temperature and time becomes more evident with results reported in Figure 17, the ANOVA of which is reported in Table 10.

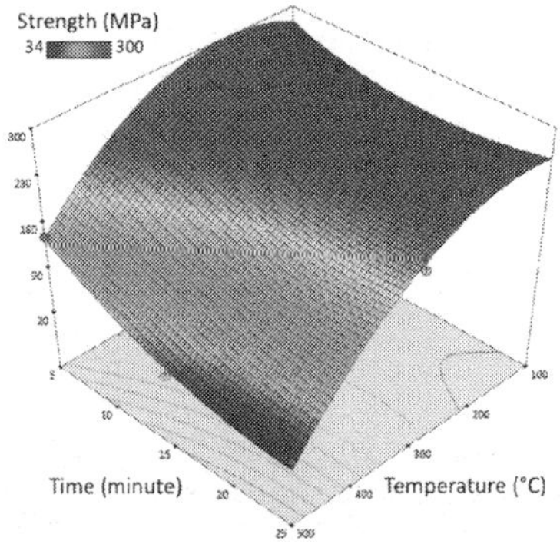

Fig. 17. Ram tensile test.

Source	SS	DF	MS	F	Prob>F
Model	50191.0	5	10038.2	20.21	0.0478
Temperature	26285.5	1	26285.5	52.91	0.0184
Time	3962.75	1	3962.75	7.98	0.1058
Temperature·Time	3391.77	1	3391.77	6.83	0.1205
Temperature^2	12417.8	1	12417.8	25.00	0.0378
Time^2	951.94	1	951.94	1.92	0.3005
Residual	993.55	7	496.77		
Total	***51184.5***	**12**			

Equation of the Response Surface

$$\text{Shear Stress}=253.66+8.388\cdot 10^{-1}\cdot \text{Temperature}-8.341\cdot \text{Time}-1.218\cdot 10^{-2}\cdot \text{Temperature}\cdot \text{Time}-1.876\cdot 10^{-3}\cdot \text{Temperature}^2+2.842\cdot 10^{-1}\cdot \text{Time}^2$$

Table 10. ANOVA – Ram tensile test values.

The justification of this fragile behaviour was that the heat treatment at high temperature promoted the inter-metallic redistribution at Al/Fe bond interface, with consequence growth of the inter-metallic thickness, causing the lowering of the mechanical properties. This hypothesis has been confirmed by previous works (*Tricarico et al., 2007*) in which the influence of the thermal loading of the Al/Fe explosion welded joints was investigated by measuring the extension and thickness of inter-metallic compounds at Al/Fe interface. The metallographic analysis measurements have pointed-out that the band extension was mainly influenced by the temperature because the increase of the band extension with increasing temperature has been recorded. Observing the evolution of the inter-metallic band thickness, it has been observed that thickness at 500°C after 5 minutes was half than that obtained after 25 minutes.

Another important aspect took into account was the shape of the fractured surfaces of the thermally loaded specimens. The fracture surfaces were characterised by a ductile fracture with very low appreciable deformation and/or absent signs of gross plastic deformation. This fracture may be justified by the presence of a continuous diffuse inter-metallic layer that triggered micro-cracks. The Figure 18 reports the fractures aspects of HTRT1, HTRT2 and HTRT3 specimens. The pattern of the HTRT1 specimen is very irregular with evident differences between peaks and voids. These difference between peaks and voids were lower in case of HTRT3 specimen, as the fragile fracture of the specimen at the end of the test shows. The surface aspect of the HTRT2 specimen seemed to be intermediate between the other two specimen surfaces, corroborating results achieved during the ram tensile test in terms of maximum load and displacement.

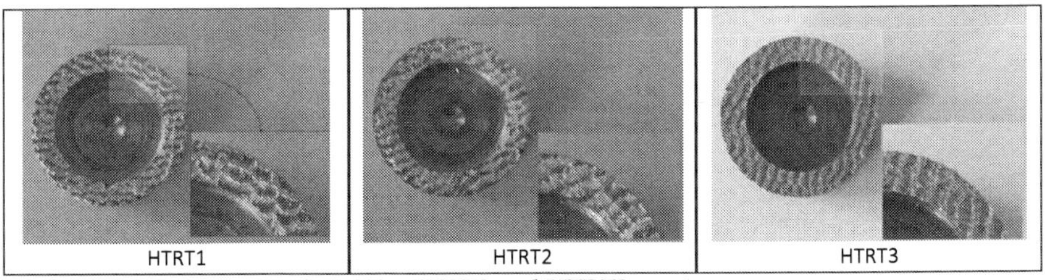

Fig. 18. Surface fractures of HTRT1, HTRT2 and HTRT3 specimens.

7. Bead on plate

Welding tests on STJs were performed on specimens machined from the same bar, which section and length were 26.3·26.0 mm^2 and 1000.0 mm respectively. The materials of the STJ bar were steel - ASTM A516 grade 55 - and aluminium alloys - AA1050 and AA5083, the thicknesses of which were 13.7±0.4 mm for steel and 11.7±0.3 mm for aluminium alloys. Bead-on-plate were carried-out by using process parameters identified in the previous preliminary analysis and varying only the joint geometry. In this way, the laser-induced thermal load was kept constant on the top surface while Al/Fe interface was subject to different thermal stresses due to the reduction of the specimen thickness. To achieve this result, upper section interacting with the laser beam was machined to reduce the distance d between the fused zone and bond interface. For this reason specimens with different thicknesses were employed. The specimen length, equal to 80.0 mm, was chosen to achieve stationary conditions during welding. Some trial tests on STJ specimens were performed to verify if the main geometrical parameters of weld sections were the same of those achieved in the preliminary analysis on the base materials. Weld geometry was characterised by the penetration depth and width equal to 5.4 and 5.0 mm, quite equal to those measured on the base materials. The small variations between results were inputted to the laser beam power fluctuation, specimen geometry (plate vs. bar) as well as materials compositions (mono-material vs. tri-materials).

The following tests on STJs were performed with the laser beam power equal to 4.3 kW and travel speed set to 1.5 or 1.0 m/min respectively for aluminium or steel top surface. The specimen LPBAL1, LPBAL2 and LPBAL3 were realised by machining the aluminium side (Table 11). The distance d between the melt zone and the Al/Fe was planned to 3.0, 1.5 and 0.0 mm, assuming that the penetration depth remained constant to 5.5 mm. Table 11 also reports cut section, details of the weld fused area and geometrical parameters such as the penetration depth h and width r. In particular, these two parameters decreased with the reduction of the distance d, in spite of the thermal load on the top surface was the same for all specimens.

Specimen	LPBAL1	LBPBAL2	LBPBAL3
Geometry	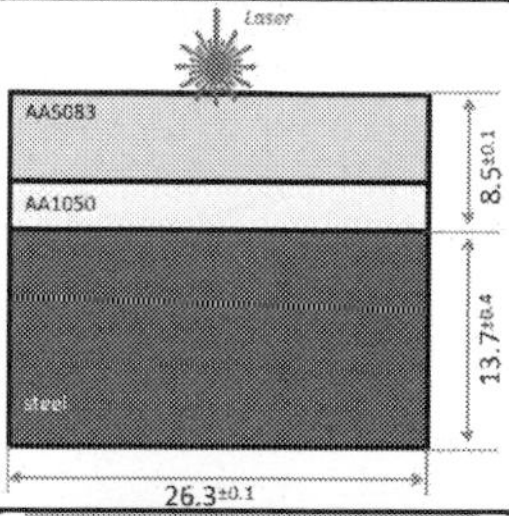	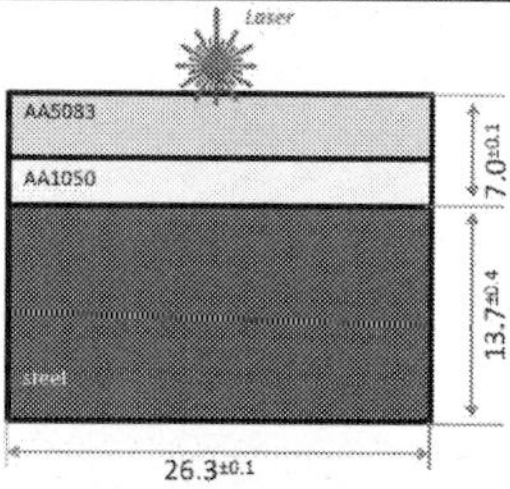	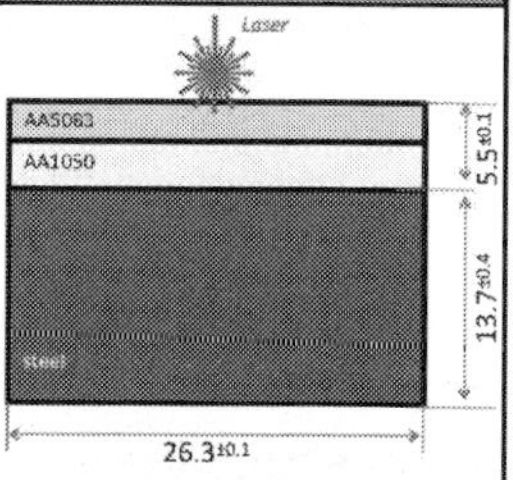
Results			
Section	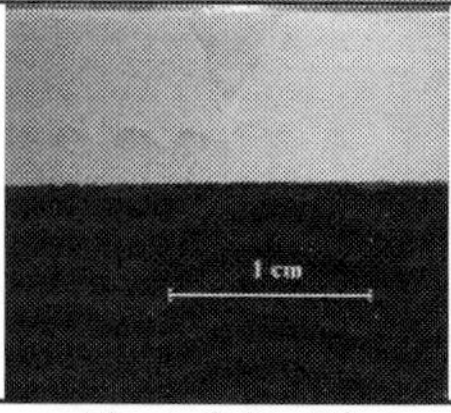	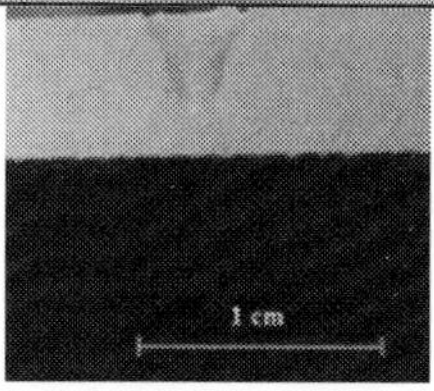	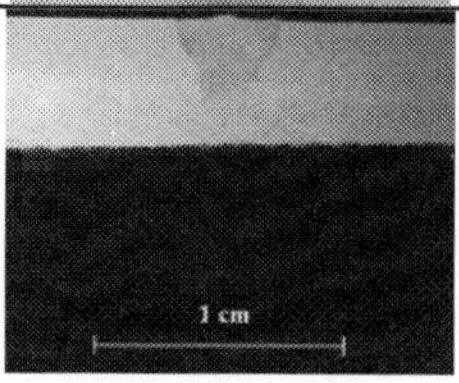
Parameters	r = 4.8 mm \ h = 5.1 mm	r = 4.7 mm \ h = 4.6 mm	r = 4.7 mm \ h = 4.5 mm

Table 11. Tests on STJs – Aluminium side.

Possible justifications of this trend were the lower conductivity of steel, that reduced the heat transmission at bond interface as well as in the steel region, and the smaller aluminium thickness, that lowered the possibility of the laser-induced heat to quickly went away from the weld fused area. The decrease of the main weld parameters become more evident by comparing the as-clad specimen with LPBAL3. The difference of the penetration depth between these two specimen was equal to 0.9 mm while the difference of the penetration width was equal to 0.3 mm.
The same experimental framework was applied for specimen welded from the steel side. The specimen LPBST1, LPBST2 and LPBST3, realised by the steel side, were machined from the same bar of the specimen processed from the aluminium side to eliminate differences due to manufacturing batch. The distance d between the melt zone and the Al/Fe was planned to 3.0, 1.5 and 0.0 mm, assuming that the penetration depth remained constant to 5.0 mm. Table 12 reports specimen dimensions, cut section, details of the weld fused area and geometrical parameters such as the penetration depth h and width r. The reduction of penetration depth h and width r with the decrease of the distance d were also detected for these tests but this reduction was considered as negligible.

Specimen	LPBST1	LPBST2	LPBST3
Geometry	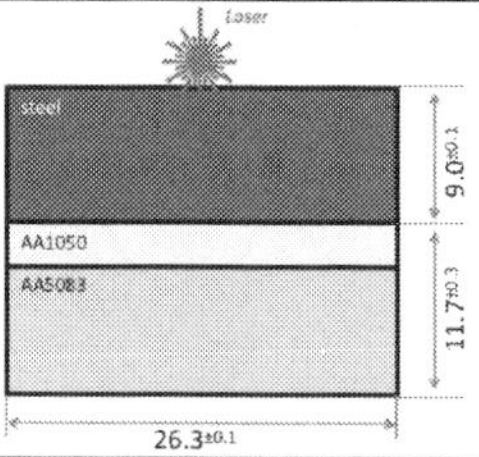	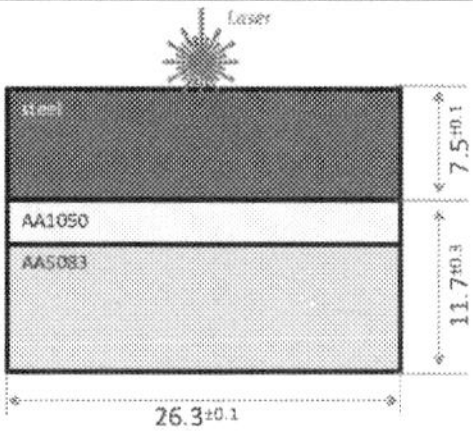	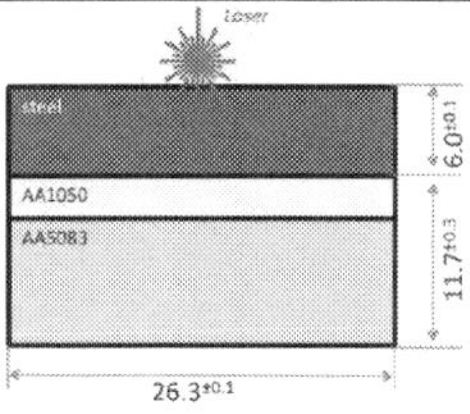
Results			
Section	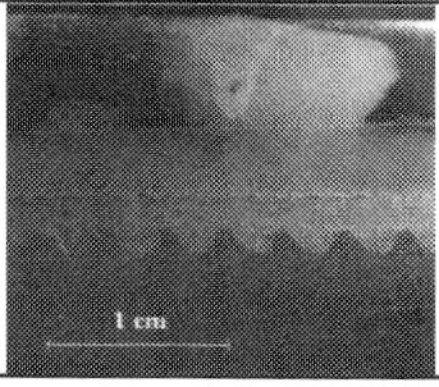	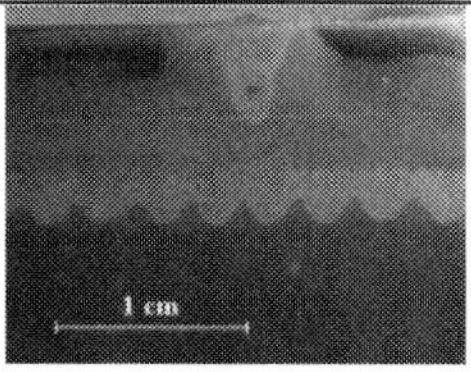	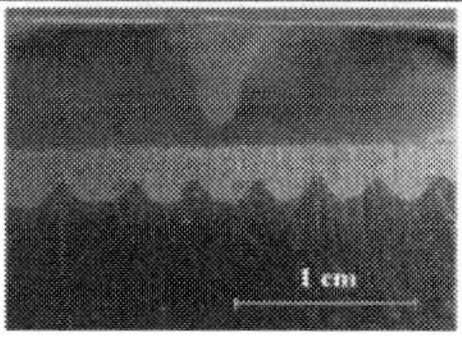
Parameters	r = 4.8 mm \ h = 5.4 mm	r = 5.1 mm \ h = 5.3 mm	r = 5.0 mm \ h = 5.2 mm

Table 12. Tests on STJs – Steel side.

8. Metallographic examination of bead on plate specimens

The study of microstructure of the bond interface was manly addressed to the qualitative and quantitative analysis of the Al/Fe inter-metallic film in terms of variation of its extension and thickness. The procedure was the same of that used for the heat treated specimens. The results of the metallographic examinations are reported in Table 13 in terms of the reference length L_{REF} and the inter-metallic extension L_{INT} as well as the ratio between these two measurements. The reference length LREF, equal to 15.0 mm, was the same for all specimens. The results pointed out that the inter-metallic extension L_{INT} was greater than 50% of the total length L_{REF} of the cut section in the as-clad specimen. The length L_{INT} increased with the decrease of the distance d between weld fused area and bond interface. This increase was more rapid for specimens welded from the steel side than those welded from the aluminium side. Another important aspect to underline was that the two main factors linked to ripples and film growth contributed to the inter-metallic extension value. Ripples existed in the as-clad specimen, as the main feature of the explosion welding process. The laser-induced heat loads influenced inter-metallic ripples, promoting their growth by inter-diffusion, but no new ripples were created during welding. The inter-metallic film was mainly promoted by laser-induced thermal loads because it also aroused in areas in which it did not exist at all. The contribution of the inter-metallic film was by made more evident by the L_{FILM}/L_{REF} ratio in Table 13. The enlargement of the inter-metallic film was also in this case higher for specimens welded from the steel side.

Specimen ID	Condition	Distance d (mm)	Length L_{REF} (mm)	Length L_{INT} (mm)	Ratio L_{INT}/L_{REF}	Ratio L_{FILM}/L_{REF}
--	As-clad	--	15.0	8.4	56.0	8.0
LPBAL1	Welded from aluminium side	3.0	15.0	8.4	56.0	8.4
LPBAL2		1.5	15.0	8.4	56.0	8.6
LPBAL3		0.0	15.0	8.9	59.3	10.9
LPBST1	Welded from steel side	3.0	15.0	8.4	56.0	8.4
LPBST2		1.5	15.0	9.5	63.3	14.7
LPBST3		0.0	15.0	10.4	69.3	16.1

Table 13. Inter-metallic extension.

The inter-metallic film thickness was evaluated in terms of average value of several random measures near the fused area. The maximum and the minimum thicknesses were also evaluated (Table 14), considering that these values were representative of local conditions. The evaluation of the average value of the film growth made observations independent of the previous state of the as-clad material, linking results to the effects of the laser induced heat.

Specimen ID	Condition	Distance d (mm)	H_{FILM} Max (μm)	H_{FILM} Avg (μm)	H_{FILM} Min (μm)	Film Growth
--	As-clad	--	10.29	6.51	3.10	--
LPBAL1	Welded from aluminium side	3.0	17.45	8.11	3.51	25%
LPBAL2		1.5	14.07	8.27	4.58	27%
LPBAL3		0.0	17.70	8.95	3.01	38%
LPBST1	Welded from steel side	3.0	11.53	7.30	3.92	12%
LPBST2		1.5	10.65	7.40	3.43	14%
LPBST3		0.0	11.05	8.21	5.53	26%

Table 14. Inter-metallic film thickness.

The average film thickness increased with the reduction of the distance d with the same trend independently form the material of the welding side. The increase of film thickness was greater for specimen welded from the aluminium side than those welded from the steel side. This behaviour, opposite to that recorded for the film extension growth, pointing out that laser induced heat remained on aluminium side of the tri-material specimen because steel created a thermal barrier with its lower thermal conductivity.

The SEM analyses were then performed from the qualitative point of view by visually inspecting morphology of the Al/Fe interface. The back-scattered electron (BSE) images near the specimen Al/Fe interface showed the AA1050 in dark grey, the ASTM A516 steel in the light grey the "wavy" interfacial area with different grey, in function existing Fe_xAl_y inter-metallics. Figure 19 is one of the acquired BSE images for the specimen LPBAL1, welded from the aluminium side. The analysis in three different positions inter-metallic compounds were more numerous and fragile, as figures show. Figure 19 also reports the results of the micro-analysis of the area indicated from the three shapes (circle, rectangle and triangle). With energy dispersive x-ray spectrometers (EDS), chemical compositions was determined quickly. Despite the ease in acquiring x-ray spectra and chemical compositions, the potentially major sources of error were minimised by optimising the operative conditions necessary to improve the statistical meaning of the electron counter. In particular, the scanning area was equal to 1 μm^2, the incident energy was 25 keV on the specimen surface with a working distance of 10 mm (in this way the x-ray take-off distance was equal to 35°), the electronic current was tuned in order to generate a X-ray counter rate of 2000 pulse per second and the effective counter time was equal to 100 s (*Capodiceci*, 2007). Figures 20-23 reports the SEM analyses for specimen LPBAL3, LPBST1 and LPBST3. It is evident that the number of grey level was higher for specimen welded form the steel side than the aluminium one. This experimental evidence was probably due by the lower conduction coefficient of steel than that of aluminium, which caused heat to be slowly removed after welding

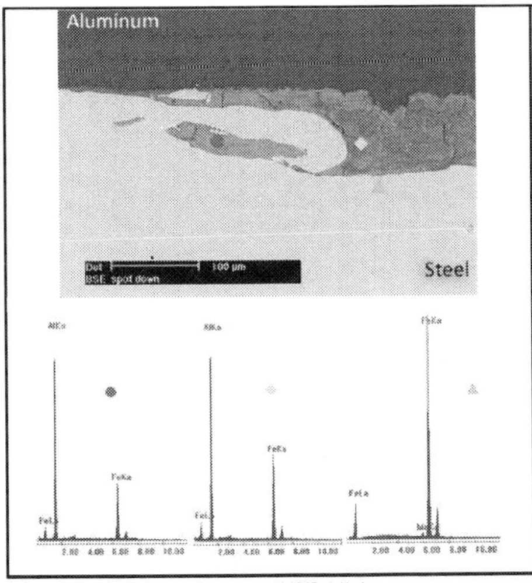

Fig. 19. SEM analysis – LPBAL1.

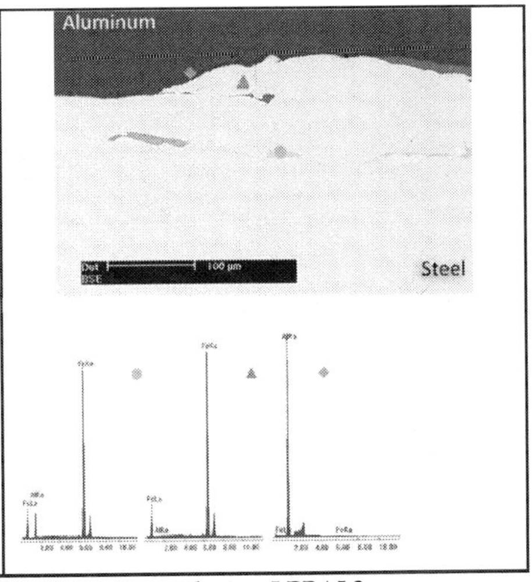

Fig. 20. SEM analysis – LPBAL3.

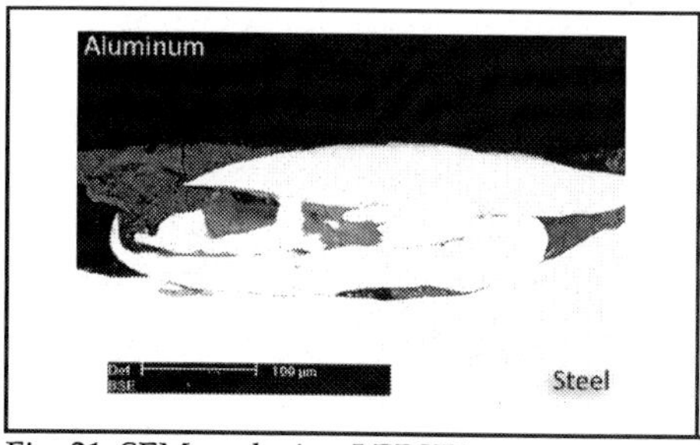

Fig. 21. SEM analysis - LPBST1.

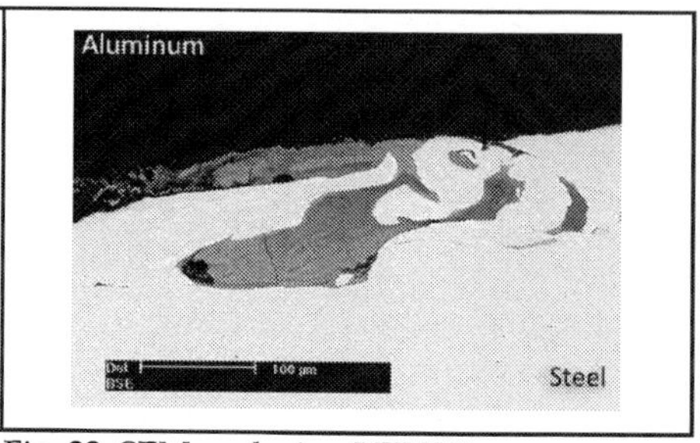

Fig. 22. SEM analysis - LPBST3.

9. Mechanical strength of laser welded specimens

The mechanical characterisation of the welded specimens allowed the modifications to the mechanical properties (shear and tensile strengths) caused by laser beam interaction to be evaluated. Shear and ram tensile samples were achieved from the same plate with the sampling scheme shows in Figure 23 in order to avoid difference in STJ lot characteristics. The laser beam passed at the centre of the small nub of the shear test specimens and sufficient far from the ram tensile specimens. The area of the small nub of the shear specimens were consequently subjected to the highest thermal stresses while the bonding area AA1050/steel of the ram tensile specimens were uniformly thermally loaded. The process parameters used for bead-on-plate welds were the same of those employed for STJ bars in the previous experimental step. Increasing thermal loads at the bond interface were achieved by removing material from the surface interacting with the laser beam. The reduction of the plate thickness required the scaling down of specimen dimensions for some samples. Table 8 and Table 9 report the dimensions of shear and ram tensile test specimens. An additional test specimen was cut at the centre of the plate to evaluate the maximum welding penetration depths in comparison with those of the welded bars as well as hardness values (*Tricarico et al, 2007*).

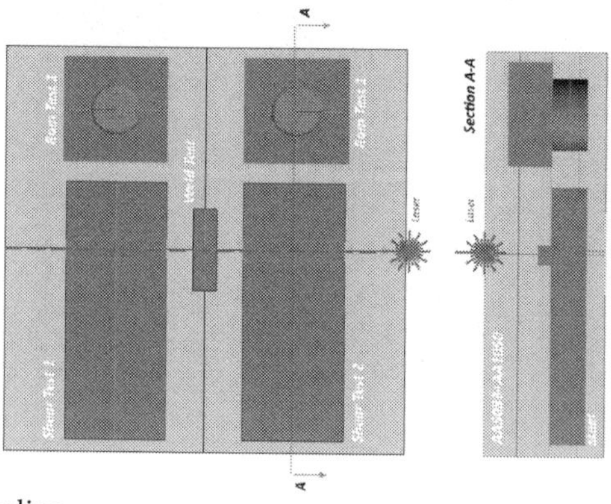

Fig. 23. Specimen sampling.

The shear test were performed with the mobile crosshead moving at 3.0 mm/min. The acquisition of several high resolution digital images during tests was useful to visually understand the mechanisms of deformation of the small nub (Figure 24), compared with numerical data.

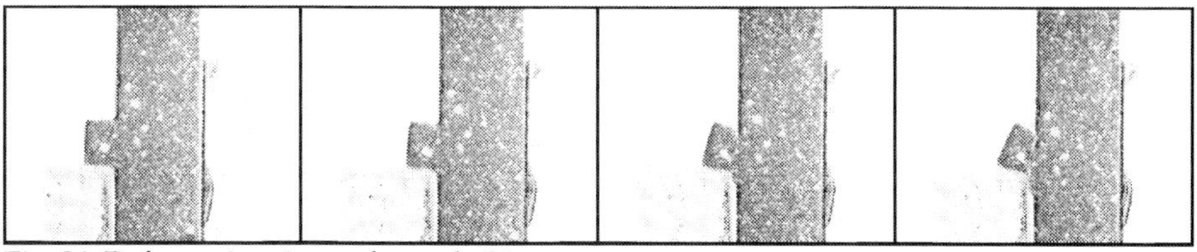

Fig. 24. Deformation times of sample B1.

The two repetitions for each welding conditions were characterised by load-displacement curves wholly overlaid, as Figure 25 shows for sample B1 and B2. The evolution of stress-displacement curve initially presented a rapid increase of the stress value, its stabilisation and finally its rapid reduction. The rupture was localised in the AA1050 and not at interface AA1050/steel, justifying the trend of this loading curve.

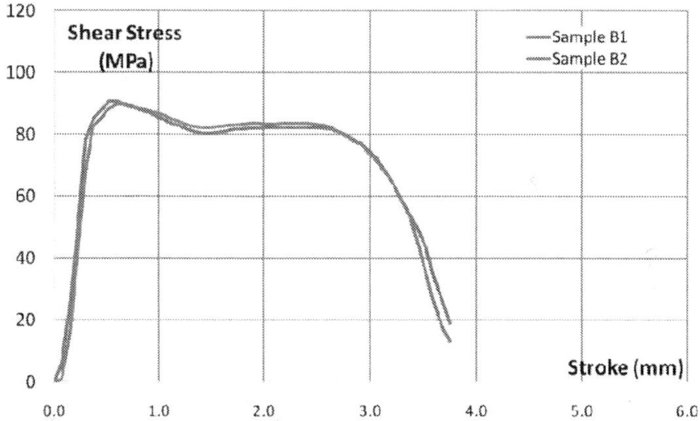

Fig. 25. Shear stress vs. punch stroke of sample B1 and B2.

Table 15 reports the final results of all tests in term of maximum shear load and stress. All stress values recorded during tests were decidedly higher than 50-60 MPa prescribed from Lloyd's Register of Shipping, revealing the good fabrication quality of the observed STJ. Results also pointed-out that the reduction of the specimen thickness and the consequent reduction of the distance between weld fused area and bond interface caused the decrease of the maximum shear strength.

Specimen ID	Condition	Al/Fe thick	Geometry			Shear	
		mm/mm	α (mm)	w (mm)	t (mm)	T (KN)	τ (MPa)
-- / A1-A2	As-clad	11.7/13.7	3.00	4.50	9.00	10.0	87.4
LPBAL1 / B1-B2	Welded from aluminium side	8.5/13.7	3.00	4.50	9.00	10.1	88.3
LPBAL2 / C1-C2		7.0/13.7	3.00	4.50	9.00	9.3	81.7
LPBAL3 / D1-D2		5.5/13.7	3.00	4.50	9.00	9.0	78.7
LPBST1 / E1-E2	Welded from steel side	11.7/9.0	3.00	4.50	9.00	10.0	87.4
LPBST2 / F1-F2		11.7/7.5	2.50	3.75	7.50	8.8	86.6
LPBST3 / G1-G2		11.7/6.0	2.00	3.00	6.00	6.3	83.0

Table 15. Shear test - Sample dimensions & results.

This decrease was more evident for specimen welded from the aluminium side than those welded from the steel side, as Figure 26 and Figure 27 show.

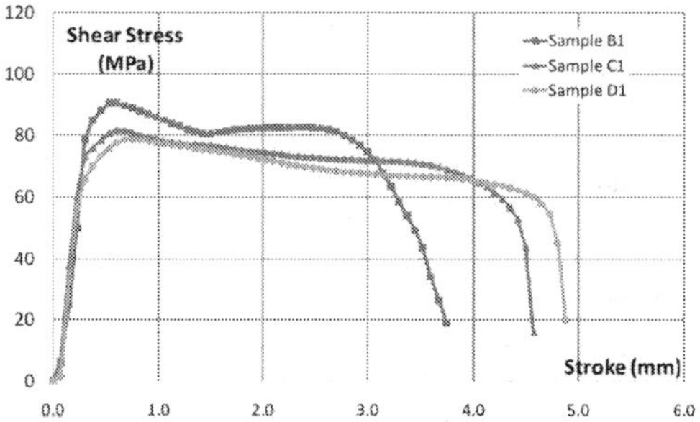

Fig. 26. Samples B1, C1 and D1.

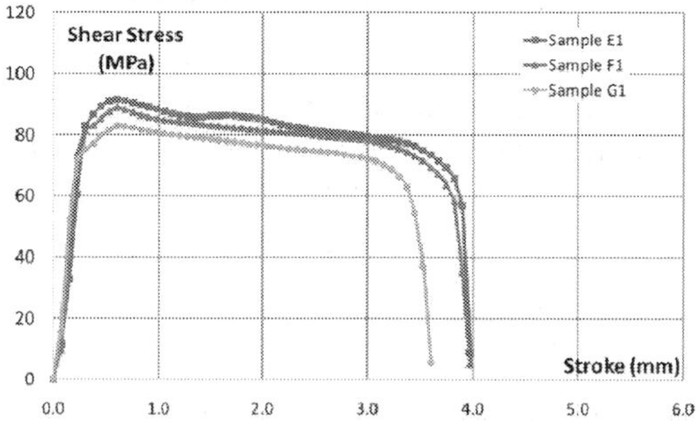

Fig. 27. Samples E1, F1 and G1.

The ram tensile test were then carried-out. Two repetitions for each welding condition was useful to assess test repeatability. Figure 28 reports results of the samples RA1 and RA2, in terms of stress-displacement in which the maximum tensile stress, equal to 235.3 MPa and corresponding to a maximum load of 29.6 KN, was equal for the two samples. The rupture was always localised at the Al/Fe interface due to the specimen shape.

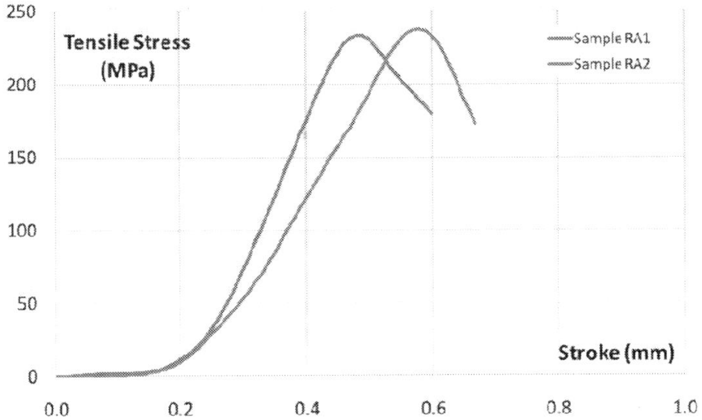

Fig. 28. Tensile stress vs. punch stroke - Samples RA1 and RA2.

Specimen ID	Condition	Al/Fe thick mm/mm	Diameter α (mm)	Tensile T (KN)	σ (MPa)
-- /RA1-RA2	As-clad	11.0/13.0	12.7	29.6	235.3
LPBAL1 / RB1-RB2	Welded from aluminium side	8.0/10.0	9.7	28.5	226.4
LPBAL2 / RC1-RC2		6.5/10.0	8.2	26.8	213.0
LPBAL3 / RD1-RD2		5.0/10.0	6.7	23.9	199.9
LPBST1 / RE1-RE2	Welded from steel side	11.0/8.5	12.7	28.1	223.7
LPBST2 / RF1-RF2		11.0/7.0	12.7	26.6	211.2
LPBST3 / RG1-RG2		11.0/5.5	12.7	26.1	206.8

Table 16. Ram tensile test - Sample dimensions & results.

Table 16 reports the final results of all tests in term of maximum tensile load and stress. The results showed the same trend detected during the shear test linked to the more evident reduction of the final strength of specimens welded from the aluminium side than those welded from the steel side. The reduction of the distance between the fused area and bond interface was less important because specimens were realised in area far from the laser beam interaction and consequently they were subjected to mild laser-induced thermal loads.
The comparison between mechanical results and inter-metallic film thickness was very interesting. The reduction of the maximum tensile and shear stresses could be inputted to the increase of the inter-metallic film thickness. In fact lower values of the mechanical strength was detected for higher values of the film thickness. This hypothesis also confirmed that specimen welded from the steel side were more critical than those welded from the aluminium side. However, the mechanical strength of the welded specimens were only blindly affected by the laser beam interaction because the measured strengths were much more higher than those normally required.

10. Mechanical strength of laser welded T-joints

Double square fillet (2F) T-joint welds of AA5083 aluminium alloy and ASTMA516 steel base materials were then produced using different welding methods (laser welding with filler wire and hybrid laser-MIG welding). T-joint welds were realised by joining two 6.0 mm thick plates. Steel (aluminium) 2SF T-joints were produced with the laser beam power equal to 5.5 (5.5) kW in continuous wave regime, travel speed set to 1.9 (1.5) m/min, filler feed equal to 0.8 (1.5) m/min. Additional process parameters kept constant during all tests were the focal length, beam focus position and Helium shielding gas flow-rate, the values of which were 300.0 mm, 0.0 mm (on the surface) and 30.0 Nl/min respectively. The laser head was angled of 51° degree respect to the Z axis to guarantee joint accessibility. The 1.2 mm diameter filler wire was used. The process parameters of the pulsed arc MIG welding during hybrid welding for steel (aluminium) were open arc voltage, peak current intensity, peak time, pulse frequency and background current intensity equal to 40 (27.8) V, 350 (380) A, 2.1 (1.7) ms, 276 (176) Hz, 80 (60) A respectively. The process evaluation in this phase was manly based on weld cross-section shape. Welds with incomplete penetration and high porosity were discarded. Figure 29 reports weld cross sections achieved during experiments. It was decide to considered anymore the hybrid welding process because of excessive undercuts detected.

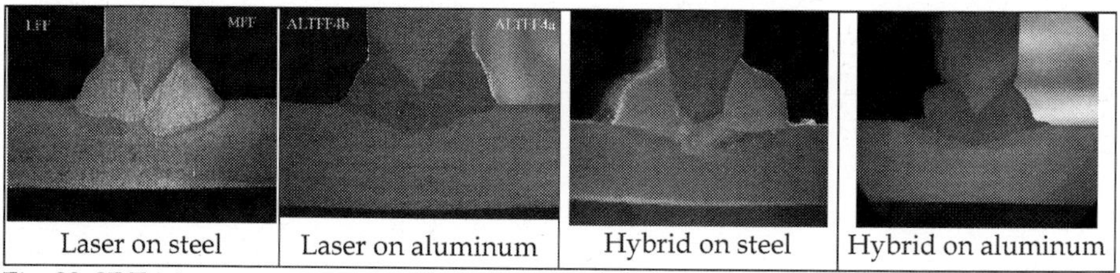

Fig. 29. 2F T-joints.

In the second experimental phase, double side/double square fillet (2S/2F) T-joints of STJ bars were realised, setting-up process parameters previously identified. The main objective was the coupling of the information of bead-on-plate with those coming from 2SF T-joints. In this way, it was possible to evaluate the effects of laser welding on the final joint geometry. Laser welding with filler wire was the only process employed and two classes of specimens were considered. In particular, the STJ were welded in as-clad condition (original height of 25.4 mm) or in condition (final height of 12.0 mm). The machined condition required that thickness of steel and aluminium alloys was reduced to 6.0 mm each by machining. Two 6.0 mm thick web plates were finally joined to STJs. Morphological and metallographic analyses were initially carried out to compare welding techniques and estimate the influence of the heat input on the Al/Fe interface. Figure 30 shows the cross sections of results of this experimental activity for hybrid and laser welding. The results were very similar to those achieved for 2F T-joints. More important was the material testing of 2S/2F T-joints. Specimen cut from centre of each joint were subject to tensile test to assess the mechanical strength (Figure 31). Length of the entire joint and of each specimen were 260.0 and 60.0 mm, according to specifications of American Bureau of Shipping.

Fig. 30. Double side/double square fillet square T-joints.

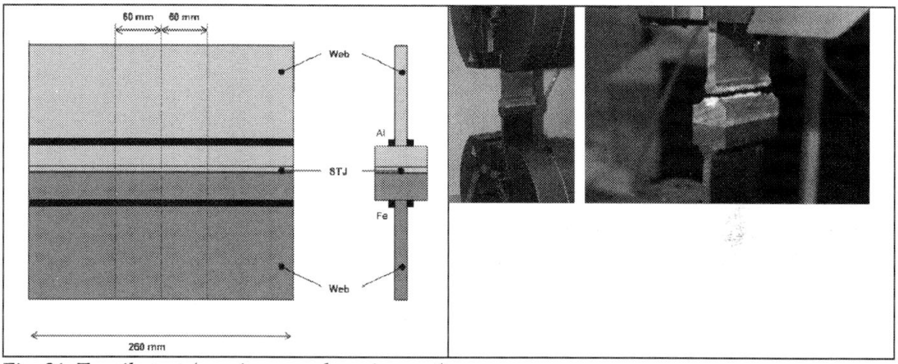

Fig. 31. Tensile test (specimen and equipment).

Tensile test were performed by grabbing both webs along the entire length and moving the crosshead of INSTRON 4485 tensile machine with a travel speed equal to 3.0 mm/min. The tensile test were stop once rupture occurred in web/STJ welds or in aluminium/steel STJ interface. Several tests were performed with several repetitions. The analysis of the results was mainly focused on mechanical strength of 2S/2F T-joints. Table 3 reports the average value between repetitions of the peak tensile load, peak tensile strength and peak tensile load on AA5083 alloy. The first two data were evaluated at bond interface while the last data was calculated at web section for the ultimate tensile strength of AA5083 alloy equal to 300 MPa. Laser beam welding led to joints with god strength in both as-clad and machined condition. All ruptures thus occurred on the aluminium side in aluminium welds or webs, satisfying the strict conditions of MILL-J-24445A (*AAVV, 1997*) for successful testing. According to this standard, each specimen tested must comply with one the following conditions for acceptance: i) failure in one of the web member and ii) failure of the bond surface at a load above that calculated to cause failure in one of the web members, based on the specific minimum tensile strength of the web material. It is possible to note these joints were also able to successfully overcome the second condition of the MIL-J-24445A, with T_{MAX} equal to 155.70 and 148.05 KN, in the hypothesis that rupture was localised in the aluminium/steel interface.

Specimen ID	Condition	Process	Tensile		Tensile	MIL-J-2445A
			T_{MAX} (KN)	σ_{MAX} (MPA)	T_{AA5083} (KN)	
2S/2F1	As-clad	Laser & Filler	155.70	93.04	108.00	YES
Machined	2S/2F2	Laser & filler	148.05	88.47	108.00	YES

Table 17. Tensile test- Sample dimensions & results.

Figure 32 shows the aspect of the rupture surfaces, organised in class of welding according to as-clad or machined state. Laser welded 2S/SF T-joints were characterised from the rupture surface positioned in the aluminium web material with no damage in the weld. The joints in the as-clad condition presented a sound weld. No particular effects were detected. Photos of Figure 33 are the most critic conditions in order to point out some specific outlines discovered during experimentation. The photo at left side represents a specimen in the machined state with aluminium alloy thickness equal to 6.0 mm. In the right side photo thickness of the specimen was further reduced before welding. In this case, the rupture occurred in the aluminium/steel interface and not in the aluminium web, pointing out that heat induced by laser beam welding seriously affected joint integrity if reduction was too high.

2S/2F1	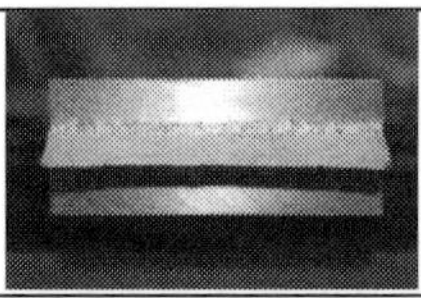	2S/2 F2	

Fig. 32. 2S/2F STJs after rupture (aluminium side).

2S/2F2		Red uced	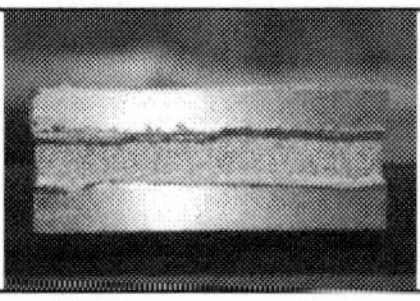

Fig. 33. 2S/2F STJs after rupture in machined state (aluminium side).

In this work the influence of different thermal loads caused by laser beam on mechanical strength of the Fe/Al explosion welded joints has been investigated. The present research has investigated the application of laser beam to weld components made of structural transition joints (STJs) between steel and aluminium base materials. The metallographic examinations and the tensile tests have pointed-out that laser-induced thermal loads strongly promoted inter-metallic growth at aluminium/steel interface. The condition has been also verified by reducing the thickness of the STJ. However, these loads have limited influence on integrity of bead-on-plate welds and double side/double square fillet T-joints because rupture mainly occurred in the web connecting the weakest material (aluminium alloy) with STJ. The condition has been also verified by reducing the thickness of the STJ.

Acknowledgement

The present research is only a part of activities performed by CALEF Consortium (http://www.consorziocalef.it/), the main members of which are the CRF, ENEA, Polytechnic of Bari and Rodriquez Cantieri Navali SPA. In particular, Rodriguez Cantieri Navali SPA directly funded the present research through the ENVIROALISWATH project. The authors wish to thank Dr. Donato SORGENTE and Dr. Gianfranco PALUMBO for their support during experimentation, Dr Marco BRANDIZZI of FIAT Research Center for his suggestions, Dr. Marco VITTORI ANTISARI and Dr. Giuseppe BARBIERI of ENEA –

Casaccia Research Center, Laura CAPODIECI of ENEA - Dept. FIM - Composites & Nanostructured materials Section, Enzo PUTIGNANO and Roberto DEBONIS of CALEF Consortium.

11. References

AAVV, (2000). *American Bureau of Shipping, Materials and Welding,* Supplementary Requirements for Naval Vessels, Part 2 Chapter 11, Section 6, Aluminum/Steel Bimetallic Transition Joints.

AAVV, (2007) *ASM Handbook Volume 6,* Welding, Brazing, and Soldering, Fundamentals of explosion welding.

AAVV, 1997, MIL-J-24445A, Joint, bimetallic bonded, aluminum to steel, Military Specifications and Standards, 25-Jul-1977.

Acarer M., Gülenç B. & Findik F., (2004). The influence of some factors on steel/steel bonding quality on there characteristics of explosive welding joints, *J. of Materials Science,* 39 , 6457-6466.

Bankers J.G., Nobili A., (2002) Aluminium-Steel Electric Transition Joints - Effects of Temperature and Time upon Mechanical Properties, *TMS 131st Annual Meeting,* Seattle (USA), February 17-21, 2002.

Borrisutthekul R. & alii, (2007). Suppression of intermetallic reaction layer formation by controlling heat flow in dissimilar joining of steel and aluminum alloy, *Materials Science and Engineering A,* 467, 108-113.

Bruckner J., Considering thermal processes for dissimilar metals - Joining steel to aluminium in heat-intensive applications, *The Fabricator,* August 28, 2003, http://www.thefabricator.com/Metallurgy/ Metallurgy_Article.cfm?ID=676.

Capodiceci L., (2007). Analysis of Explosion welding steel /aluminum transition joint with Scanning Electronic Microscope and X-Ray Microanalysis, *ENEA Internal Technical Report,* RT/ENEA/06/11 of FIM COMP Dept.

Chao R.M., Yang J.M. & Lay S.R., (1997). Interfacial toughness for the shipboard aluminium/steel structural transition joint, *Marine Structures,* 353-362.

Chen C.M., Kovacevic R., (2004). Joining of Al 6061 alloy to AISI 1018 steel by combined effects of fusion and solid state welding, *Int. J. of Machine Tools & Manufacture,* 44, 1205–1214.

Chen L. & alii, Clad metals, roll bonding and their applications for SOFC interconnects, *Journal of Power Sources,* 152, 40–45.

Deqing W., Ziyuan S. & Ruobin Q., (2007). Cladding of stainless steel on aluminium and carbon steel by interlayer diffusion bonding, *Scripta Materialia,* 56, 369–372.

Durgutlu A., Gulenc B. & Findik F., (2005). Examination of copper/stainless steel joints formed by explosive welding, *Materials and Design,* 497–507.

Kobayashi S. & Yakou T., (2002). Control of intermetallic compound layers at interface between steel and aluminum by diffusion-treatment, *Materials Science and Engineering A,* 338, 44-53.

Jindal V., Srivastava V.C., (2008), Growth of intermetallic layer at roll bonded IF-steel/aluminum interface, *J. of Materials Processing Technology,* 195, 88–93.

Lee J.E. et alii, (2007). Effects of annealing on the mechanical and interface properties of stainless steel/aluminum/copper clad-metal sheets, *J. of Materials Processing Technology*, 187–188, 546–549.

Li S.-X. et al., (2008). Fatigue damage of stainless steel diffusion-bonded joints, *Materials Science and Engineering A*, 480, 125-129.

Missori S., Murdolo F. & Sili A., (2004), Single-Pass Laser Beam Welding of Clad Steel Plate, *Welding Journal*, 02, 65-71.

Montgomery D.C., 2000, *Design and Analysis of Experiments,* 5th Ed., J.Wiley and sons, 1 (2000) 274-275.

Mousavi S.A.A.A., Al-Hassani S.T.S. & Atkins A.G., (2008). Bond strength of explosively welded specimens, *Materials & Design*, 29 , 1334-1352.

Mousavi S.A.A.A., Barrett L.M. & Al-Hassani S.T.S., (2008). Explosive welding of metal plates, *J. of Materials Processing Technology*, 1-3, 224-239.

Mousavi S.A.A.A. & Sartangi P.F., (2008). Effect of post-weld heat treatment on the interface microstructure of explosively welded titanium–stainless steel composite, *Materials Science and Engineering A*, 494, 329-336.

Naoi D. & Kajihara M., (2007), Growth behavior of Fe_2Al_5 during reactive diffusion between Fe and Al at solid-state temperatures, *Materials Science and Engineering A*, 459, 375-382.

Peyre P. & alii, (2007). Generation of aluminium–steel joints with laser-induced reactive wetting, *Materials Science and Engineering A*, 444, 327-338.

Sierra G. & alii, (2007). Steel to aluminium key-hole laser welding, *Materials Science and Engineering A*, 447, 197-208.

Tricarico L. & alii, (2007). Effects of laser welding on properties of Al/Fe explosionwelded structural transition joints, *ATA Proc. of 2nd Int. Conf. on "Advanced Materials and Technologies for Transportation Industry*, ELASIS in Pomigliano D'Arco (Italy), October 18-19 2007.

Young G.A., Banker J.G., Explosion Welded, Bi-Metallic Solutions to Dissimilar Metal Joining, *13th Offshore Symposium*, February 24, 2004, Houston (USA).

Watanabe T., Takayama H., Yanagisawa A. (2006). Joining of aluminium alloy to steel by friction stir welding, *Journal of Materials Processing Technology*, 178, 342–349.

Vollertsen F. (2005). Developments and trends in laser welding of sheet metal, *Advanced Materials Research*, 6-8,59-70.

5

Laser welding application in crashworthiness parts

Nuno Peixinho
University of Minho - Dept. Mechanical Engineering
Portugal

1. Introduction

This chapter presents applications and research results of laser welding in crashworthiness parts. The term "structural crashworthiness" is used to describe an investigation into the impact performance of a structure when it collides with another object. This type of study is required in order to calculate the forces during a collision, which are needed to assess the damage of structures and the survivability of passengers in vehicles, for example. An important aspect of crashworthiness studies deals with impact energy absorption since the main purpose of vehicle crashworthiness is the dissipation of energy in specially designed zones while maintaining a survival space for passengers in stiff zones.

Several important technologies have been used to improve the crashworthiness of structures. Particularly in the transportation industry, the conflicting requirements of weight reduction, emissions reduction, improvement of energy absorption and reduction of intrusion and force levels, has led to advances in different areas: materials (increased use of light-alloys, polymers and high-strength steels), technologies (laser welding, tailor-welded blanks, hydroforming), design tools (computational power, advanced software, analytical tools).

The improvement of crashworthiness and reduction of weight using high strength steels (or other alternative materials) cannot be applied directly without considering other features. Thickness reduction cannot be performed without taking into account other design restrictions, particularly stiffness, but also dynamic stability i.e. resistance to the transition from progressive folding to global buckling. For this purpose design simulation tools such as the finite element method can make an important contribution. But also a conjugation with developments in technology processing can be adopted since laser welding and hydroforming have the potential to increase stiffness.

The interest of laser welding is related to the concept of intelligent design wherein the manufacturing process can bring advantages to both energy absorption and stiffness thereby increasing the potential advantage of using higher strength materials. Of particular importance has been the use of laser welding in tailor welded blank technology, which further allows the use of material in a more efficient way. For example, using higher thickness sheet metal where it is more needed. Or using higher strength materials in areas where more strength is needed. The application of such concept is currently in use but can become more widespread with more extensive research and driven by the need to reduce weight in automobile structures. Joining of dissimilar materials has been given much attention in recent years due to their superior functional capabilities. The welding in tailor welded blanks of different high strength steels has already been accomplished and tested by the author. The application in beam structures of Dual-Phase DP600 and DP1000 steel materials with different thicknesses allowed for improvements in specific energy absorption in impact loading.

This chapter presents a literature review of the application of laser welding in crashworthiness parts, including a description of relevant background information, major issues, current trends and future challenges. Additionally several research results and applications from the author's experience are presented. These include:

- Laser welded crashworthiness parts: impact energy absorption of tubular structures and use of tailor welded blank technology in beam structures made of dissimilar materials (Figure 1);
- Application of laser welding in the development of components with localized thermal triggers.

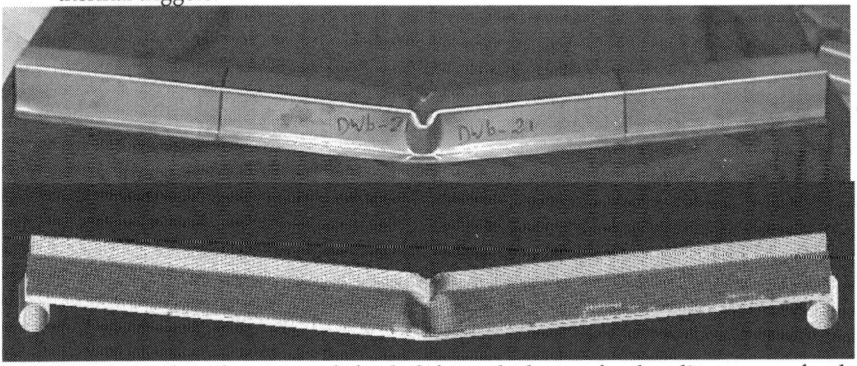

Fig. 1. Experimental and numerical final deformed shapes for bending tests of tubes manufactured using laser welding and tailored blanks (DP600 + DP800)

2. Laser Welding process and applications

For several years car body assembly techniques were dominated by spot-welding techniques. Resistance spot-welding is an affordable technology to join steel sheets and other metals. It can also be effectively automated while not requiring special preparation of the parts before joining. However, it has some disadvantages: spot connection system can cause high stress concentration near its location that can lead to local failure; there is a heat-affected zone and therefore a local modification in the material strength and relevant residual stresses; spot welding cannot be used to join different materials, unless using costlier special procedures. Laser-welding, as an alternative procedure, is a continuous joining technique, yet a very flexible process that can also be easily made automatic. Recent developments in high power CO2 lasers and robotic control have accelerated the application of laser-welding to vehicle structure fabrication and assembly in the automotive industry. It has also been shown that laser-welding gives many advantages if compared with other welding processes: a low heat input; a small heat-affected zone (Figure 2); low welding distortion, welding in an exact and reproducible manner, and welding at high speed. With the introduction of new laser technology, such as high power Nd:YAG and CO2 lasers and fibre-optic beam delivery systems, the automotive industry is re-evaluating assembly systems for body manufacture.

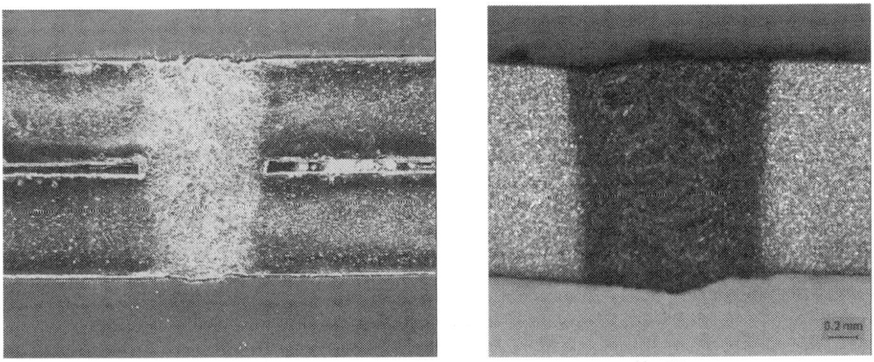

a) b)

Fig. 2. Examples of laser welds: a) lap weld; b) butt weld

Welding occurs when materials are heated to a molten state and fused together. Lasers generate light energy that can be absorbed into materials and converted to heat energy. By employing a light beam in the visible or infrared portion of the electromagnetic spectrum, energy can be transmitted from its source to the material using delivery optics which can focus and direct the energy to a very small, precise point. Since the laser emits coherent radiation, the beam of energy has minimal divergence and can travel large distances without significant loss of beam quality or energy.

Laser welding techniques take advantage of the capability for applying thermal energy to small areas in an efficient manner. This feature offers some distinctive metallurgical advantages in some welding applications, but also creates some unique problems. Since the surface heating generated by the laser light relies upon the material's heat conductivity to produce the weld, penetration is usually limited to less than 2 millimeters. However, using a technique known as "keyholing," higher power lasers (>106 W/cm^2) can make deeper penetrations: by heating the spot of laser focus above the boiling point, a vaporized hole is formed in the metal. This is filled with ionized metallic gas and becomes an effective absorber, trapping about 95 percent of the laser energy into a cylindrical volume, known as a keyhole. Temperatures within this keyhole can reach as high as 25,000 °C, making the keyholing technique very efficient. Instead of heat being conducted mainly downward from the surface, it is conducted radially outward from the keyhole, forming a molten region surrounding the vapor. As the laser beam moves along the work-piece, the molten metal fills in behind the keyhole and solidifies to form the weld.

Generally, there are two types of lasers that are being used for welding operation: CO2 and Nd:YAG. Both CO2 and Nd:YAG lasers operate in the infrared region of the electromagnetic radiation spectrum, invisible to the human eye. The Nd:YAG provides its primary light output in the near-infrared, at a wavelength of 1.06 microns. This wavelength is well absorbed by conductive materials, with a typical reflectance of about 20 to 30 percent for most metals. The near-infrared radiation permits the use of standard optics to achieve focused spot sizes as small as 0.025mm diameter. CO2 lasers have an output wavelength of 10.6 micron and an initial reflectance of about 80 percent to 90 percent for most metals thus requiring special optics to focus the beam to a minimum spot size of 0.08mm to 0.1mm diameter. However, whereas Nd:YAG lasers have power outputs up to 500 watts, CO2 systems can easily supply 10,000 watts and greater. As a result of these differences, the two laser types are usually employed for different applications. The powerful CO2 lasers overcome the high reflectance by keyholing, wherein the absorption approaches blackbody. The reflectivity of the metal is only important until the keyhole weld begins. Once the material's surface at the point of focus approaches its melting point, the reflectivity drops within microseconds.

One of the most promising technologies enabled by laser welding is tailor welded blanks. The application of such concept is becoming more widespread, driven by the need to reduce weight in automobile structures. Several challenges related to tailor welded blank technology have been addressed in recent years. Gaied and co-authors (Gaied et al., 2009) performed an experimental and numerical assessment of the formability of tailor welded blanks, being possible to predict the behavior of forming limit curves numerically and match with experimental results. Another recent study (Sheng, 2008) was focused on the forming of tailor welded strips with the interest of application in progressive forming. Dome tests were conducted and compared with simulation results allowing the validation of the approach for progressive forming. The forming parameters and processing dial-up for tailor welded blanks presents its own challenges (Panda & Kumar, 2001) as described in a study on the improvement of formability through application of counter pressure in biaxial stretch forming of tailor welded parts.

Joining of dissimilar materials has been given much attention in recent years due to their superior functional capabilities. The application in tailor welded blanks of different high strength steels has already been accomplished and tested by the author (Peixinho, 2004). The application in beam structures of Dual-Phase DP600 and DP1000 steel materials with different thicknesses allowed for improvements in specific energy absorption in impact loading. Another combination of dissimilar materials is of aluminium with steel due to its potential application in aerospace and automotive industries. Joining material combinations such as aluminium and steel poses a number of problems including formation of brittle intermetallic compounds, poor wetting behaviour of aluminium and differences in physical and chemical properties of the base metals. A recent numerical study by Padmanabhan and co-authors (Padmanabhan et al., 2008) determined the formability characteristics of aluminium–steel tailor-welded blanks. Aluminium (AA6016-T4) blank sheet was combined with a range of steel blank sheets namely, mild-steel (DC06) and high strength steels (AISI-1018, HSLA-340, and DP600) to form four different Al–steel tailor-welded models.

Other numerical studies on tailor welded blanks (Liu et al., 2007) and (Qiu & Chen, 2007), have clarified the need for reliable and numerical simulations in stamping and hydroforming processing of tailor welded blanks. The reliability of numerical simulations is a resultant of mesh quality, finite element formulation but also the availability of material properties that describe adequately material behaviour. For example, (Cheng et al., 2007) presented an experimental method of analysis to determine the tensile properties of welds of the heterogeneous tailor-welded blank (TWB) and its base metal. A real-time microscopic recording system was developed to acquire the true stress–strain data of the weld during tensile testing. A characterization of tensile properties of tailor welded IF steel sheets was

also presented by Panda and co-authors (Panda et al., 2007). Efficient numerical models can also be used for design optimization of welded blanks, as presented by (Kim et al., 2000).
The design of vehicle front structures for crashworthiness is commonly based on a series of rigid subsystems that constitute a nearly undeformable survival cell for the passengers, and deformable subsystems able to efficiently dissipate the vehicle kinetic energy. During frontal crash the front rails represent the main deformable components aimed to dissipate the kinetic energy of the vehicle, therefore their behaviour is crucial to obtain good vehicle performance, with stable and controlled energy dissipation. The design of the front rail, usually consisting of a thin-walled prismatic column, requires definition of materials and geometry: shape and dimensions of the cross section, thickness of the material.
An important issue in automotive design the trend for adopting materials allowing more efficiency. Widely used materials like deep-drawing steels are being substituted by high strength steels (dual-phase, TRIP steels.), aluminium alloys, magnesium alloys, and various types of polymeric materials and composites. There are several reasons for this change: the structure weight reduction that allows for more accessories and safety components, the need for higher stiffness and strength of the car body structure and cost reduction. Several problems are associated to the introduction of new materials: their properties are still not completely known, the usually adopted technologies are sometimes not usable anymore, and new environmental and protection problems arise. Additional problems are associated to the joining techniques.
Regarding laser-welded tubular structures subjected to impact loading there are not extensive results available in the literature. Radlmayr and co-authors (Radlmayr et al., 1993) examined octagonal sections, whereas (Geoffroy et al., 1993) reported that laser-welded crash boxes have slightly improved or similar energy absorption characteristics, as compared to spot-welded ones, but a consistent stabilization effect of the plastic folding was obtained. The author (Peixinho et al., 2006) presented experimental and numerical results for impact testing of thin-walled structures made of high strength steels and using spot-welding, laser welding and tailor welded blanks techniques. The results highlighted the advantages of continuous joints in thin-walled structures and improvements in energy absorption using laser welding and tailor welded blank technology. Peroni and co authors (Peroni et al., 2009) presented results of an experimental program aimed to study the progressive collapse behaviour of some thin-walled closed-section structural sections made from deep-drawing steels and joined with different joining. Solutions characterized by different continuous joining technologies were examined and compared to the usual spot-welding solution. The obtained results indicate that continuously joined structures are at least equivalent to and generally better than spot-welded structures, and have further advantages typical of these joining solutions (higher stiffness and fatigue strength, improved vibration response, especially in the case of adhesive joints). Stability of folding was reported as much improved with laser welding compared to the other joining solutions.

3. Research work in laser welding applications

3.1 Laser welded crashworthiness parts

This section presents results from an experimental program that included quasi-static and dynamic testing of tubes made of high-strength steels, as described in (Peixinho, 2004). These included short tubes, with a length of 250 mm for axial crush testing, and tubes with a length of 1000 mm used in three-point bending tests. The selection of the geometries was motivated by practical loading situations and structural sections commonly used in crash-absorbing structures and components. This was the case with the top-hat sections. Also the hexagonal section was examined since it should be more efficient than the traditional top-hat and double-hat sections.

The materials examined are Dual-Phase and TRIP steels, having commercial designations: DP600; DP800; TRIP600. These designations reflect the nominal Ultimate Tensile Strength of the material. All investigated steels were industrially melted, cold-rolled, annealed and temper rolled to a final thickness between 1.0 and 1.4 mm. They were tested in the as-shipped condition. Table 1 shows the nominal chemical compositions of the investigated steels, the elements not mentioned are within the usual limits for cold rolled steels. The chemical composition was obtained recurring to mass spectrometers: Philips model X'Unique II and Spectrolab model M5. Typical microstructures of the tested steels are shown in figures 3 and 4, obtained by optical microscopy and Scanning Electron Microscopy.

Steel grade	C (%)	Si (%)	Mn (%)	P (%)	S (%)	Nb (%)	Al (%)
DP600	0.11	0.31	0.78	0.015	0.01	-	0.04
DP800	0.13	0.21	1.48	0.015	0.01	0.02	0.04
TRIP600	0.23	0.33	1.49	0.011	-	-	0.88

Table 1. Nominal chemical composition, mass contents in %

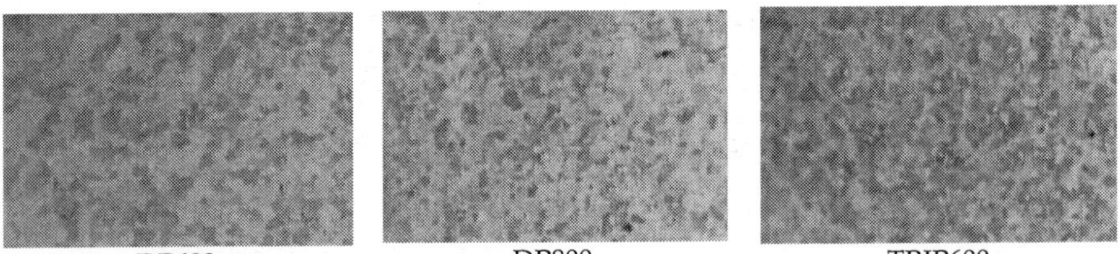

DP600 DP800 TRIP600

Fig. 3. Optical micrographs of steel samples (200x)

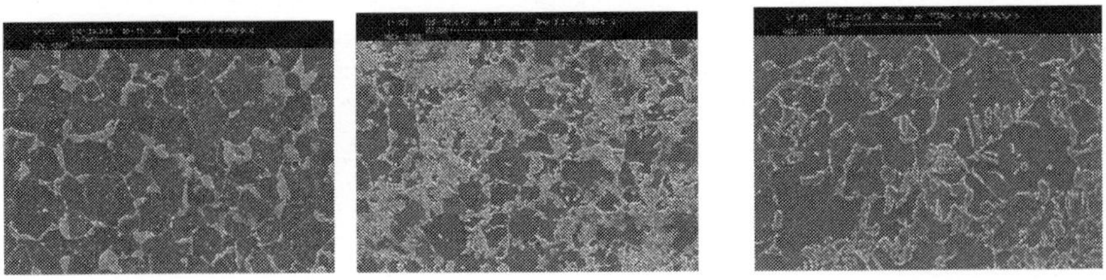

DP600 DP800 TRIP600

Fig. 4. SEM images of microstructures (1500x)

Besides the three types of steels, top-hat and hexagonal geometries, also spot welding and laser welding technologies were used in the manufacturing of tubes. Laser welds were performed continuously at the centre of the joining flanges of the tubes. After specimens had been cut to length, the end faces were ground to ensure the faces were square and parallel to each other. Generally the tubes used for dynamic testing were manufactured with triggers. These are indentations in the tubes used to initiate the folding process. The crush and bend tests were performed using three different test equipments to cater for different test speeds. A summary of the experimental program is shown in table 2. In figure 5 the nominal dimensions of the studied sections are presented.
The bending tests were performed on tubes with a top-hat geometry and having dimensions of the section presented in figure 5.a). Some of these tubes were manufactured using two different steel grades: DP600 and DP800 with the manufacturing process making use of tailor welded blank construction. In this case, the central section of the tubes was manufactured with DP800 and the extremities using DP600. This type of solution is currently of great interest for the automotive industry since it allows the manufacture of components having a variable thickness. In this way greater weight efficiency can be achieved since higher thickness can be used in specific locations, for example in the impacted area of a energy-absorbing beam. In this case case, besides different thickness, also different materials were used which makes the welding process more difficult. The set-up for the bending tests is presented in figure 6. The details of the tubes manufactured using tailor-welded blanks are presented in figure 7.

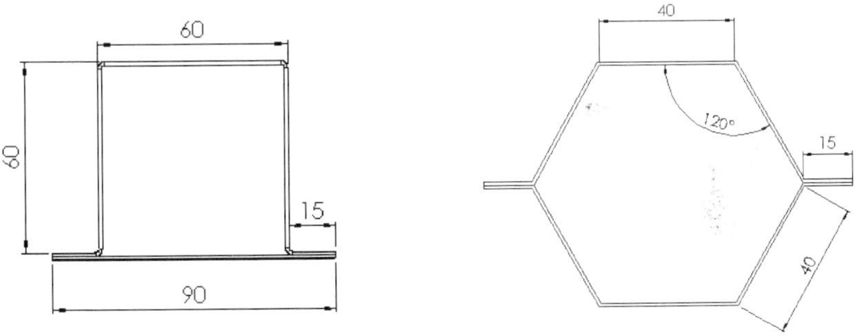

a) b)

Fig. 5. Nominal dimensions of sections: a) Top-hat; b) Hexagonal

Material	Geometry		Welding technique		Trigger	Reference of tests
	Top-hat	Hexagonal	Laser	Spot-weld		
Axial Crush tests						
DP600	V		V		V	DW1; DW2; DW3; DW4 Dtc7; dtc8
	V		V			DW17 QS4; QS5; QS6
	V			V		DW5; DW6; DW7; DW8 dtc3; dtc4 QS1; QS2; QS3

TRIP600	V		V		V	DW9; DW10; DW11; DW12 dtc1; dtc2
	V		V(1)		V	DW22; DW23 QS13; QS14
		V	V		V	DW13; DW14; DW15; DW16 dtc5; dtc6
	V		V			DW18 QS7; QS8; QS9
		V	V			QS10; QS11; QS12
	V			V		DW24; DW25; DW25; DW27 QS15; QS16
Bending tests						
DP800	V		V			DWb19; DWb20 QSb1; QSb2
DP800+DP600	V		V(2)			DWb21 QSb3; QSb4

(1) Laser-welding using two parallel welds
(2) Tube manufacturing using tailor-welded blanks

Legend of test nomenclature:
DW: drop-weight crush tests
DWb: drop-weight bending tests
dtc: crush tests at 250 mm/s
dtcb: bending tests at 250 mm/s
QS: quasi-static crush tests
QSb: quasi-static bending tests

Table 2. Summary of experimental program

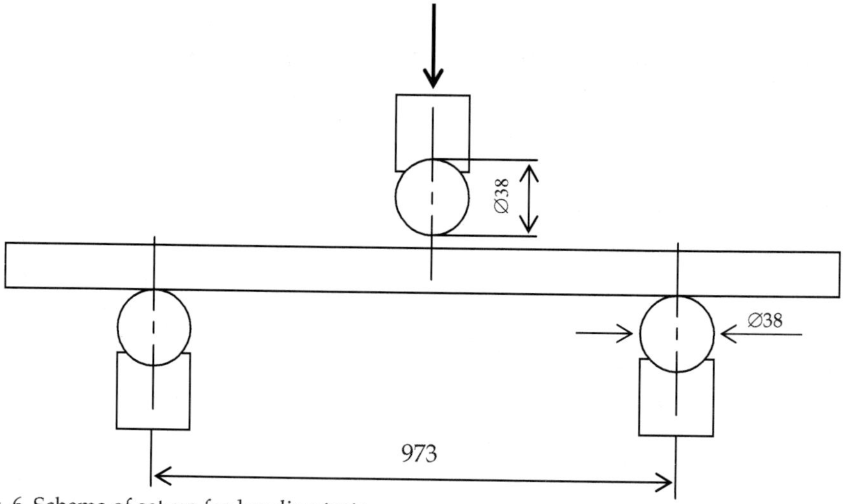

Fig. 6. Schema of set-up for bending tests

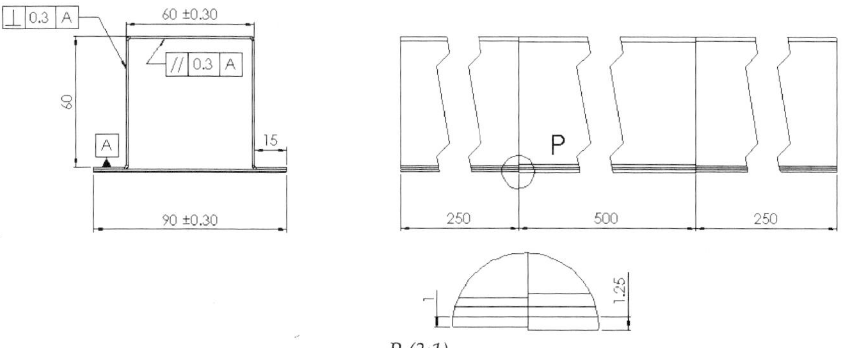

P (3:1)

a)

b)

Fig. 7. a), b). Details of tubes manufactured using tailor welded blanks

Quasi-static tests on thin-walled tubes were performed on a DARTEC M1000 machine with a 600kN capacity. The DARTEC machine was operated at a constant cross-head speed of 0.1 mm/s. During the tests, the compressive load and displacement were measured using a strain-gauge load-cell and a LVDT. The machine was controlled by a PC that also recorded and processed the measured data from the test machine. The entire crushing process was composed of individual strokes of 90 mm displacement, as the test machine was only capable of performing strokes to a maximum of 100 mm extension. The bending tests were performed with only one stroke of 90 mm displacement.

Several tests were performed at intermediate speeds of approximately 250 mm/s, using a DARTEC testing machine with a load capacity of 250kN. The control of the test machine and recording of data also made use of a PC. In this equipment the specimens were placed centrally and upright between two endplates but without any further support.

The impact tests were conducted on a drop hammer. The crush tubes were impacted at their top by a falling mass, which was laterally guided by rails. The specimens were placed vertically on an anvil and hit by the impactor. No end constraints were provided, however special care was taken with the surfaces of the anvil, impactor and test specimens in order to obtain parallel faces. This included machining the top ends of the tubes as well as the anvil and the impacting face of the falling mass. The impactor used in the dynamic bending tests

had a cylindrical end with a 38mm diameter and a support for the tubes as presented in figure 6.
The dynamic tests were carried out at test energies ranging from 0.575 to 14.270 kJ. Different test energies were obtained changing the drop height and the impact mass. Figure 8 shows the drop hammer rig as well as associated instrumentation, test supports and specimens. A Laser-Doppler velocimeter was used to obtain the velocity-time history during the dynamic tests. It was then possible to obtain the load-time, displacement-time and load-displacement histories. From these data, the axial displacement, or crushing distance, as well as the displacement averaged mean load values may be calculated.

a)

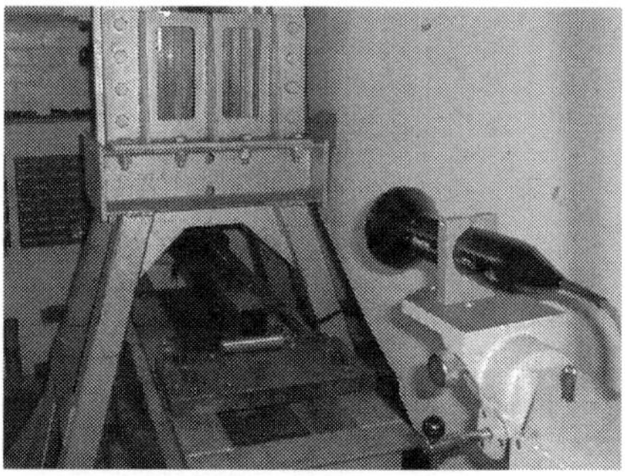

b)

Fig. 8. a) Drop-hammer rig and instrumentation (recording camera on the left); b) Image of drop-hammer rig with Laser-Doppler velocimeter in the foreground.

The crushing tests of tubes were used to determine of maximum crushing force $P_{máx}$, mean crushing force P_m, absorbed energy E_a, as well as to perform a qualitative analysis of the crushing behaviour that included the number of lobes formed, types of lobes, and collapse type. The specimens were accurately measured prior to and after testing. The total crushing distance δ was measured as the difference of the height of the specimen before and after testing. The recorded force-displacement curves obtained in the DARTEC tests were integrated with respect to the deflection δ to determine the mean crushing force. The mean load P_m was then calculated using the expression:

$$P_m = \frac{E_a}{\delta_f} \tag{1}$$

where δ_f is the final deflection. The mean load is an indication of the energy-absorbing ability of a structure, when compared to the axial displacement required to absorb that energy. Subsequently, the mean load and absorbed energy were also calculated for prescribed displacement values. The maximum crushing force was determined from the load curves. However, this value is only reliably obtained in the quasi-static tests since inertia effects and fluctuations in the initial load peak exist in the dynamic tests which makes accurate recording difficult.

In the dynamic tests the velocity-time readings obtained with the Laser-Doppler velocimeter were differentiated and integrated to obtain the load-time, displacement-time and load-displacement histories. From these data, the axial displacement, or crushing distance, as well as the displacement averaged mean load values may be calculated using the absorbed energy in the same manner as with the quasi-static tests.
In general, the spot-welds resisted well the loading and deformations. Besides localised material fracture, only in a few tubes and in a few locations, spot-welds were halfway torn apart. Laser welds only presented problems for the TRIP600 steel. Only in a few of the top-hat tubes manufactured with this material it was possible to obtain regular progressive folding without separation of the hat-section and closeout panel. However, the hexagonal laser-welded sections and the spot-welded tubes manufactured with TRIP600 did not present that problem.
The analysis of results of energy absorption properties should consider the folding behaviour and its initiation. Generally, the dynamic tube crushing tests made use of initiators or triggers in the form of indentations in the tubes. These worked satisfactorily in the dynamic tests, providing an efficient initialisation of the crushing process near the top of the specimen (proximal face to the impact mass). This feature could be observed from the camera recordings. Figures 9 and 10 present examples of the initiation of folding. The images were obtained with the recording camera rotated for best resolution within the test area.

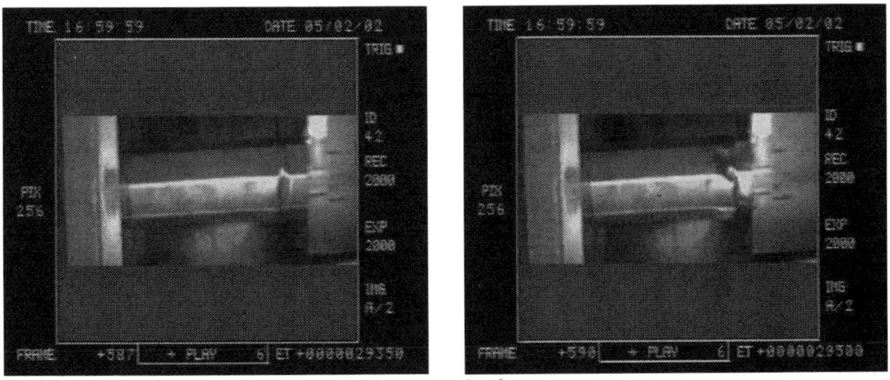

Fig. 9. Initial sequence of crushing of a hexagonal tube

Generally, buckling was initiated at the proximal face of the specimens and progressed towards the distal end. However, in some cases, there was a simultaneous initiation of folding at both ends with a plastic buckle being developed near the distal end of the specimen. This buckle generally remained stable during further deformation of the specimen, which could be attributed to the contribution of the triggers at the opposite end of the specimens. In some of the tests with spot-welded tubes this buckle caused a near-simultaneous progression of the crushing process from both ends, or also instability towards the end of the deformation process. Since the spot-welded tube did not have triggers this occurrence is attributed to the competition between both ends in the contribution to the deformation process. In figure 10 this occurrence is also observed.

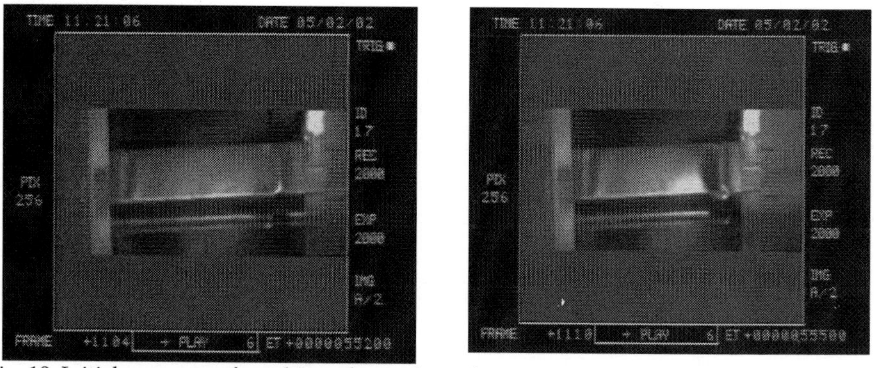

Fig. 10. Initial sequence of crushing of a top-hat tube

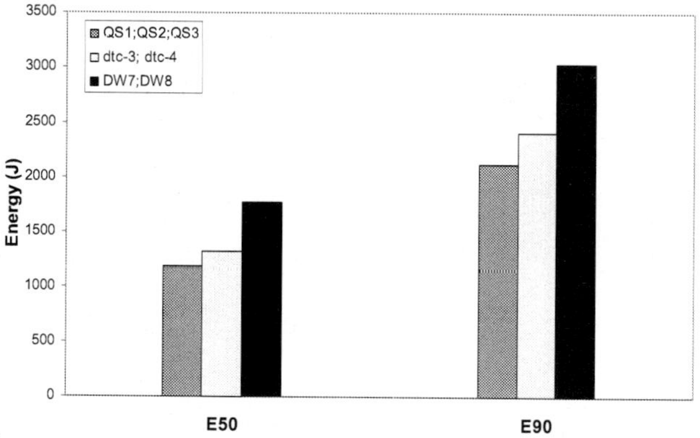

Fig. 11. Absorbed energies for DP600, top-hat geometry, spot welding

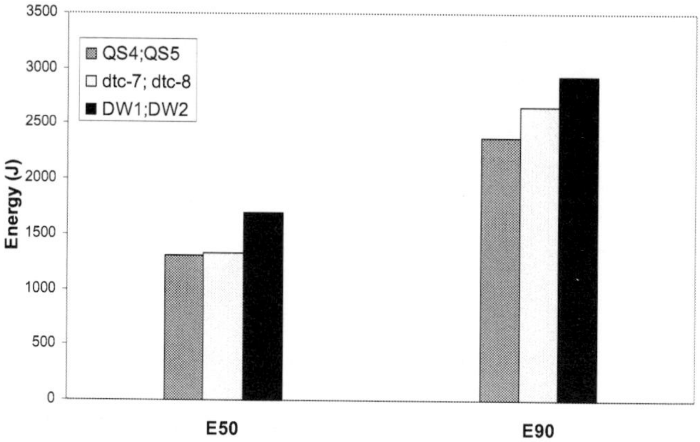

Fig. 12. Absorbed energies for DP600, top-hat geometry, laser welding

Several features can be observed from the results that allow a comparison of different materials, geometries and welding processes. This analysis can be performed by comparing the absorbed energies at prescribed displacements, in this case energies at 50mm and 90mm of crushing length. This analysis is important since the absorption of energy and its management are critical to obtain crashworthy structures. In figures 11 to 13 examples of absorbed energies at different crushing lengths (E_{50}; E_{90}) and different test velocities are presented. In these cases an increase of absorbed energies for impact loading is observed which was expected when considering inertia and strain rate effects.

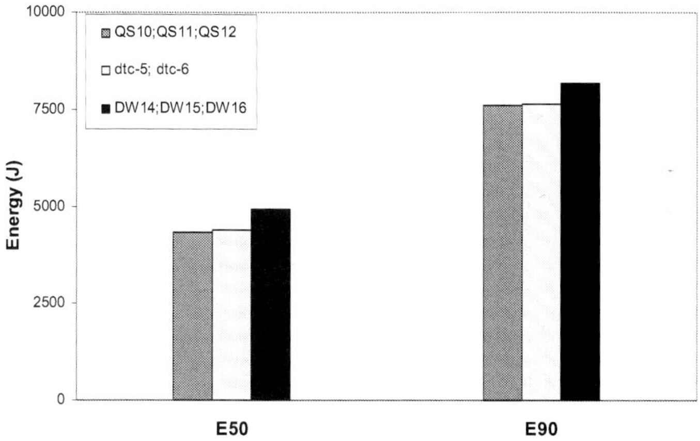

Fig. 13. Absorbed energies for TRIP600, hexagonal geometry, laser welding

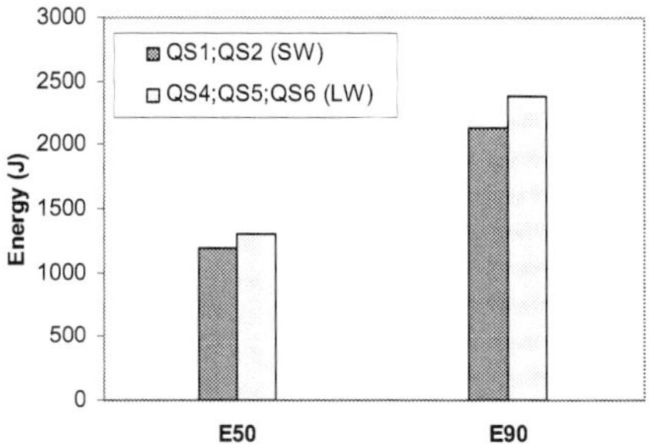

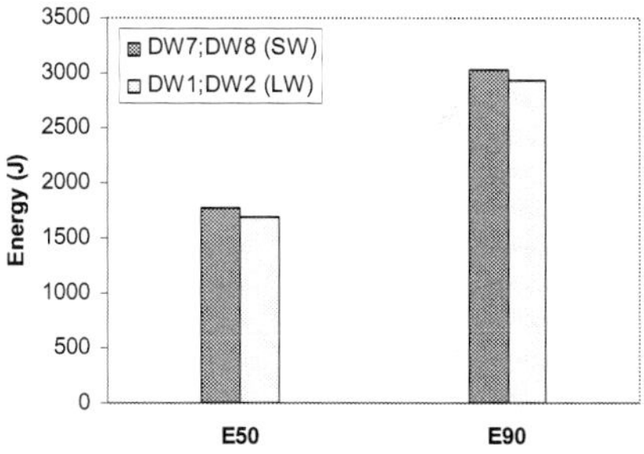

a) quasi-static crush tests b) dynamic crush tests

Fig. 14. Comparison of absorbed energies for spot-welded (SW) and laser welded(LW) top-hat tubes (DP600)

One of the observed characteristics in this study was the differences between spot-welded and laser welded connections used in the manufacturing process of the tubes. Figures 14 and 15 present a graphical comparison of absorbed energies in tubes manufactured using the two processes. The moderate increase in the amount of absorbed energy for a given crush distance in laser welded connections was expected, considering previously published results. However, in figure 14-b) it is observed that at higher impact speeds the spot-welded tubes absorbed a higher amount of energy. This was not observed for TRIP600 steel, although with this material the difference in absorbed energies between spot-welded and laser welded tubes in dynamic crush testing was very small. It is possible that at impact loading the continuous connection obtained using laser welds has undergone some local separation although this was not observed in the tests considered for this analysis.

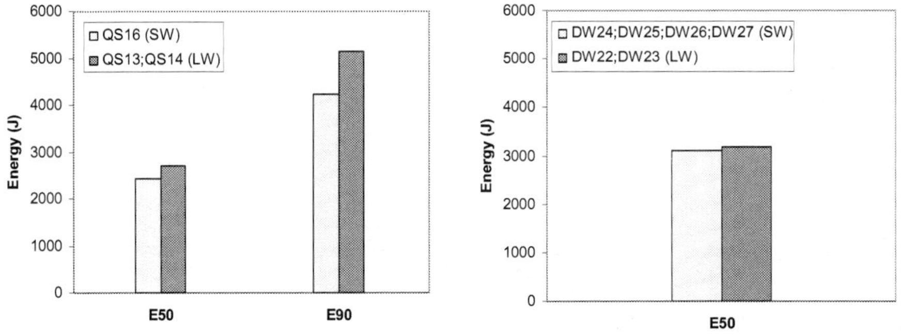

a) quasi-static crush tests b) dynamic crush tests

Fig. 15. Comparison of absorbed energies for spot-welded (SW) and laser welded (LW) top-hat tubes (TRIP600)

Another observed feature in the experimental tests was the efficiency of different sections for the purpose of energy absorption. This was possible in the tests of the TRIP600 material where the specific absorbed energies of top-hat and hexagonal sections were compared. Figure 16 presents results of that comparison. A remarkable increase in absorbed energy per unit weight is observed for hexagonal sections. This was expected considering existing results in the available literature (Auto/Steel Partnership, 1998) where the difference in the average static crush force between top-hat and hexagonal tubes having the same mass was of approximately 40%. In the present tests the increase in the average static crush force was of approximately 32% with the increase in the absorbed energies E_{50} and E_{90} ranging from 32.9 to 37.4 % in the quasi-static tests and 29.6 to 35.5% in the dynamic tests. This increase in the efficiency of the energy absorption is expected considering that thin-walled cylindrical shells have more efficient folding modes and that octagonal and hexagonal thin-walled sections are closer to the more efficient circular shape than top-hat sections.

In figure 17 a comparison of specific absorbed energies of DP600 and TRIP600 is presented, based in tests using the same geometry (top-hat). A noticeable increase in specific absorbed energy is observed for the TRIP600 material, in both quasi-static and dynamic tests. This difference can be attributed to the higher strain hardening and strength properties and also the higher elongation to fracture that implies a higher area under the stress-strain curve, which is directly related with energy absorption. However, it should be noted that the tests were performed in tubes manufactured using steel sheets with different thicknesses, which might induce differences in the folding process with consequences in the absorbed energy.

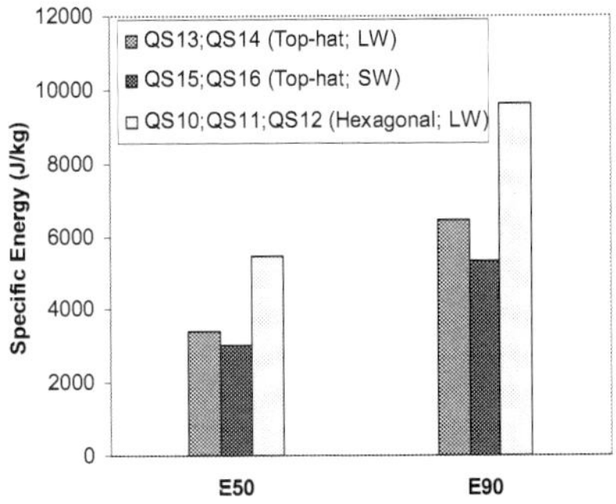

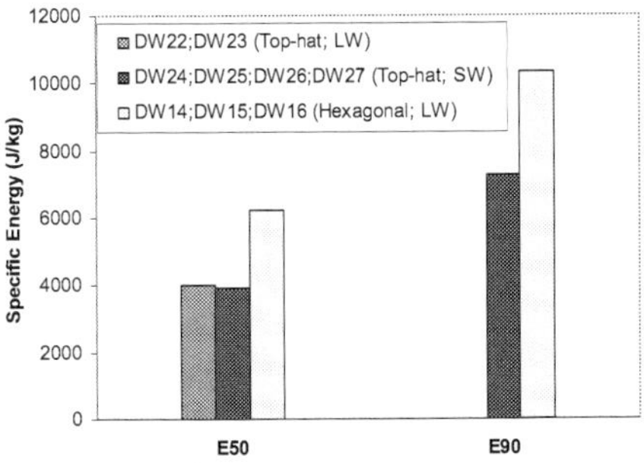

a) quasi-static crush tests b) dynamic crush tests

Fig. 16. Comparison of specific absorbed energies for top-hat and hexagonal tubes (TRIP600)

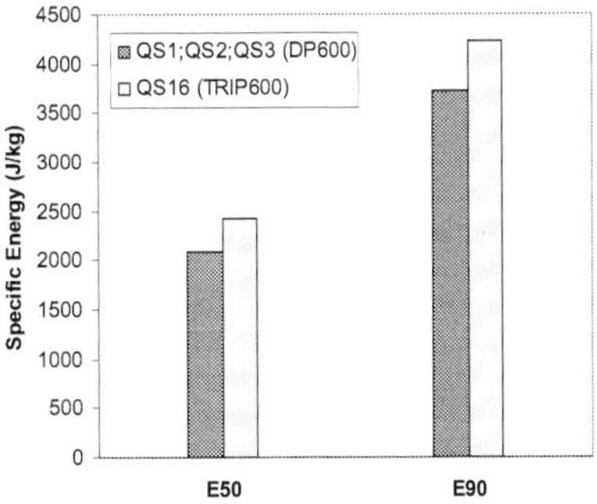

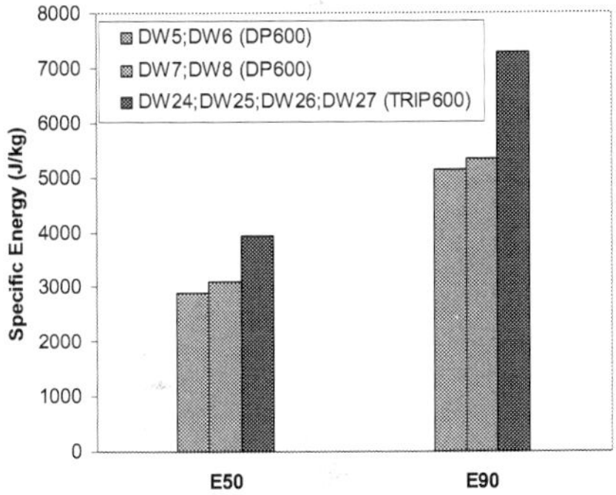

a) quasi-static crush tests b) dynamic crush tests

Fig. 17. Comparison of specific absorbed energies for DP600 and TRIP600 steels using top-hat geometry

The available data for bending tests allows the evaluation of some features. In figure 18 a comparison of quasi-static and dynamic absorbed energies is presented for the tubes manufactured using tailor-welded blanks. As expected a slight increase is observed for the dynamic case. Figure 19 presents a comparison of specific absorbed energies (E_{50} and total absorbed energy) between the tubes made of DP800 steel and the ones manufactured using tailor welded blanks (that use DP600 and DP800 steel grades). The tubes manufactured using tailor-welded blanks are more efficient because the plastic deformation is localized in the central area where the striker impacts the tube.

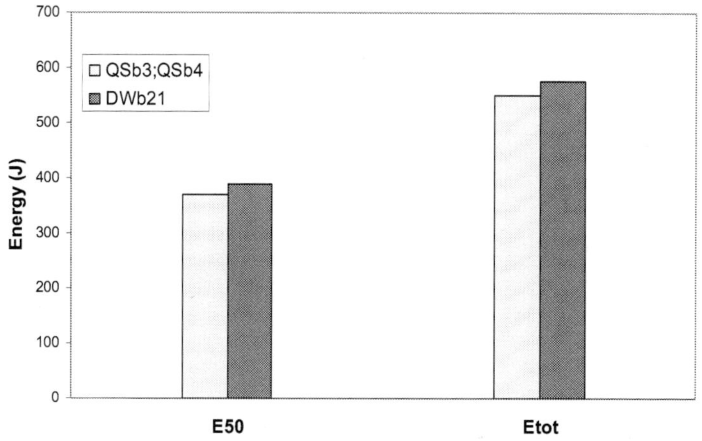

Fig. 18. Comparison of absorbed energies for bending tests of tailor welded tubes tested quasi-statically and dynamically.

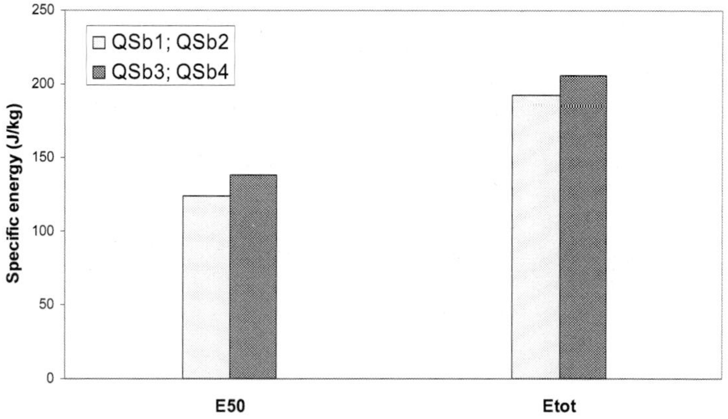

Fig. 19. Comparison of specific absorbed energies in bending tests of tubes manufactured using DP800 steel and tailor-welded blanks (DP600 and DP800 steel).

3.2 Application of laser welding in the development of components with localized thermal triggers

This section presents results of a study aimed at developing an approach consisting of local heating of aluminium alloy structures with the purpose of introducing a local modification of material properties. The main objective of this approach is the management of crash-energy absorption in a cost effective manner through the introduction of triggers: by local heating in areas chosen for triggers, local softening of aluminium can be induced thus

forcing the tubular structure to initiate deformation in prescribed locations and assure deformation in the mode of highest energy absorption.
Research studies have reported attempts to improve energy absorption of extruded aluminium tubing by artificially introducing various types of triggering dents (Kim, 2002); (Lee et al., 1999). The absorbed energy and crushing morphology were analyzed depending on number, shape, and location of triggering dents by using computer simulation.
The concept of using thermal modification of an aluminium alloy in localized areas can provide for a larger global deformation of a part and higher energy absorption before failure. Thus fracture in critical regions can be delayed and the total energy absorption can be accordingly increased. Such design features are also highly cost-effective in implementation compared to the alternative process of geometric redesign. This advantageous use of aluminium is therefore possible by applying "local material design", which in the present context is defined as controlled manipulation of material properties like strength, work hardening and ductility by means of non-homogenous heating, as originally presented (Bjørneklett & Myhr, 2003).
In particular, the buckling of crash boxes during a crash situation may be controlled by deliberately imposing local soft zones (i.e. thermally induced triggers). For the impact event simulation tools can be used to assess crashworthiness performance and even enable a combined simulation of the thermal processing and subsequent response in the final component subjected to dynamic loading.
This study presents preliminary results of temperature and heating cycle influence in material properties and microstructure of a selected 6060-T5 aluminium alloy. The objective of this research work is to improve the crushing stability and the absorption of energy originated from impact in tubular components. The improvement of the deformation is done by CO2 laser welding technology applied as a local heat treatment. This process will induce a micro structural modification caused by the heating in predefined zones that act as triggers of the folding process in the progressive impact energy absorption of tubular structures. It is well known that the 6060-T5 aluminum alloy suffers modifications in microstructure with heat-treatment. Technical literature presents different diagrams that show the behavior of this material at different temperatures and heat-cycle duration. It is shown that with temperature between 250 ° C and 550 ° C there is a significant modification in the microstructure with decrease on hardness. It should be mentioned that the time is also an important factor being the temperature and time interdependent

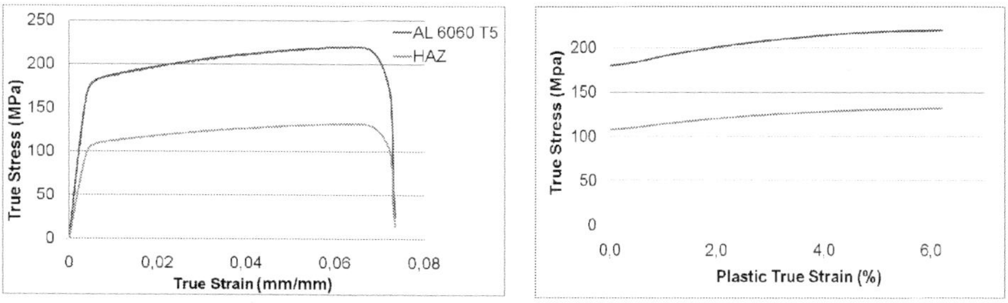

a) b)

Fig. 20 - a) AA 6060 T5 True stress–strain curve and on the heat affected zone; b) Model of plastic behaviour used in the numerical simulations.

The mechanical properties of the aluminium alloy 6060-T5 were obtained by static tensile tests, and the properties of the heating affected zone (HAZ) are about 60% less than the base material, according to the Vickers micro-hardness test, as one can see in Figure 20.
The aluminium alloy studied suffers modifications in microstructure for certain parameters, such as temperature and heating cycle. The significant changes in the microstructure of the alloy occur for temperatures between 250°C and 550° C where a decrease in hardness occurs. This is attributed to the dissolution of copper rich precipitates due to the imposed thermal cycle. Appropriate choice of heating cycle parameters is also important because the alloy may not need very long temperature cycles for full transformation, or very high temperatures, and these two factors vary depending on another, being at this moment the objective of the heat treatment the highest softening possible of the alloy.
For carrying out the furnace heat treatment, several samples were cut from aluminium alloy sheet (average thickness 1.5mm). The cut samples were then subjected to heat treatment: each sample placed in the central zone of a furnace for prescribed temperature and time.
For the laser heat treatment a CO_2 laser welding machine was used (Trumpf - 4000W). This was found suitable for the local softening approach. The density of energy could be regulated from laser power and feed rate thus varying material parameters and the heat affected zone.

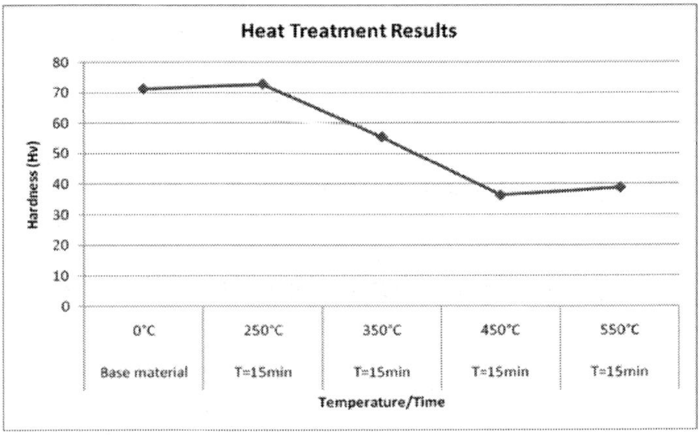

Fig. 21. Hardness results for furnace heat treatment.

Figure 21 presents results of Vickers micro-hardness test (with 100gf load) for the samples with furnace heat-treatment. Temperature and time are presented for the furnace tests.
The laser was used with 4 kW power and different feed rates. The hardness results and sample superficial aspect, presented in Figures 22 and 23, show a significant increase of the HAZ with the feed rate of 5 m/min. It is also possible to see that the minimum hardness, in the HAZ, is similar with the obtained in the bulk treated specimens (furnace heat treatment).

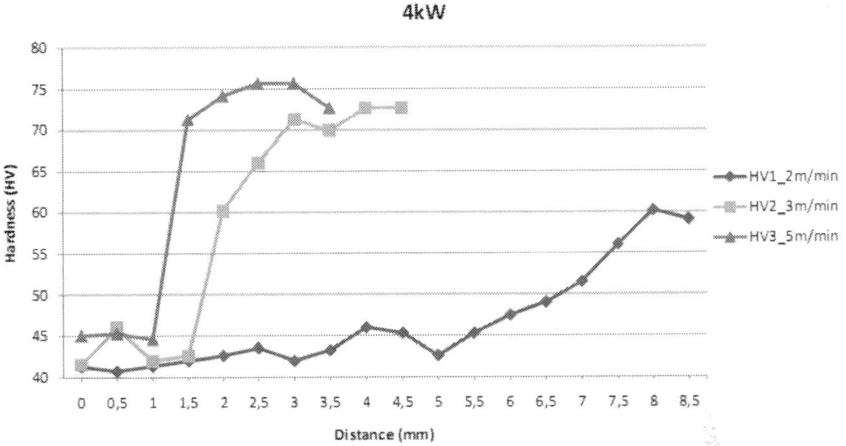

Fig. 22. Hardness results for laser heat treatment at center of HAZ (0 mm) and distance from center of HAZ.

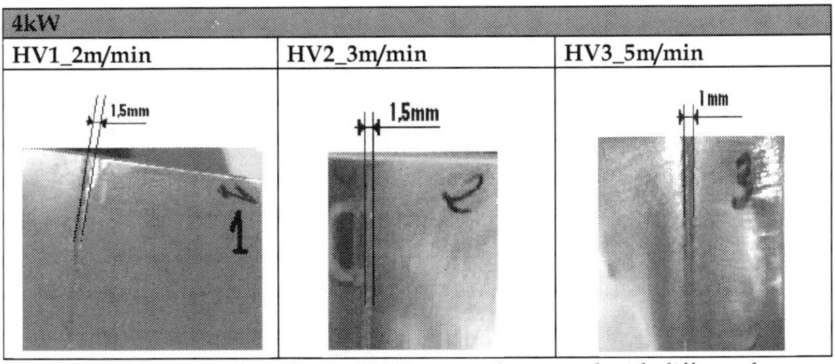

Fig. 23. Images of the heat affected zone HAZ in samples treated with different laser speeds.

The structure considered in this study is a prismatic column with square cross-section of aluminium 6060-T5. The dimension of the cross-section is 75x75 mm with 1.5 mm wall thickness, and the length of the column is 300mm. The local heating in areas chosen for triggers will be modelled in the numerical simulations through the modification of the mechanical properties, as shown in figure 20.b). The location of these triggers on aluminium alloy will be precisely induced thus forcing the column to deform in that zone.

The mechanical properties considered on the numerical simulations are Young's modulus $E=69\times10^3$ MPa, Poisson's ratio $\nu=0.3$, density $\rho=2700Kg/m^3$ and the initial yield stress $\sigma_y=180MPa$ for the base material and $\sigma_y=108MPa$ for the heat affected zone (HAZ). The complete true stress-strain relation used in the simulations is shown in 20-b). As the aluminium is insensitive to the strain rate effect, this is neglected in the finite element modelling.

The present simulations were performed with the commercial software LS-DYNA that is appropriate for non-linear explicit dynamic simulation for large deformations. The loading condition is the impact of a rigid mass of 70kg at an initial speed of 45km/h on the top of the model, as shown in Figure 24.a), being the lower part of the model clamped.
The elements used in this type of modelling need a good bending capacity and membrane behaviour for large in-plane deformations allowing for axial loads. With these requirements the chosen element is a Belytschko-Lin-Tsay shell element of four nodes, which is commonly used in crash simulations. This element type is suitable for the large deformations which occur in the folding process. Five integration points were used in the thickness direction.

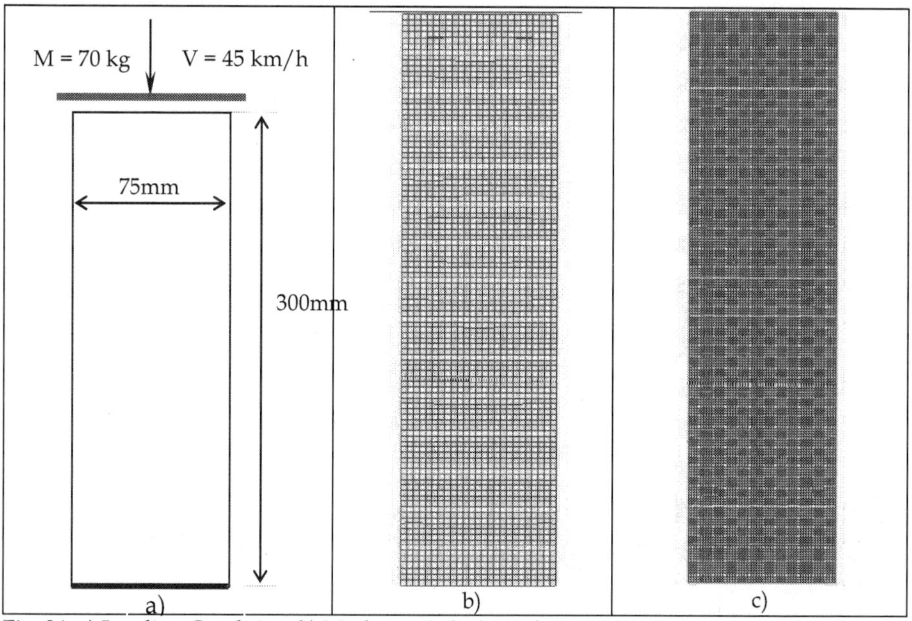

Fig. 24. a) Loading Condition; b) Mesh size 3x3; c) Mesh size 1.5x1.5

The contact between the rigid wall and the model is defined as surface-surface interaction with a friction coefficient equal to 0.1. Besides, self-contact with a friction coefficient 0.1 is defined on the model walls and gravitational acceleration is applied to the whole model.
In the numerical simulations the focus of the laser heat treatment was chosen for trigger dimension, and appropriate mesh size triggers are also chosen, as shown in Figure 24 b) and c). The studies are based on the 3mm and 1.5mm width of the laser focus, that is, the weld3 and weld1.5 as indicated in the figures.
A total of six triggered configurations were defined, depending on number, width, and location. The number of triggers can be largely divided into three types, i.e. without trigger, triggers in opposite sides (2 sides) and triggers around of the model (4 sides of the model), and their width is also varied either 3 or 1.5 mm, as shown in Table 3.
In all models the triggers are referenced to the top of the numerical model. For example, in reference 14x20 it is meant that the triggers are inserted in up to intervals of 20 mm, fourteen

triggers in along of the model. When the reference is 9x30 and 6x40 the same process is done, inserted at even intervals of 30mm/40mm with nine/six triggers in along of the model, respectively. For models with the reference 4x20, 4x30 and 4x40, only the initial four triggers are introduced in up to intervals of 20mm, 30mm and 40mm, respectively.
Plastic folds are initially formed in the upper part of the smart models, and continue to develop gradually down into the lower parts. Besides, as soon as the folds consist in a side of the model, they develop in the side opposed in turns. These folds are facilitating a mechanism to absorb the energy on the compressive deformation, therefore the tendency of formation of folds fulfils an important role in the absorption of energy.
The numerical results of some smart models are shown in Figures 25-27, where it is possible to observe that under dynamic loading models generally had a regular progressive folding, but some of them exhibited irregular plastic folding during the terminal crushing stages, as observed in model Weld3 4sides 4x20 (Figure 27), where the folds are well induced at the trigger sites in the initial phase of deformation, but showing quite unstable deformation later on. In both models without triggers, in the middle of the plastic deformation phase the folds are quite irregular inducing to a structural instability.

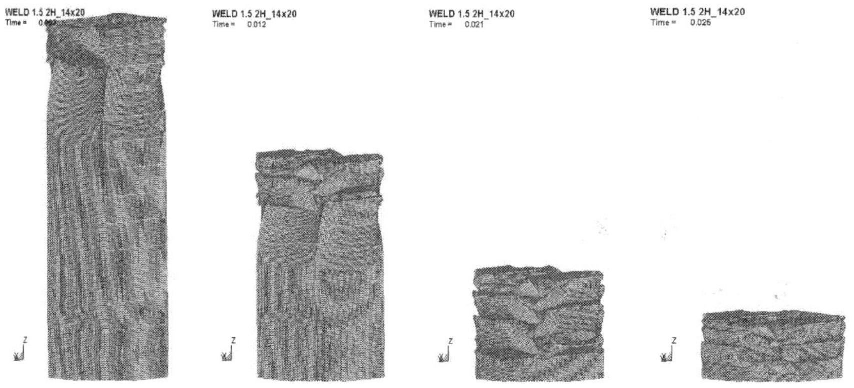

Fig. 25. Deformed shape along of the model (14x20) with 1.5mm width of the HAZ triggers.

Crash energy absorption in the axially loaded model proceeds by the folding process. The elements compressed by the axial compression at the critical load loose the stability of the equilibrium configuration of the structure. Figure 28 shows through the force-displacement curves where it is folding outward (A), contact outward (B), folding inward (C) and contact inward (D). Through the deformed shape of the model the last statement can be confirmed.
When the first fold is forming, the model reaches the maximum force capacity, which represents the first peak and is referred to as the maximum peak force. The load decrease as the first fold is being developed where the folding outward is started. After the completion of the first fold, the force reduces to the first lowest point where the contact outward happened. The further deformation causes the load to increase until the next peak is formed with the formation of the second fold. The process repeats with the folding the third, forth, and fifth folds until the kinetic energy of the striking mass has been reduced to zero, as shown in Figures 29-33.

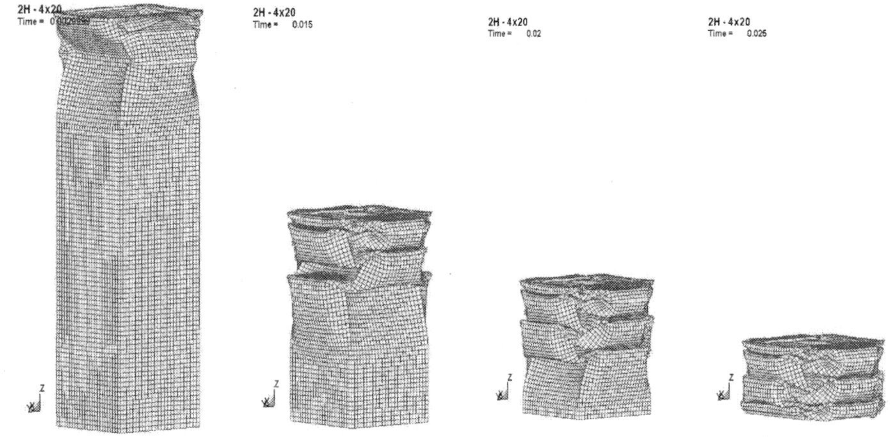

Fig. 26. Deformed shape along of the model (4x20) with 3mm width of the four HAZ triggers.

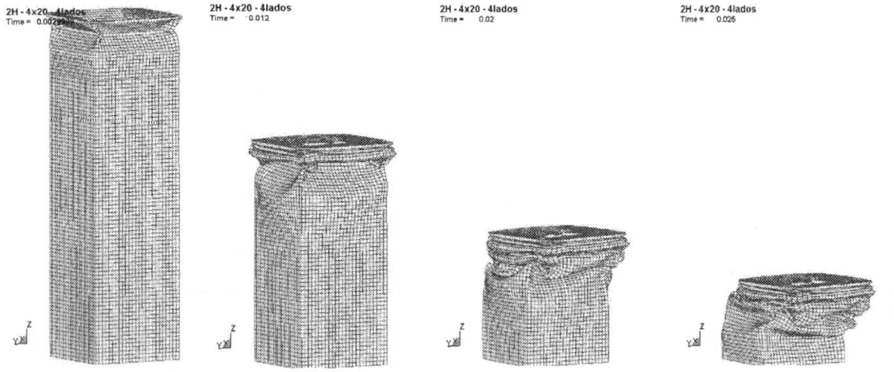

Fig. 27. Deformed shape on model with four HAZ triggers around the model with 3mm width (4sides 4x20).

Figure 29 shows the force-displacement and absorbed energy-displacement curves of the models with triggers along of the model and for the distance 40mm the results are quite different than the other ones because the first fold is forming in the top of the model but the second fold is being started at the bottom of the model. The absorption energy during the crushing process for the same displacement than the others is increasing. This model is an exception when compared to the others smart models studied here.

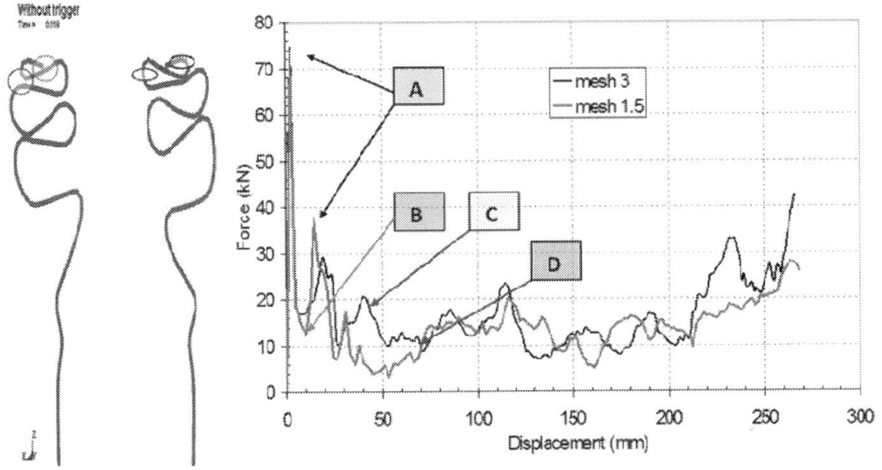

Fig. 28. Deformed shape of the model without trigger with mesh size at 3mm and force-displacement curves of the models without triggers for both mesh size.

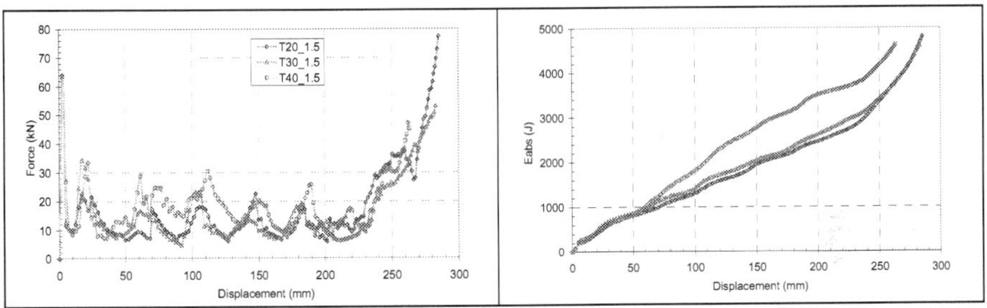

Fig. 29. Force-displacement and absorbed energy-displacement curves of the models with 1.5m wide of the HAZ triggers, on the adjacent sides at the same distance along of the model.

In the case of models without triggers, the maximum force exerted goes through a sharp maximum significantly higher than the other models, as seen in Table 3. This maximum force is effectively reduced about 14% for almost the models by the imposed HAZ triggers.

The highest efficiency of energy absorption during the crushing is found in models that have four triggers in opposite sides with 1.5mm width, and in models that have four triggers around the model with 3mm width. In these cases the folds present a regular shape.

This section presented research work regarding experimental results of heat-treatment of aluminium alloys with the purpose of inducing local modification of material properties. This was achieved using laser heat-treatment and in furnace tests. It was verified that it is possible to change the local hardness in a controlled way, i.e. by the copper rich precipitates dissolution effect in the sample, with a laser treatment, by changing the feed rate.

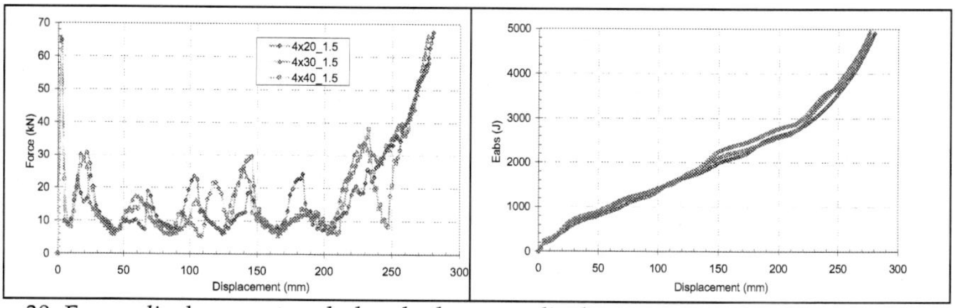

Fig. 30. Force -displacement and absorbed energy-displacement curves of the models with four HAZ triggers on the opposite sides with 1.5mm of the width.

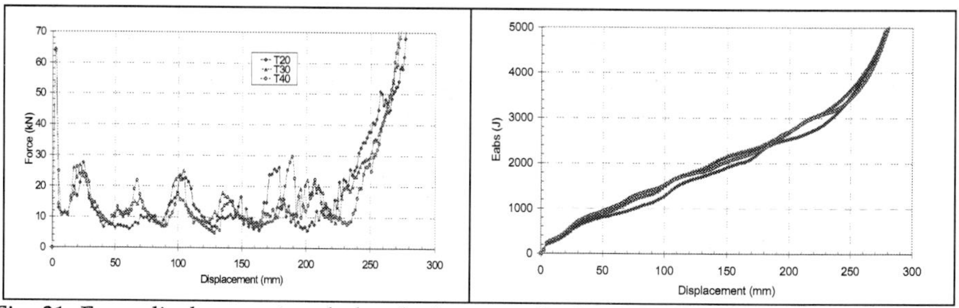

Fig. 31. Force-displacement and absorbed energy-displacement curves of the models with 3mm wide of the HAZ triggers, on the opposite sides at the same distance along of the model.

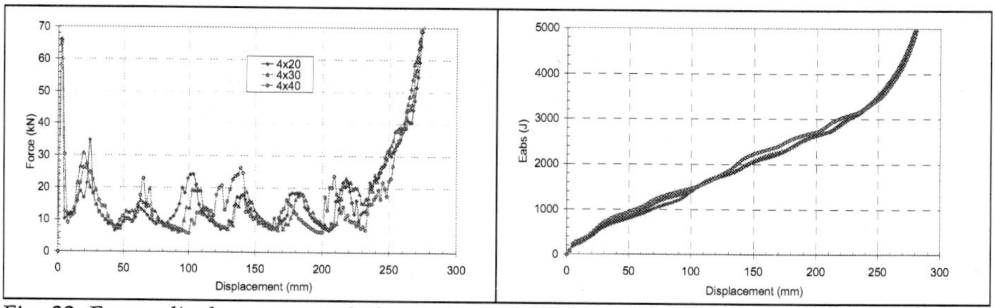

Fig. 32. Force -displacement and absorbed energy-displacement curves of the models with four HAZ triggers on the opposite sides with 3mm of the width.

Numerical simulations of crushing behaviour of aluminium tubes with local triggers obtained through heat treatment were performed. The highest efficiency of absorption energy during crushing is found in models that have four triggers in opposite sides with 1.5mm wide, and in models with four triggers around the model and 3mm width.

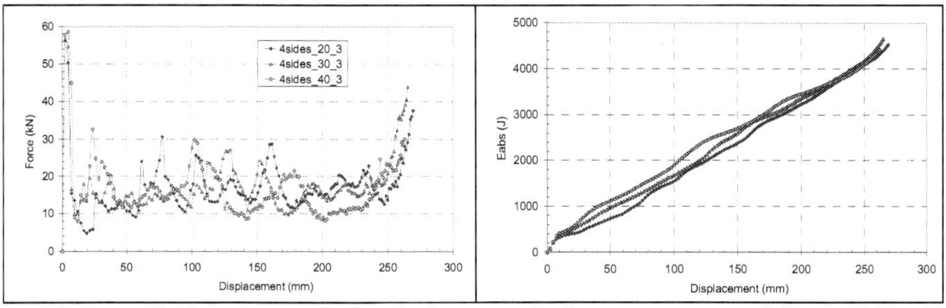

Fig. 33. Force-displacement and absorbed energy-displacement curves of the models with four HAZ around the entire model with 3mm of the width.

ITEM			Δl	E_{int}	F_{peak}	F_{med}	Folds
			mm	J	kN	kN	
Without trigger		mesh 3	262.0	4508	74.6	17.6	7
		mesh 1.5	268.0	4466	74.2	15.1	10
WELD 1.5	2 sides	14x20	277.3	4332	63.9	16.1	8
		9x30	277.6	4364	63.8	15.6	8
		6x40	253.6	4282	64.0	16.4	8
		4x20	274.1	4562	65.2	17.3	7
		4x30	270.5	4631	65.5	17.9	7
		4x40	273.3	4593	64.8	16.4	8
WELD 3	2 sides	14x20	275.9	4690	64.3	19.5	7
		9x30	258.3	4394	56.5	17.0	7
		6x40	259.7	4314	57.7	16.5	8
		4x20	273.4	4396	65.9	17.7	7
		4x30	271.7	4457	66.2	18.2	7
		4x40	271.5	4255	64.2	16.6	8
	4 sides	4x20	267.4	4455	56.2	16.5	10
		4x30	258.4	4394	56.4	18.0	8
		4x40	259.7	4313	58.5	16.5	5

Table 3. Numerical results

The research revealed that, by using a thermal trigger, a reduction of 15% of the initial crushing force is achievable. It is also found that this thermal trigger can not only reduce the initial maximum force but also ensure stable and uniform absorbed energy at most smart models.
The concept of using thermal modification of an aluminium alloy in localized areas for providing a larger global deformation of a part and higher energy absorption before failure appears as possible and effective in the experimental work presented and numerical simulations.

4. References

Auto/Steel Partnership, (1998). *Automotive design manual*, version 5.1, edited by American Iron and Steel Institute - Auto/Steel Partnership, 1998.

Bjørneklett, B. ; Myhr, O., (2003). Materials Design and Thermally Induced Triggers in Crash Management, *Proceedings IBEC Conference*, 2003.

Cheng, C. ; Jie, M. ; Chan, L. ; Chow, L., (2007). True stress–strain analysis on weldment of heterogeneous tailor-welded blanks—a novel approach for forming simulation, *International Journal of Mechanical Sciences*, Volume 49, Issue 2, 2007, Pages 217-229.

Gaied, S. ; Roelandt, J-M ; Pinard, F.; Schmit, F. ; Balabane, L., (2009). Experimental and numerical assessment of Tailor-Welded Blanks formability, *Journal of Materials Processing Technology*, Volume 209, Issue 1, 2009, Pages 387-395.

Geoffroy, J. ; Cambien, I. ; Jouet, A., (1993). Contribution of high strength steels to the absorption of impact energy. *La metallurgia Italiana* 1993;85(6):377–82.

Kim, H-S, (2002). New extruded multi-cell aluminium profile for maximum crash energy absorption and weight efficiency, *Thin-Walled Structures*, 40, pp. 311-327, 2002.

Kim, J. ; Kim, N. ; Huh, M., (2000). Optimum blank design of an automobile sub-frame, *Journal of Materials Processing Technology*, Volume 101, Issues 1-3, 2000, Pages 31-43.

Lee, S.; Hahn, C. ; Rhee, M. ; Ohd, J., (1999). Effect of triggering on the energy absorption capacity of axially compressed aluminum tubes, *Materials and Design*, 20, pp.31-40, 1999.

Liu, G. ; Yuan, S. ; Chu, G., (2007). FEA on deformation behavior of tailor-welded tube in hydroforming, *Journal of Materials Processing Technology*, Volumes 187-188, 2007, Pages 287-291.

Padmanabhan, R.; Oliveira, M.; Menezes, L., (2008). Deep drawing of aluminium–steel tailor-welded blanks, *Materials & Design*, Volume 29, Issue 1, 2008, Pages 154-160.

Panda, S.; Kumar, D. ; Kumar, H. ; Nath, A., (2007). Characterization of tensile properties of tailor welded IF steel sheets and their formability in stretch forming, *Journal of Materials Processing Technology*, Volume 183, Issues 2-3, 2007, Pages 321-332.

Panda, S.; Kumar, D., (2001). Improvement in formability of tailor welded blanks by application of counter pressure in biaxial stretch forming, *Journal of Materials Processing Technology*, Volume 204, Issues 1-3, 2008, Pages 70-79.

Peixinho, N., (2004). Study of viscoplasticity models for the prevision of the mechanical behaviour of high-strength steels subjected to impact, PhD thesis, University of Minho, 2004.

Peixinho, N.; Pinho, A., (2007). Study of viscoplasticity models for the impact behaviour of high-strength steels, *Journal of Computational and Nonlinear Dynamics*, Vol. 2, pp. 114-123, 2007.

Peroni, L.; Avalle, M.; Belingardi, G., (2009). Comparison of the energy absorption capability of crash boxes assembled by spot-weld and continuous joining techniques, *International Journal of Impact Engineering*, 36 (2009) 498–511.

Qiu, X.; Chen, W., (2007). The study on numerical simulation of the laser tailor welded blanks stamping, *Journal of Materials Processing Technology*, Volumes 187-188, 2007, Pages 128-131.

Radlmayr, K-M.; Ponschab, H. ; Stiaszny, P.; Till, E., (1993). Comparative behaviour of safety structures from soft and higher-tensile qualities as well as aluminium alloys in crashes. In: *Proceedings to the twenty-sixth ISATA conference on road and vehicle safety*. Aachen (Germany); 1993. Paper 93SF061.

Sheng, Z., (2008). Formability of tailor-welded strips and progressive forming test, *Journal of Materials Processing Technology*, Volume 205, Issues 1-3, 2008, Pages 81-88.

6

Computational modelling of conduction mode laser welding process

S. Bag[1] and A. De[2]
[1]*CEMEF-Mines ParisTech, 06904 Sophia Antipolis Cedex, France*
[2]*IIT Bombay, Powai, Mumbai 400076 India*

1. Introduction

Laser welding has the advantage of localised heat, low distortion and rapid solidification, and is used in wide variety of material joining applications. Laser welding is performed either in conduction or in keyhole mode. In conduction mode, the applied power density is smaller and vaporization of work piece material is absent. Keyhole mode laser welding involves the application of very high power density creating a vapour filled cavity into the work piece that also helps in greater absorption of beam energy (Benyounis et al., 2005; Liu et al., 1993; Tzeng, 1999). An appropriate design for welding procedure requires a-priori knowledge of the peak temperature, weld thermal cycle and cooling rate. Due to high peak temperature and small weld pool size, real-time measurements of temperature and velocity fields, and the growth of weld pool are difficult in laser welding (DebRoy & David, 1995; Zhao et al., 1999; Pitscheneder et al., 1997; Lhospitalier et al., 1999; Lee & Kim, 2004). Thus, the computational models, which can simulate temperature and velocity field in laser welding, is in ever demand.

Conduction heat transfer based models are simpler, computationally inexpensive and yet can provide fairly reliable results in several simple welding systems (Trivedi et al., 2007; Goldak et al., 1984; Frewin & Scott, 1999; De et al., 2003). The transport phenomena based heat transfer and fluid flow analysis involves larger physical attributes, generally complex and can be computationally expensive especially for large and complex weld joint geometry (Bag & De, 2008; Bag & De, 2009; Mackwood & Crafer, 2005). Thus, conduction heat transfer based models are often preferred to the convective heat transport based weld pool simulations for smaller weld pool sizes and joining processes involving rapid melting and solidification. The conduction heat transfer based weld pool models also find tremendous application in the calculations of weld distortion and residual stress (Teng et al., 2001; Jung & Tsai, 2004; Deng et al., 2007; Cho & Kim, 2002; Trivedi et al., 2007; Deng, 2009), where the temperature field over a very large domain is of greater importance in comparison to its local variation in weld pool. We present here both of these modelling approaches and a comparison of the relative error in respective predictions.

In both the conduction and convective transport based models, the laser beam is considered as a surface heat source with Gaussian energy distribution. In conduction based heat transfer analysis, a volumetric heat source is often used further to numerically compensate the influence of convection heat transport in weld pool. The existing approach to define a volumetric heat source needs a-priori definition of its shape and size that restricts the use of the same (Goldak et al., 1984; Frewin & Scott, 1999; De et al., 2003). We have introduced an adaptive volumetric heat source term to make it more general and close to the reality. The adaptive volumetric term is defined by mapping the instantaneous value of the computed weld dimensions (length, width and depth) with respect to time step (transient) or load step (steady-state) and thus, the requirement of a-priori definition of the source geometry is avoided. Lastly, both conduction and convective heat transport based simulations of laser welding needs a number of parameters, which are required for model calculations and cannot be defined by scientific principles alone. Absorption coefficient, effective thermal conductivity and viscosity of molten material in weld pool, and parameter defining the nature of energy distribution are examples of such parameters in laser welding simulations (Chande & Mazumder, 1884; Tanriver et al., 2000; De & DebRoy, 2005). Here we show that a robust optmization algorithm integrated with the numerical process models can help in identifying suitable values of the uncertain model parameters and provide reliable computed results.

The present work includes a finite element based three-dimensional transient and quasi-steady heat transfer and fluid flow model for the prediction of temperature and velocity field, and weld dimensions in laser welding process. The novel feature introduced in the conduction model is the adaptive volumetric heat source term that is used to account for the energy absorbed by the molten weld pool in conduction based analysis. Temperature dependent material properties and the latent heat of melting and solidification are considered. The transport phenomena based heat transfer and fluid flow model considers effective values of thermal conductivity and viscosity to account for the effect of high momentum transport within small weld pool. The numerical heat transfer models are integrated with a differential evolution (DE) based optimization tool to estimate the value of uncertain parameters in an inverse manner. The predicted weld pool dimensions from the overall integrated model are validated successfully against similar experimentally measured results for laser spot and linear welding. The comparative results in terms of weld pool shape and size for both conduction and transport phenomena based models are presented.

2. Finite element based numerical model

The finite element formulation of 3D conduction mode heat transfer model using adaptive volumetric heat source both for spot and linear welding is described first. The discretization of transport phenomena based heat transfer and fluid flow model is presented next. Various issues regarding computational difficulties are also pointed out in this section.

2.1 Heat transfer model with adaptive volumetric heat source

The conduction mode heat transfer in transient state is governed by the following equation in 3D Cartesian coordinate system.

$$\frac{\partial}{\partial x_m}\left(k\frac{\partial T}{\partial x_m}\right)+\dot{Q}=\rho C_p\frac{\partial T}{\partial t} \tag{1}$$

where x_m is the distance along the m = 1, 2 and 3 (same as x, y and z) orthogonal directions and ρ, C_p and k refer respectively to density, and temperature dependent specific heat and thermal conductivity of the work piece material. The term $\dot{Q}$ refers to the rate of internal heat generation per unit volume and t refers to time variable. Physically, the internal heat generation due to joule heating is neglected in the present model. However the volumetric heat source is mathematically incorporated through the term $\dot{Q}$. In steady state analysis, the transient growth of temperature field is transformed to special distribution by considering that the laser beam is moving with a constant velocity (V_w) (say, in x_2 i.e. y-direction). Hence, eq. (1) is rewritten for moving heat source with reference to the moving coordinate system (x_1, x_2, x_3) as

$$\frac{\partial}{\partial x_m}\left(k\frac{\partial T}{\partial x_m}\right)+\dot{Q}=-\rho C_p\left(-V_w\frac{\partial T}{\partial x_2}\right) \tag{2}$$

It is to be noted from the nature of eq. (2) that the temperature is distributed specially and it is termed as quasi-steady state analysis. The boundary interaction along with the solution domain is described in Fig. 1. Mathematically, the heat balance along with the surface is expressed as

$$-k\frac{\partial T}{\partial n}=Q_c+Q_r-Q_s \tag{3}$$

where Q_c and Q_r represent heat loss by convection and radiation from the surface, respectively, and Q_s is the heat flux added by laser on the surface. The first term in eq. (2) indicates the heat conducted normal to the surface. On the symmetric surface, the temperature gradient normal to the surface is zero. The initial temperature of whole solution domain is considered as ambient temperature (T_0). To avoid the computational difficulty of the radiation term, an effective heat transfer coefficient (combined effect of convection and radiation) is considered as (Frewin & Scott, 1999)

$$h_{eff}=2.4\cdot 10^{-3}\cdot\varepsilon\cdot T^{1.61} \tag{4}$$

where ε is the emissivity of the work piece material. Hence, the convection and radiation term of eq. (3) is expressed in modified form as

$$Q_{c'}=Q_c+Q_r=h_{eff}(T-T_0) \tag{5}$$

The distribution of laser energy on the top surface follows Gaussian distribution which is mathematically expressed as (Goldak et al., 1984)

$$Q_s = \frac{P\eta_G d_l}{\pi r_{eff}^2} \exp\left(-\frac{d_l}{r_{eff}^2} \sum_{m=1}^{2} x_m^2 \right) \tag{6}$$

where P refers to laser power, η_G and r_{eff} the absorption coefficient and effective radius of laser beam on the work piece surface, respectively, and d_l the power density distribution factor of heat source. Figure 2 describes the typical simulation of laser heat flux distribution on the substrate surface. The nature of distribution with respect to maximum heat flux is mainly extended by the distribution factor (d_l) which is typically ~ 3.5 for laser welding process (Frewin & Scott, 1999).

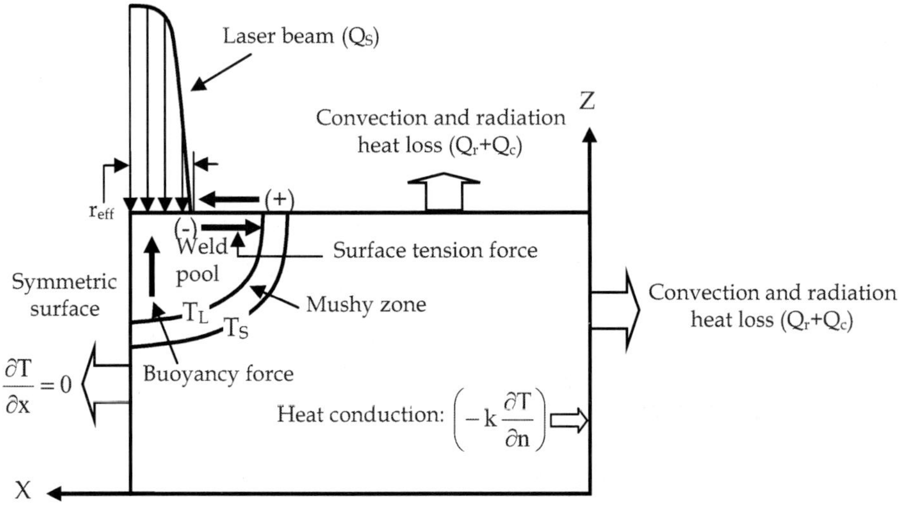

Fig. 1. Boundary conditions applied in numerical modelling of laser welding process.

Figure 3 schematically represents the shape of the volumetric heat source and it is mathematically represented as

$$\dot{Q} = \frac{6\sqrt{3} f P \eta_v}{\pi\sqrt{\pi}\, a_1 a_2 a_3} \exp\left(-3 \sum_{m=1}^{3} \frac{x_m^2}{a_m^2} \right) \tag{7}$$

where a_1, a_2 and a_3 represent the computed values of the weld dimensions obtained iteratively after every time-step or load-step, and η_v refers to volumetric efficiency. The volumetric efficiency dictates the actual amount of volumetric heat that is utilised to develop the weld pool. The distribution of heat is uneven in the case of moving heat source (linear welding). The convective transport in weld pool gets lesser time to develop in the front of the heat source in comparison to the rear resulting in an asymmetric weld pool with respect to the centre of the source in the direction of weld velocity. In stationary welding, however, the weld pool remains symmetric owing to a symmetric energy distribution. To incorporate the asymmetric effect in linear welding, eq. (7) is multiplied with an empirical constant f with f = 0.6 for the front and 1.4 for the rear of the weld pool (Bonifaz, 2000) with the dimensions of double ellipsoidal $2a_1$, a_3, a_2^1 and a_2^2 ($a_2 = a_2^1 + a_2^2$ and $a_2^1 < a_2^2$) as shown in Fig. 3(a). In laser spot welding, f remains as unity with a symmetrical pool shape with respect to the centre of the heat source (Fig.3b).

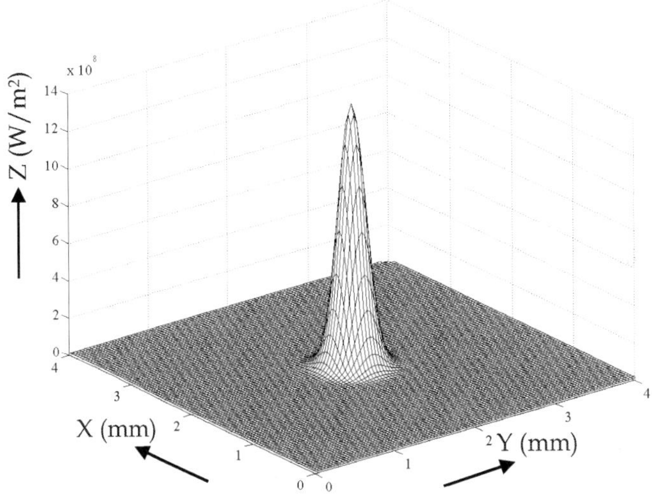

Fig. 2. Distribution of surface heat flux on work piece following Gaussian distribution.

The numerical calculations are performed through a number of small load steps in steady state analysis (linear welding) and time steps for transient analysis (spot welding). These small steps facilitate incorporation of the temperature dependence of materials properties through iterative procedures. The volumetric heat source term is activated when a finite size of molten weld pool forms. A direct iteration scheme is adopted to get a converged solution of temperature field by minimizing the error between the adaptive weld pool size of current load step and the previous load step.

The solution domain is discretized using eight nodded isoparametric element with linear variation of temperature. The temperature variable, T, within the element is expressed in terms of nodal temperatures as

$$T = \sum_{i=1}^{8} [N_i]\{T_i\} \tag{8}$$

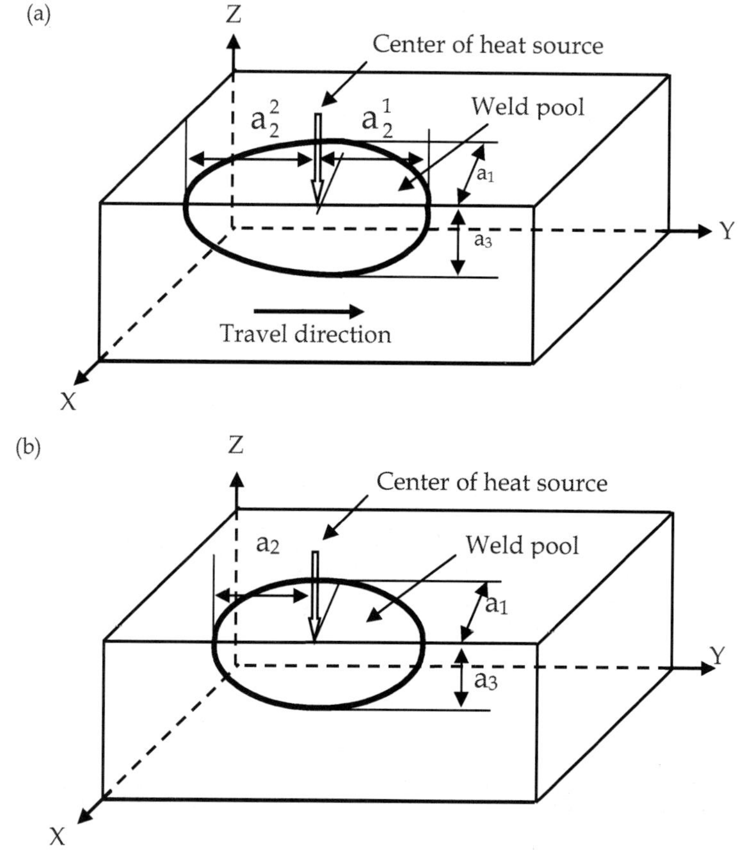

Fig. 3. Adaptive volumetric heat source in (a) linear welding and (b) transient spot welding.

The governing equation along with the boundary conditions is discretised with Galerkin's weighted residue technique (Gupta, 2002). By using Gauss theorem and boundary conditions described in eq. (3), the discretised form of the governing equation can be rewritten in matrix form for any specific element 'e' as

$$[K^e]\{T\}+[S^e]\left\{\frac{\partial T}{\partial t}\right\}+[H^e]\{T\}-\{F_v^e\}-\{F_s^e\}-\{F_{c'}^e\}=0 \quad (9)$$

where

$$[K^e]=\int_{\Omega^e} k\sum_{m=1}^{3}\left(\frac{\partial N_i}{\partial x_m}\frac{\partial N_j}{\partial x_m}\right)d\Omega \quad (10)$$

$$[S^e]=\int_{\Omega^e}\rho C_p N_i N_j d\Omega \;\; ; \quad [H^e]=\int_{\Gamma^e} h_{eff} N_i N_j d\Gamma \quad (11, 12)$$

$$\{F_s^e\} = \int_{\Gamma_s^e} N_i Q_s d\Gamma \;;\; \{F_{c'}^e\} = \int_{\Gamma_c^e} N_i h_{eff} T_0 d\Gamma \tag{13, 14}$$

$$\{F_v^e\} = \int_{\Omega^e} N_i \dot{Q} d\Omega \tag{15}$$

$$i, j = 1, 2, 3, 4, 5, 6, 7, 8 \quad \text{for 8 noded brick element} \tag{16}$$

Here, $[K^e]$ represents the overall heat conduction within solution domain, $[S^e]$ represents the heat capacity of domain, $[H^e]$ defines the heat loss by convection and radiation from the surface, and $\{F_s^e\}$, $\{F_v^e\}$ and $\{F_{c'}^e\}$ represent heat added by laser beam through surface, volumetric heat added to the domain, and heat added through surface due to reference temperature, respectively. Considering the contribution from all the elements, the final set of equation in matrix form is written as

$$[\overline{K}]\{T\} + [S]\left\{\frac{\partial T}{\partial t}\right\} = \{F\} \tag{17}$$

where

$$[\overline{K}] = [K] + [H] \;\;;\;\; \{F\} = \{F_v\} + \{F_s\} + \{F_{c'}\} \tag{18, 19}$$

However, to get the temperature distribution over time, eq. (17) is further discretised linearly in the time domain following Galerkin's scheme which is unconditionally stable (Reddy & Gartling, 2000). Assuming that the temperature field at the beginning of the time interval, Δt, is known, the same at the end of the time interval is calculated as

$$\{T^2\} = -\left[\frac{2}{3}[\overline{K}] + \frac{1}{\Delta t}[S]\right]^{-1}\left\{\left[\frac{1}{3}[\overline{K}] - \frac{1}{\Delta t}[S]\right]\{T^1\} - \{F\}\right\} \tag{20}$$

In similar fashion, the matrix form of the equation for pseudo-steady state heat transfer analysis can be written as

$$[K^e]\{T\} + [S^e]\{T\} + [H^e]\{T\} = \{F_v^e\} + \{F_s^e\} + \{F_{c'}^e\} \tag{21}$$

where

$$[S^e] = -\int_{\Omega^e} \rho C_p V_w N_i \frac{\partial N_j}{\partial x_2} d\Omega \tag{22}$$

The expressions of other terms are already described in transient heat transfer analysis. Considering the contribution from all the elements within the solution domain the final set of equation for pseudo-steady state heat transfer analysis is expressed as

$$[\overline{K}]\{T\} = \{F\} \tag{23}$$

where

$$[\overline{K}] = [K] + [S] + [H] \;\;;\;\; \{F\} = \{F_v\} + \{F_s\} + \{F_{c'}\} \tag{24, 25}$$

2.2 Transport phenomena based heat transfer and fluid flow model

In addition to the energy equation, heat transfer and fluid flow analysis requires the solution of the momentum conservation equation expressed as (Reddy & Gartling, 2000)

$$\rho\left(\frac{\partial u_m}{\partial t}+u_n\frac{\partial u_m}{\partial x_n}\right)=F_m+\frac{\partial}{\partial x_m}\left[-P\delta_{mn}+\mu_{eff}\left(\frac{\partial u_m}{\partial x_n}+\frac{\partial u_n}{\partial x_m}\right)\right] \tag{26}$$

where u_m is the velocity in respective direction and m, n = 1, 2, 3, μ_{eff} is the effective viscosity, P the modified pressure obtained by subtracting hydrostatic pressure from local pressure, F_m the body forces in respective directions, ρ the density of the material, δ_{mn} is knocker delta. The governing equation for quasi-steady state analysis with respect to moving coordinate system (x_1, x_2, x_3) is expressed as

$$\rho\left(u_n\frac{\partial u_m}{\partial x_n}\right)=F_m+\frac{\partial}{\partial x_n}\left[-P\delta_{mn}+\mu_{eff}\left(\frac{\partial u_m}{\partial x_n}+\frac{\partial u_n}{\partial x_m}\right)\right]-\rho V_w\frac{\partial u_n}{\partial x_2} \tag{27}$$

Figure 1 describes various driving forces and corresponding boundary interactions during the heat transfer and fluid flow analysis. The body force in laser welding consists of buoyancy force only. The buoyancy force acts in x_3 i.e. z-direction and is expressed considering Boussinesq approximation as (Reddy & Gartling, 2000)

$$F_3=-\rho\beta(T-T_{ref})g \tag{28}$$

where β is the coefficient of thermal expansion, g the gravitational acceleration, and T_{ref} the reference temperature. The continuity equation (conservation of mass) for incompressible fluid is expressed as

$$\frac{\partial u_m}{\partial x_m}=0 \tag{29}$$

The energy equation of transient state is expressed as (Reddy & Gartling, 2000)

$$\frac{\partial}{\partial x_m}\left(k_{eff}\frac{\partial T}{\partial x_m}\right)+\dot{Q}=\rho C_p\left(\frac{\partial u_n}{\partial t}+u_n\frac{\partial T}{\partial x_m}\right) \tag{30}$$

where k_{eff} refer to effective thermal conductivity of liquid metal. The last term of the right hand side indicates the energy transport within the melt pool due to movement of liquid metal which was absent in conduction heat transfer analysis. The equation for the conservation of energy in pseudo-steady state is stated in 3D Cartesian coordinate system as

$$\frac{\partial}{\partial x_m}\left(k_{eff}\frac{\partial T}{\partial x_m}\right)=\rho C_p\left(u_n\frac{\partial T}{\partial x_m}-V_w\frac{\partial T}{\partial x_2}\right) \tag{31}$$

The solution boundaries for both of mass and momentum equations are defined by the solid-liquid interface and the free surface of the weld pool which is assumed flat to avoid extra computational effort. A no-slip boundary condition for laminar flow at the solid-liquid interface is expressed as

$$u_m = 0 \tag{32}$$

A slip boundary condition is expressed along the symmetric plane as

$$u_1 = 0\,,\ \frac{\partial u_2}{\partial x_1} = 0 \text{ and } \frac{\partial u_3}{\partial x_1} = 0 \tag{33}$$

The free surface of weld pool is subjected to surface tension force and the corresponding boundary conditions are expressed as

$$\mu\frac{\partial u_1}{\partial x_3} = f_L\frac{\partial \gamma}{\partial T}\frac{\partial T}{\partial x_1}\quad ;\ \mu\frac{\partial u_2}{\partial x_3} = f_L\frac{\partial \gamma}{\partial T}\frac{\partial T}{\partial x_2}\quad ;\ u_3 = 0 \tag{34}$$

where γ is the temperature dependent surface tension coefficient and f_L is the volume fraction of liquid metal along the weld pool top surface.
The penalty finite element method is designed in the present case to solve momentum equations by linking the continuity equation as constraint with the pressure and is expressed as

$$P = -\lambda\left(\frac{\partial u_m}{\partial x_m}\right) \tag{35}$$

where λ is the penalty parameter that is set as equal to a large number so that it can satisfy the continuity equation (Reddy & Gartling, 2000). To avoid nonlinearity due to presence of velocity term in the convective term, the velocities (u_1^0, u_2^0, u_3^0) are made independent from the nodal velocity variables. Hence, u_1^0, u_2^0 and u_3^0 for an element are calculated as average of the corresponding nodal velocity components. The velocity variable within the element is expressed

$$u_m = \sum_{i=1}^{8}[N_i]\{u_m^i\};\qquad \text{where } m = 1, 2, 3 \tag{36}$$

Equation (26) for a specific element 'e' can be written in a matrix form as

$$[M^e]\frac{\partial\{U\}}{\partial t} + \left\{[\bar{C}^e] + [\hat{K}^e] + [K^e]\right\}\{U\} = \{F^e\} \tag{37}$$

where

$$[M^e] = \int_{\Omega^e} \rho N_i N_j d\Omega ; \qquad [\hat{K}^e_{mn}] = \int_{\Omega^e} \lambda \frac{\partial N_i}{\partial x_m}\frac{\partial N_j}{\partial x_n} d\Omega \tag{38, 39}$$

$$[K^e_{mn}] = \int_{\Omega^e} \mu \frac{\partial N_i}{\partial x_m}\frac{\partial N_j}{\partial x_n} d\Omega ; \; [\bar{C}^e] = \int_{\Omega^e} \rho \sum_{m=1}^{3} \left(u^0_m N_i \frac{\partial N_j}{\partial x_m} \right) d\Omega ; \tag{40, 41}$$

$$\{F^e\} = \int_{\Omega^e} N_i \dot{F} d\Omega ; \quad \{U\} = \{\{u^i_1\} \;\; \{u^i_2\} \;\; \{u^i_3\}\}^T \tag{42, 43}$$

$$i, j = 1, 2, 3, 4, 5, 6, 7, 8; \qquad m, n = 1, 2, 3 \tag{44}$$

Similarly, eq. (27), in the case of quasi steady state analysis is expressed as

$$\{[M^e] + [\bar{C}^e] + [\hat{K}^e] + [K^e]\}\{U\} = \{F^e\} \tag{45}$$

where

$$[M^e] = -\int_{\Omega^e} \rho V_w N_i \frac{\partial N_j}{\partial x_2} d\Omega ; \tag{46}$$

All other terms of eq. (45) are already defined. For all the elements in the solution domain, the assembly form of momentum equations for quasi-steady state analysis is further written as

$$[\bar{M}]\{U\} = \{F\} \tag{47}$$

where

$$[\bar{M}] = [M] + [\bar{C}] + [\hat{K}] + [K] \tag{48}$$

However, the integral term involving the penalty function i.e. $[\hat{K}]$ matrix in the derivation of eq. (37) or (48), should be under-integrated (one point less) than the viscous and the convective terms i.e. $[K]$ and $[\bar{C}]$ matrices (Reddy & Gartling, 2000).
By similar mathematical treatment of momentum equations, the energy equation for quasi-steady state analysis can be represented in matrix form for any specific element 'e' as

$$[H^e]\{T\} + [C^e]\{T\} + [S^e]\{T\} + [\bar{H}^e]\{T\} = \{f^e_Q\} + \{f^e_q\} + \{f^e_h\} \tag{49}$$

where

$$[C^e] = \int_{\Omega^e} \rho C_p \sum_{m=1}^{3} \left(u^o_m N_i \frac{\partial N_j}{\partial x_m} \right) d\Omega \tag{50}$$

All other terms of energy equation are already defined in conduction heat transfer analysis. The matrix equation of transient energy equation follows similar procedure described above.
The presence of surface active elements such as sulfur and oxygen play an important role in the formation weld pool geometry since the surface tension fore (i.e. surface tension coefficient) differs considerably in molten material which is a function of weight percent of

surface active elements and temperature. The detailed variation of surface tension gradient as a function of temperature and activity of solute is represented as (Sahoo et al., 1988)

$$\partial\gamma / \partial T = -A - R_g \Gamma_s \ln(1 + C_s b_i) - \frac{C_s b_i}{1 + C_s b_i} \frac{\Gamma_s \Delta H^\circ}{T} \tag{51}$$

where $\partial\gamma / \partial T$, A, Γ_s, C_s, b_i, ΔH° and R_g referred to the surface tension gradient, adsorption coefficient, surface excess at saturation, segregation coefficient, activity of the 'ith' species, heat of adsorption and characteristic gas constant respectively. Their computed results showed that the surface tension decreased linearly with temperature when the sulfur content in the weld pool was negligible. At a constant temperature the surface tension showed an upward curvature with increase in sulfur content. When the sulfur content in the weld pool was significant, surface tension first increased and then decreased with the increase in temperature as surface active elements tended to segregate at higher temperature. Present numerical model of heat transfer and fluid considers the formation of weld geometry due to the effect of surface active element present in the parent material.

3. Inverse modelling approach

The reliability of numerical model intuitively depends on correct representation of several input parameters that is essential in modelling calculations. A number of inverse methods have recently been used in conjunction with numerical models for determining suitable values of required uncertain model parameters (De & DebRoy, 2005; Mishra & DebRoy, 2005; Bag & De, 2008; Bag et al., 2009; Bag & De, 2010). This is achieved in the present work by integrating a real number based differential evolution (DE) algorithm for the numerical models. The link between the numerical model and optimization algorithm as well as search direction towards the optimum conditions are evaluated through the formation of a suitable objective function which is defined as

$$O(\theta) = \sum_{l=1}^{M} \left[\frac{w_l^c - w_l^{exp}}{w_l^{exp}} \right]^2 + \left[\frac{p_l^c - p_l^{exp}}{p_l^{exp}} \right]^2 = \sum_{l=1}^{M} [w_l^* - 1]^2 + [p_l^* - 1]^2 \tag{52}$$

where 'c' refers to computed value from numerical model, 'exp' refers to experimentally measured values and '*' refers to non-dimensional form of the variables that indicate the extent of over or under-prediction of variables. The subscript l refers to specific data points in a series of M number of total data points. In equation (52), θ stands for the independent variable set which is unknown. The objective function, $O(\theta)$, depicts the error between the estimated from numerical model and the corresponding measured values with similar process conditions with M number of observations. However, it is pointed out that this integrated modelling calculation considers few experimental results. Differential evolution (DE), proposed by Storn (Storn, 1997), is a derivative of genetic algorithm (GA). The algorithm is described elsewhere (Price et al., 2005).

Figure 4 describes the overall solution algorithm of the integrated model. The algorithm starts with the creation of large volume of discrete data sets that referred to as population. Each individual in this population is a possible solution and consists of assumed values of the uncertain parameters, to begin with. The numerical solutions are carried out using all the individuals and the corresponding error in prediction ($O(\theta)$) for each individual is computed. In case the minimum value of the error is beyond a pre-defined value of tolerance in prediction, further iteration or calculations are discouraged. The choice of the best solution is dictated by the numerical value of objective function after N iterations.

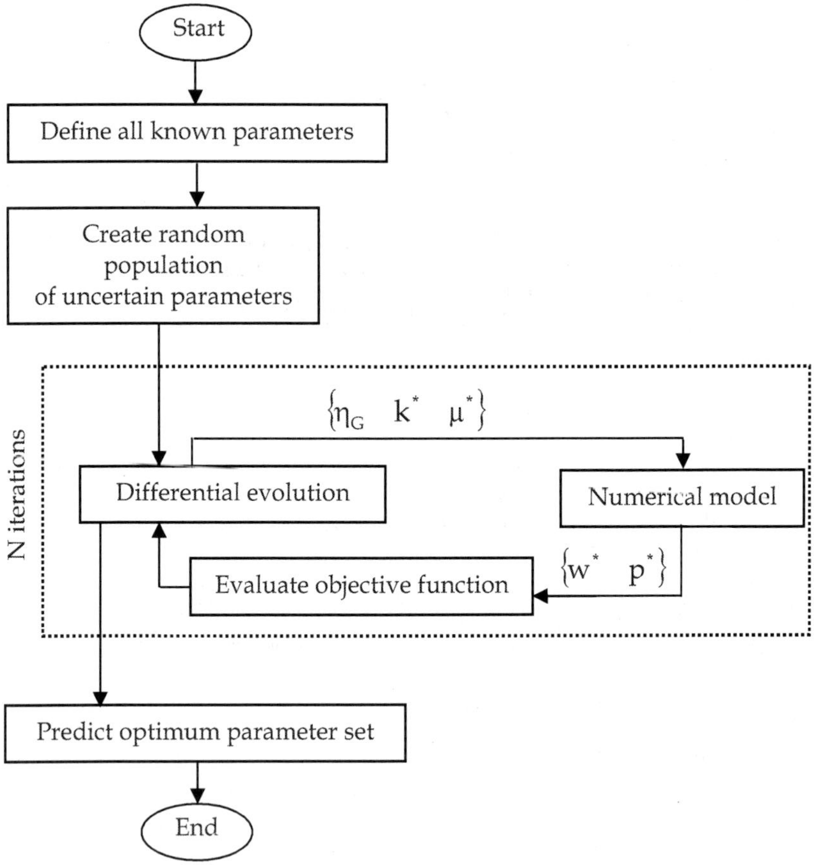

Fig. 4. Overall solution algorithm of integrated model.

4. Results and discussions

The overall results of the model calculations are arranged in the following sequence. First, the identification of uncertain parameters is done through inverse modelling approach. A detail sensitivity analysis of these parameters on weld geometry is performed to get the range of physically meaningful values and to identify less or more sensitive parameters related to the process modelling. Next, the comparison between experimental and numerically calculated weld geometry are described in various process conditions using the optimized set of uncertain parameters. The influence of surface active elements on the shape of weld pool geometry is highlighted here. The relative importance of various driving forces during momentum transport within weld pool is performed next. Finally, the numerically computed weld geometry is compared using conduction only heat transfer and transport phenomena based heat transfer and fluid flow analysis.

To validate the numerical model, several laser weld experiments are conducted both for spot and linear welding using pulsed Nd:YAG laser. In addition to that some of the experimental results are considered from independent literature (Pitscheneder et al., 1996; Tzeng, 2000). The experiments are conducted at two cases: one for spot welding at average laser power 1.0 kW at several on-times varying from 0.5 to 2.5 s and another for linear welding at average power of 1.2 kW at travel speed varying from 5 to 10 mm/s with different weight percent of sulfur present in low carbon steel. The effective beam radius of the incident laser beam is measured as 1.20 mm for all the experimental conditions. The weld samples are prepared on 2.0 mm thick low carbon steel sheet. The chemical composition of the material is described elsewhere (Frewin & Scott, 1999). Table 1 describes the various welding conditions (six cases) and corresponding parameters used to conduct experiments. Figure 5 depicts the measured weld dimensions corresponding to all six cases used to validate the numerically computed results.

[a, b] considered from literature

Case	Type of welding	Material	Sulfur content (wt %)	Laser power (kW)	Effective beam radius (mm)	Power density (kW/mm^2)	Laser on-time (s) /travel speed (mm/s)
i	Spot welding	High-speed steel[a]	0.002	5.2	1.4	845	0.1 ~ 0.75
ii			0.015				
iii		Low carbon steel	0.002	1.0	0.6	885	0.5 ~ 2.5
iv	Linear welding	Bare steel[b]	0.010	0.4	0.4	796	6 ~ 8
v		Low carbon steel	0.002	1.2	0.6	1062	5 ~ 10
vi			0.012				5 ~ 7

Table 1. Experimental conditions of laser welding used in present work.

It is evident from Fig. 5(a) that with the increase of laser on-time the weld pool dimensions increases since it absorbs more energy at higher weld time. On the other hand, the weld dimensions decreases with increase in travel speed (Fig. 5(b)) since the heat input per unit length decreases with increase in travel speed. However, it is also observed from Fig. 5 that the weld pool aspect ratio (penetration to width) varies from 0.14 ~ 0.60 and the power density varies from 796 to 1062 W/mm^2, which indicates typical weld pool shapes in conduction mode laser welding process. The weld dimensions differ considerably between case-i and case-ii, and between case-v and case-vi irrespective of similar experimental conditions except the weight percent of sulfur content in the parent material. This manifests the predominance effect of convective heat transfer in weld pool in the presence of considerable amount of surface active elements. Table 2 describes the material properties used in present numerical calculation and to calculate various non-dimensional numbers.

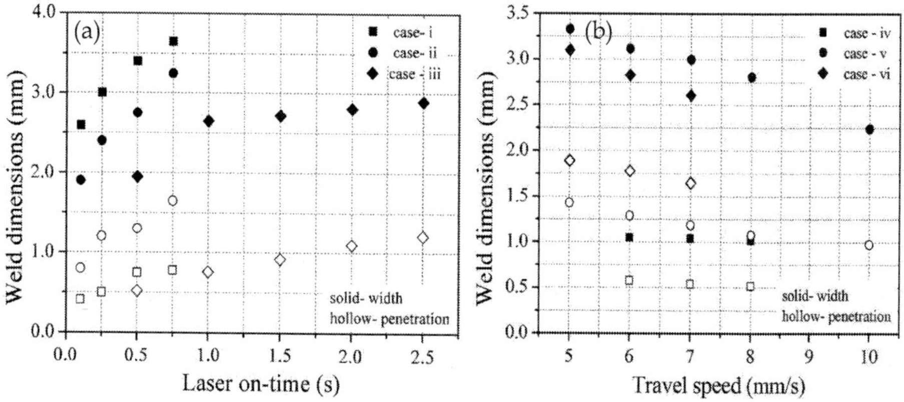

Fig. 5. Experimentally measured weld pool dimensions for (a) laser spot welding and (b) linear welding.

4.1 Identification of uncertain parameters

It is evident from theoretical formulation of finite element based numerical model that it involves various uncertain parameters such as absorption coefficients (η_G, η_V), distribution coefficient (d_l), effective radius of laser beam (r_{eff}), coefficient of uneven heat distribution in linear welding (f) and material properties at high temperature (k_{eff}, μ_{eff}). Hence, a detailed sensitivity analysis of the computed weld dimensions on these uncertain parameters is performed by using 3D numerical model at various welding conditions. It is observed that with the increase of absorption coefficients, the weld pool dimensions increase with the weld penetration being more sensitive than weld width. This is primarily due to smaller thickness dimension in comparison to the length and width of the plate. An increase in the value of effective thermal conductivity and viscosity reduces weld width and increases weld penetration. Higher values of thermal conductivity reduce the surface temperature gradient and hence the convective transport of heat towards the periphery that reduces weld width. Greater values of effective viscosity also reduce the convective transport of heat leading to the smaller weld width. A secondary sensitivity analysis of weld dimensions is also performed on d_l and f. It is realised that these parameters have negligible influence over a wide range of changes in the process parameters. Hence these are considered as known parameters to make the inverse approach tractable with the minimum number of uncertain parameters. The effective radius of laser beam is considered as certain parameter. The uncertain parameter sets of conduction heat transfer analysis and transport phenomena based heat transfer and fluid flow analysis are considered as

$$\{\theta^{cond}\} = \{\eta_G \quad \eta_V\} \tag{53a}$$

$$\{\theta^{conv}\} = \{\eta_G \quad k_{eff}/k_s \quad \mu_{eff}/\mu\} \equiv \{\eta_G \quad k^* \quad \mu^*\} \tag{53b}$$

where k^* and μ^* is non dimensional form with respect to k_s (conductivity at room temperature) and μ (molecular value at room temperature), respectively.

Parameters	**Value**	
	Low carbon steel	Bare steel
Density (ρ) - kg m^{-3}	7.8 x 10^3	8.0 x 10^3
Melting temperature (T_m) - K	1790	1800
Specific heat (C_p) - J kg^{-1}	755.0	745.8
Latent heat (L) - J kg^{-1} K^{-1}	2.45 x 10^5	2.50 x 10^5
Thermal conductivity (k_s) - J s^{-1} m^{-1} K^{-1}	32.4	31.2
Coefficient of thermal expansion (β) - K^{-1}	2.0 x 10^{-5}	2.0 x 10^{-5}
Temperature coefficient of surface tension $\left(\frac{d\gamma}{dT}\right)$ of pure iron - N m^{-1} K^{-1}	-0.5 x 10^{-3}	-0.5 x 10^{-3}
Molecular viscosity (μ) - kg m^{-1} s^{-1}	6.0 x 10^{-3}	6.0 x 10^{-3}

Table 2. Material properties used in numerical calculation.

To start with the optimization calculation, the feasible ranges of uncertain parameters are defined first. A correct choice of the parameter ranges also influence the overall computational time. Table 3 reports the feasible range of parameters used for optimization calculation and the optimum values of uncertain parameters derived from the integrated model. However, the choice of the initial range of parameters is based on literature reported results (Benyounis et al., 2005; Liu et al., 1993; Tzeng, 1999; Tanriver et al. 2000) as well as the experience gained from several numerical experiments. It is to be noted that the parameters for case- i and case - ii are considered from independent literature (Pitscheneder et al., 1996) without performing any optimization calculation. Moreover, the estimation of the optimum parameters is independent of the presence of surface active elements i.e. the optimum uncertain parameter set is similar for a laser process irrespective of the surface active elements presents in the material. Table 4 describes the typical values of the objective function and the corresponding uncertain parameter set after 23 iterations for case - iii. Further improvement in the value of the objective function was not possible corresponding to predefine values of control parameters (crossover constant, mutation factor and number of initial population) of DE. The optimum set of parameter is chosen from table 4 corresponding to the minimum value of objective function.

Type of welding	Mode of analysis	Uncertain parameter	Range of parameter	Optimum parameter
Spot welding (case - iii)	Conduction heat transfer	η_G	0.20 ~ 0.50	0.37
		η_V	0.20 ~ 0.70	0.48
	Heat transfer and fluid flow	η_G	0.20 ~ 0.50	0.36
		k^*	1.0 ~ 10.0	5.2
		μ^*	1.0 ~ 10.0	4.5
Linear welding (case -iv, v, vi)	Conduction heat transfer	η_G	0.20 ~ 0.50	0.36
		η_V	0.20 ~ 0.70	0.51
	Heat transfer and fluid flow	η_G	0.20 ~ 0.50	0.38
		k^*	1.0 ~ 10.0	6.0
		μ^*	1.0 ~ 10.0	7.0

Table 3. Optimum calculation of uncertain parameters by selecting range of parameters.

Individual index	$O(\theta^{conv})$ X10-3	η_G	k^*	μ^*
1	4.2	0.385	5.01	4.81
2	3.6	0.391	5.42	4.53
3	4.7	0.392	5.32	5.02
4	1.7	0.382	5.21	4.53
5	2.3	0.374	5.91	4.66
6	2.2	0.365	5.65	4.12
7	3.9	0.331	5.78	4.44
8	**1.2**	**0.381**	**5.22**	**4.51**
9	2.6	0.394	5.10	4.78

Table 4. Optimum set of uncertain parameters in laser spot welding using DE corresponding to case - iii using only three experimental data sets.

4.2 Prediction of weld geometry

Figure 6 describes the comparative study for the prediction of a target weld pool dimensions using three different heat source models: surface heat flux without any volumetric heat source, volumetric heat source with predefined heat source term and adaptive volumetric heat source. It is evident from the figure that surface only heat flux is not always satisfactory to predict the weld dimensions whereas volumetric heat source models are more reliable to such prediction. Volumetric heat with predefined heat source terms predicts the target weld dimensions with a-priori knowledge of weld dimensions. However, adaptively defined volumetric heat source predicts target weld geometry without the knowledge of target weld dimensions since the growth of source terms evolves with time as weld pool grows. Hence, the adaptive nature of volumetric heat source term essentially enhances the robustness and applicability of conduction mode laser welding process.

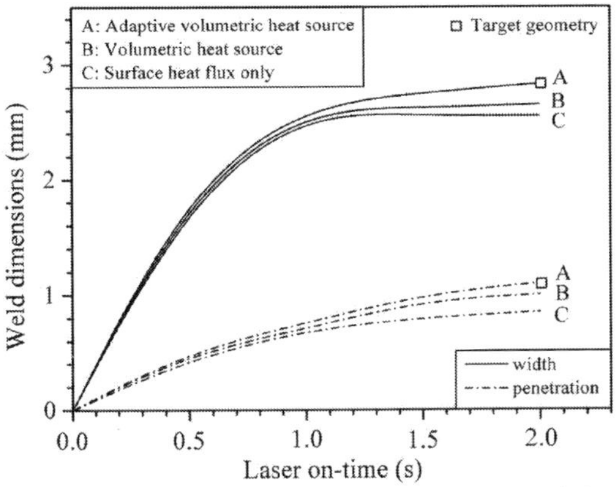

Fig. 6. Prediction of weld dimensions using various heat source models.

The experimentally measured weld dimensions for laser welding process is used to compare the numerically computed results using the optimum set of uncertain parameters of heat transfer and fluid flow analysis. Figure 7 describes such comparison both for spot and linear welding. The relative error in this case is defined by

$$w_R = \left|\frac{w^c - w^{exp}}{w^{exp}}\right|; \qquad p_R = \left|\frac{p^c - p^{exp}}{p^{exp}}\right| \tag{54}$$

where w_R or p_R represents the deviation of the calculated dimensions with reference to the corresponding experimental values. It is evident from Fig. 7(a) that the relative errors for most of the cases are smaller than 0.10. However, the transport phenomena based heat transfer and fluid flow model is more proficient for relatively bigger weld pool (higher on-time) for all the three cases. It is evident from Fig. 7(a) that the relative error for low laser on-time (case - i and case - iii) is high and this is possibly due to the lack of reaching the fully developed flow. In this case the conduction heat transfer model may be appropriate to predict weld dimensions. Figure 7(b) depicts the relative errors for linear welding for three cases. It is evident from this figure that the relative errors are high for comparatively smaller weld dimensions. However, the overall relative error is below 0.07. This indicates that the numerical model is robust enough to predict weld dimensions over a range of variable process parameters (travel speed = 5 ~ 10 mm/s, sulfur content = 0.002 ~ 0.012 wt %).

Figure 8 describes the 3D computed temperature and velocity profile for laser spot and linear welding corresponding to laser on-time of 1.0 s (case - iii of Table 1) and travel speed 10 mm/s (case - v of Table 1), respectively. The temperature and velocity profile for spot welding is symmetric with respect to XZ plane (Fig. 8a). Sine the surface tension coefficient is negative (for low sulfur content), the material moves from centre of laser beam towards the periphery and the buoyancy force acts in upward direction (Z direction). The combined effect of these two driving forces makes circulation loop in clockwise direction.

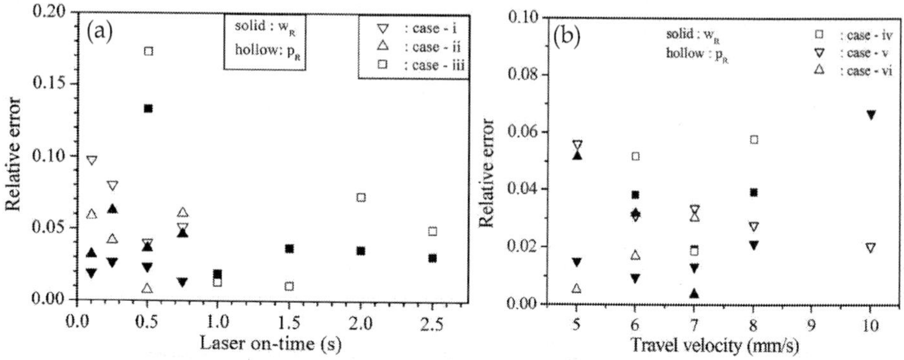

Fig. 7. Comparison between computed and experimentally measured weld dimensions in case of (a) spot welding and (b) linear welding.

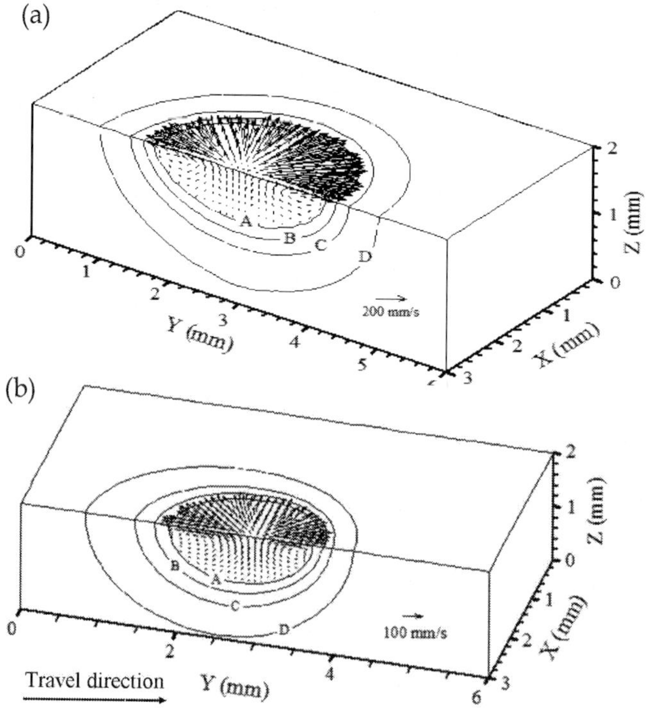

Fig. 8. A 3D computational view of temperature and velocity distribution in (a) spot welding and (b) linear welding (A – 1780 K, B – 1473 K, C – 1203 K, D – 993 K).

This nature of material movement in weld pool enhances weld width and decreases the weld penetration. Similar nature of material movement is also observed in the case of linear welding (Fig. 8b). However, in this case the temperature and velocity distribution is asymmetric due to the linear weld velocity along Y-direction. Figure 9 depicts the comparison between the experimentally measured macrograph (left side) with the computed weld geometry (right side) corresponding to travel speed of 7 mm/s for case - v. The cross-section of computed weld geometry is extracted from XZ plane on the location of the centre of laser beam. A good agreement of the shape and size of computed weld pool with the corresponding experimentally measured result is observed in Fig. 9.

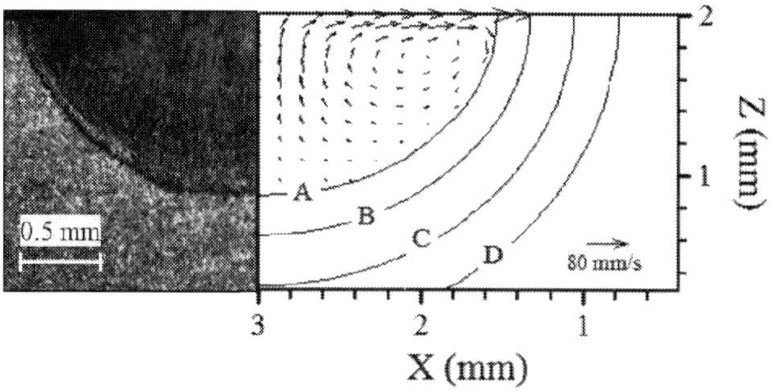

Fig. 9. Comparison between computed and experimentally measured weld macrograph (A - 1780 K, B - 1473 K, C - 1203 K, D - 993 K).

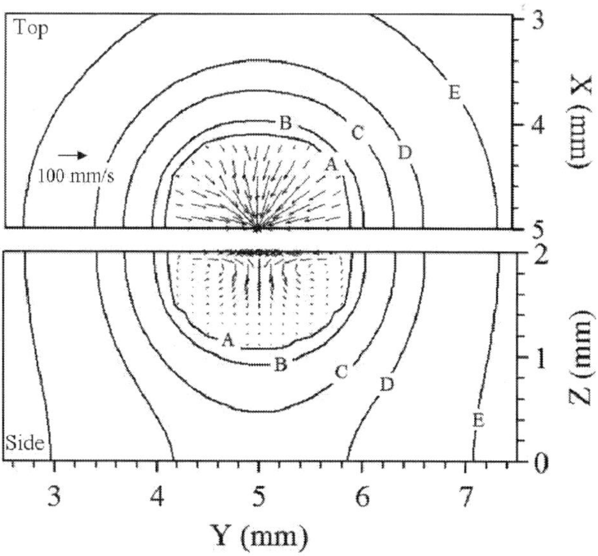

Fig. 10. Distribution of temperature and velocity profile in laser spot welding at on-time 0.1 s for case – ii (A - 1620 K, B - 1500 K, C - 1200 K, D - 1000 K, E - 773 K).

Figure 10 shows the nature of fluid flow in laser spot welding for the material having 0.015 weight percent of sulfur. It is evident from the figure that the liquid metal flows from periphery to the centre of heat source. This nature of flow is due to the positive surface tension coefficient at this high percentage of surface active elements present in molten material. It is also obvious that the reversal nature of flow reduces the weld width and increases the penetration as compared to molten martial having small percentage of sulfur for similar welding conditions. This trend is also observed in experimental results.
Figure 11 describes the shape of weld geometry in similar welding conditions and material except having different quantities of sulfur. The resultant velocity direction is completely opposite in these two cases. This clearly indicates the importance of the coupled heat transfer and fluid flow simulation for the prediction of weld pool geometry in the presence of surface active elements within parent material. To achieve similar trend with single set of uncertain parameters is nearly impossible for conduction heat transfer analysis alone.

4.3 Relative importance of driving forces

The validation of numerical heat transfer and fluid flow model in fusion welding process is extremely difficult or nearly impossible by means of experiments (Pitscheneder et al., 1997). Hence, a relatively simple approach is followed by researchers (Bag et al., 2009; Oreper & Szekely, 1987; Hong et al., 2003; Zhang et al., 2003; He et al., 2003; He et al., 2005). The relative importance of driving forces for liquid metal motion is quantitatively analyzed using dimensionless numbers and an order of magnitude analysis is followed to estimate the expected velocity of liquid metal (Bag et al., 2009). A comparison of the quantative values obtained from the order of magnitude analysis and the numerical model stands to validate the fluid flow analysis in fusion welding process.
The relative importance of the mode of heat transfer within weld pool is evaluated by using Peclet number (Pe) which is defined by (He et al., 2003)

$$Pe = \frac{U_{av}\rho C_p L_{av}}{k_{eff}} \tag{55}$$

where U_{av} and L_{av} are the average velocity and average length on the top of weld pool surface. It is evident from eq. (48) that the liquid metal convection affects the heat transfer when Pe is large whereas small Pe indicates the heat dissipation mainly by conduction. Hence, the fluid flow analysis of molten weld pool is significant when Pe is more than one or well above one. Table 5 describes the computed values of Peclet number for various welding conditions (case - iii and case - v). It is evident from the quantative values of Peclet number that convection within weld pool is significant as compared to only conduction analysis. Hence the transport phenomena based heat transfer analysis is necessary to predict weld dimensions for these welding conditions.

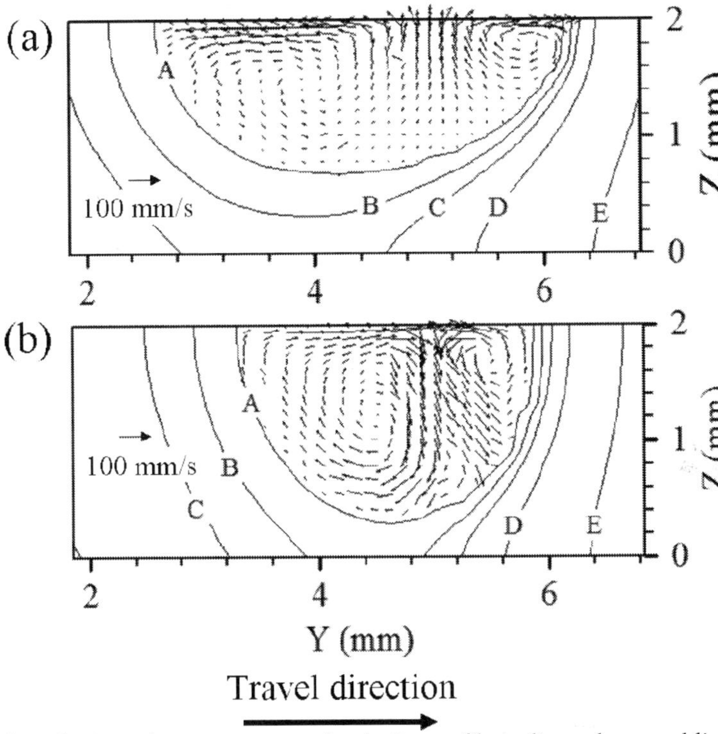

Fig. 11. Distribution of temperature and velocity profile in linear laser welding at travel speed of 6 mm/s with different sulfur content in steel: (a) 0.002 wt % sulfur and (b) 0.012 wt % sulfur (A - 1780 K, B - 1473 K, C - 1203 K, D - 993 K, E - 773 K).

The transport phenomena in laser welding is characterised by three driving forces such as surface tension force, buoyancy force and viscous force. The relative importance of surface tension force is described by surface tension Reynolds number (R_{ST}) which is the ratio of surface tension gradient force to viscous force, and is expressed as (He et al., 2005)

$$R_{ST} = \frac{\rho L_{av} \Delta T \left| \frac{\partial \gamma}{\partial T} \right|}{\mu_{eff}^2} \tag{56}$$

where $\frac{\partial \gamma}{\partial T}$ the temperature coefficient of surface tension and ΔT is is the mean temperature difference between peak pool temperature and solidus temperature on the top of weld pool. The Grashof number (He et al., 2003) is defined by the ratio of buoyancy force to viscous force and is represented as

$$Gr = \frac{g\beta L_{CR}^{3}\Delta T\rho^{2}}{\mu_{eff}^{2}} \tag{57}$$

where g is gravitational acceleration, β is the thermal expansion coefficient, and L_{CR} is the characteristics length. To understand the relative importance between surface tension force and buoyancy force, the dimensionless number is defined by the ratio of surface tension Reynolds number and Grashof number and is expressed as

$$R_{ST/B} = \frac{R_{ST}}{Gr} \tag{58}$$

From the order of magnitude analysis, the maximum velocity under surface tension force, U_{ST}, can be done assuming a boundary layer develops due to Marangoni shear stress and the maximum velocity occurs at a location approximately halfway between the heat source and weld pool edge (DebRoy & David, 1995),

$$U_{ST}^{3/2} \approx \left|\frac{\partial\gamma}{\partial T}\right|\left|\frac{dT}{dx}\right|_{x=w/4} \frac{w^{1/2}}{0.664\rho^{1/2}\mu_{eff}^{1/2}} \tag{59}$$

where $\frac{dT}{dx}$ is average temperature gradient on the top of weld pool at the position of w/4.
An order of magnitude analysis of the maximum velocity due to buoyancy driven flow is estimated as (He et al., 2003)

$$u_g = \sqrt{g\beta\Delta T p} \tag{60}$$

where p is the depth of weld pool.
The quantative estimation of various dimensionless numbers for present welding conditions is reported in Table 5 by using material data depicted in Table 2. It is obvious from the tabulated data of R_{ST} that the viscous force is less significant as compared to surface tension force. However, the computed values of Gr indicate that viscous force is more significant as compared to buoyancy force. Overall, the analysis on the quantative values of all driving force within molten pool indicates that surface tension force acts as the main driving force for the liquid metal movement in laser welding process. Hence the maximum magnitude of velocity is observed on the top of the weld pool (Fig. 8) due to the surface tension force. Figure 12 describes the comparison of maximum magnitude of expected velocity between order of magnitude analyses and predicted from numerical model. The relatively small deviation between these values indicates that the numerical model predicts the velocity distribution well.

Type of welding	On-time (s)/Travel speed (mm/s)	Dimensionless numbers			
		Pe	R_{ST} (x10^3)	Gr (x10^{-2})	$R_{ST/B}$ (x10^4)
Spot welding (case - iii)	0.5	3.04	0.36	0.20	18.0
	1.0	6.41	1.1	1.20	9.5
	1.5	7.67	1.4	1.60	9.0
	2.0	8.30	1.6	1.90	8.5
	2.5	10.5	2.3	2.90	8.0
Linear welding (case - v)	5	9.91	0.14	2.20	6.0
	6	8.71	0.11	1.60	6.9
	7	7.59	9.10	1.20	7.4
	8	6.20	6.70	0.79	8.5
	10	5.03	4.90	0.37	13.0

Table 5. Quantative estimation of dimensionless numbers in fluid flow analysis of laser welding.

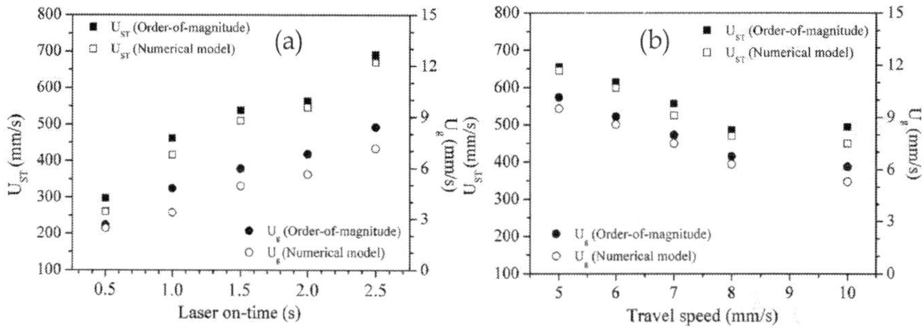

Fig. 12 Comparison of maximum magnitude of velocity between numerical model results and calculated from order of magnitude analysis in case of (a) spot welding (case - iii) and (b) linear welding (case - v).

4.4 Comparative study between conduction and transport phenomena based model

Figure 13 describes the comparison between computed weld dimensions using both conduction heat transfer and transport phenomena based model in laser spot welding. This comparison is also performed with reference to experimentally measured results for similar welding conditions. It is obvious from Fig. 13(a) that the conduction model predicts weld geometry well in case of small geometry (low on-time) and material having low weight percent of sulfur whereas the transport phenomena based heat transfer and fluid flow model predicts bigger weld pool (high on-time) better. However, the conduction based model fails to predict the weld geometry for the material having considerable amount of surface active elements (0.015 wt % of sulfur). Figure 13(b) indicates that both the models predict weld geometry well since the surface active elements is less in this case (0.002 wt % sulfur). However, the transport phenomena based model predicts weld penetration well as compared to the conduction based model. Hence, it is concluded that the transport phenomena based model is suitable for wide range of process capability i.e. longer laser on-time and presence of surface active elements. Figure 14 depicts a comparative study of weld dimensions in linear welding between conduction and convection based model with reference to experimentally measured results. It is obvious from Fig. 14 (a) that the convection based model predicts better than conduction based model results. This possibly due to fact that the material contains 0.010 weight percent of sulfur that changes the shape of weld geometry considerably as compared to material having low sulfur (0.002 wt %).

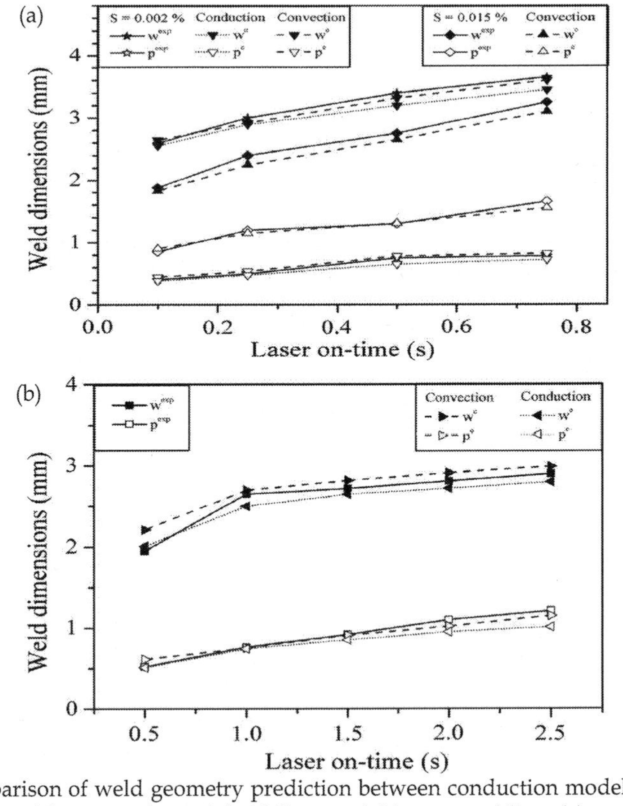

Fig. 13. Comparison of weld geometry prediction between conduction model and transport phenomena based heat transfer and fluid flow model in spot welding: (a) case - i and case - ii) and (b) case - iii.

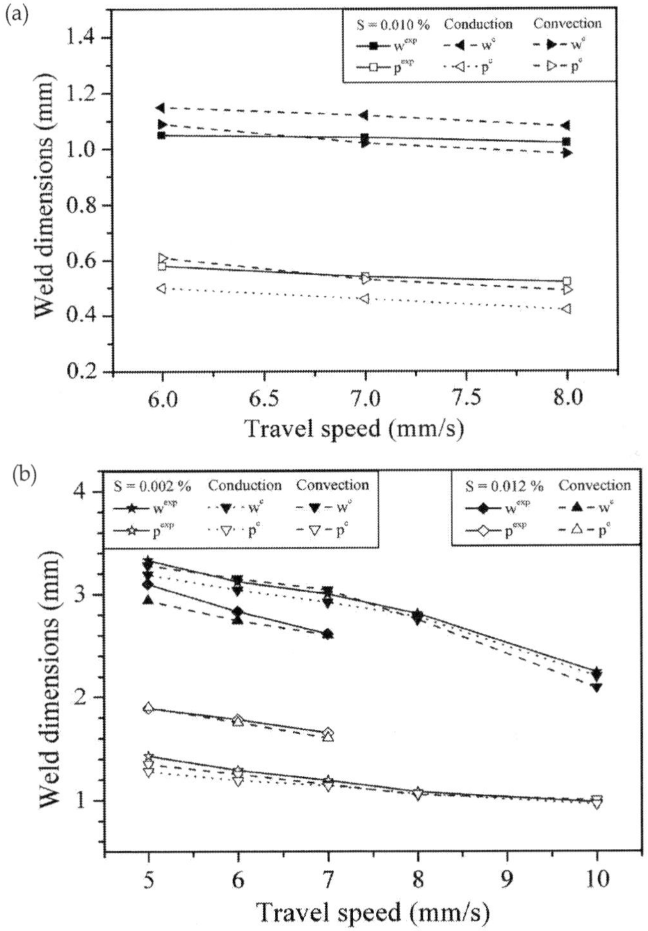

Fig. 14. Comparison of weld geometry prediction between conduction model and transport phenomena based heat transfer and fluid flow model in linear welding: (a) case - iv and (b) case - v and case vi.

5. Conclusions

An integrated model of conduction mode laser welding process is depicted in present work that is capable of undertaking 3D transient and pseudo-steady state heat conduction as well as transport phenomena based heat transfer and fluid flow analysis in weld pool using finite element method. The real parameter based differential evolution (DE) assists the numerical process model to predict uncertain parameters in an inverse manner. Conduction heat transfer based numerical models are important when weld geometry is small and, fast and repetitive calculation is of primary interest. The proposed adaptively defined volumetric heat source term in the frame of conduction heat transfer analysis is successfully demonstrated for a number of laser spot and linear welds. Transport phenomena based heat transfer and fluid flow analysis enhances the reliability of computed temperature field of comparatively bigger weld pool and is essential for material having considerable amount of surface active elements. The quantitative estimation of the fluid velocity is validated through order of magnitude analysis. The significant quantative knowledge extracted from this work in laser welding is expected to improve the physical understanding of laser welding process and serve as a basis for the design of welding process.

6. References

Bag, S. & De, A. (2008). Development of a three-dimensional heat transfer model for GTAW process using finite element method coupled with a genetic algorithm based identification of uncertain input parameters. *Metallurgical and Materials Transactions A*, **Vol. 39A**(No. 11), 2698-2710.

Bag, S. & De, A. (2009). Development of an efficient numerical heat transfer model coupled with genetic algorithm based optimization for the prediction of process variables in GTA spot welding, *Science and Technology of Welding and Joining*, Vol. **14**, 333-345.

Bag, S.; De, A. & DebRoy, T. (2009). A genetic algorithm assisted inverse convective heat transfer model for tailoring weld geometry. *Materials and Manufacturing Processes*, **Vol. 24**(No. 3), 384-397.

Bag, S. & De, A. (2010). Probing reliability of transport phenomena based heat transfer and fluid flow analysis in autogeneous fusion welding process, *Metallurgical and Materials Transactions A*, Vol. **41A**(No. 9), 2337 - 2347.

Benyounis, K. Y., Olabi, A. G. & Hashmi, M. S. J. (2005). Effect of laser welding parameters on the heat input and weld-bead profile. *Journal of Materials Processing Technology*, **Vol. 165**, 978-985.

Bonifaz, E. A. (2000). Finite element analysis of heat flow in single-pass arc welding. *Welding Research Supplements*, **Vol. 79**(No. 5), 121s-125s.

Chande, T. & Mazumder, J. (1984). Estimating effects of processing conditions and variable properties upon pool shape, cooling rates, and absorption coefficient in laser welding. *Journal of Applied Physics*, **Vol. 56**(No. 7), 1981-1986.

Cho, S. H. & Kim, J. W. (2002). Analysis of residual stress in carbon steel weldment incorporating phase transformation. *Science and Technology of Welding and Journal*, **Vol. 7**, 212-216.

De, A. & DebRoy, T. (2005). Reliable calculations of heat and fluid flow during conduction model laser welding through optimization of uncertain parameters. *Welding Journal*, **Vol. 84**(No. 7), 101s-112s.

De, A.; Maiti, S. K.; Walsh, C. A. & Bhadeshia, H. K. D. H. (2003). Finite element simulation of laser spot welding. *Science and Technology of Welding and Joining*, **Vol. 8**, 377-384.

DebRoy, T. & David, S. A. (1995). Physical processes in fusion welding. *Reviews of Modern Physics*, **Vol. 67**, 85-112.

Deng, D. (2009). FEM prediction of welding residual stress and distortion in carbon steel considering phase transformation effects. *Materials and Design*, **Vol. 30**, 359–366.

Deng, D.; Murakawa, H. & Liang, W. (2007). Numerical simulation of welding distortion in large structures. *Computational Methods in Applied Mechanics and Engineering,* **Vol. 196**, 4613–4627.

Frewin, M. R. & Scott, D. A. (1999). Finite element model of pulsed laser welding. *Welding Journal,* **Vol. 78**, 15-22.

Goldak, J. A.; Chakravarti, B. & Bibby, M. J. (1984). A new finite element model for welding heat sources. *Metallurgical and Materials Transactions B,* **Vol. 15B**, 229-305.

Gupta, O. P. (2002). Finite and boundary element methods in engineering. Oxford and IBH Publications, New Delhi, India.

He, X.; Fuerschbach, P. W. & DebRoy, T. (2003). Heat transfer and fluid flow during laser spot welding of SS 304 stainless steel. *Journal of Physics D: Applied Physics,* **Vol. 36**, 1388-1398.

He, X.; Elmer, J. W. & DebRoy, T. (2005). Heat transfer and fluid flow in laser micro welding. *Journal of Applied Physics,* **Vol. 97**, 084909:1-9.

Hong, K.; Weckmann, D. C.; Strong, A. B. & Zheng, W. (2003). Vorticity based turbulence model for thermo fluids modeling of welds. *Science and Technology of Welding and Joining,* **Vol. 8**(No. 5), 313-324.

Jung, G. H. & Tsai, C. L. (2004). Plasticity based distortion analysis for fillet welded thin plate t-joint. *Welding Journal,* **Vol. 83**, 177-187.

Lee, M. Y. & Kim, J. W. (2004). On-line penetration depth measurement system using infrared temperature sensing in CO_2 laser welding. *Advances in Non Destructive Evaluation,* **Vol. 270-273**, 2308-2314.

Lhospitalier, S.; Bourges, P.; Bert, A.; Quesada, J. & Lambertin, M. (1999). Temperature measurement inside and near the weld pool during laser welding. *Journal of Laser Applications,* **Vol. 11**, 32-37.

Liu, J. T.; Weckman, D. C.; & Kerr, H. W. (1993). The effects of process variables on pulsed Nd:YAG laser spot welds: Part I. AISI 409 stainless steel. *Metallurgical and Materials Transactions B,* **Vol. 24**, 1065-1076.

Mackwood, A. P. & Crafer, R. C. (2005). Thermal modelling of laser welding and related processes: a literature review. *Optics & Laser Technology,* **Vol. 37**, 99-115.

Mazumder, J. & Steen, W. M. (1980). Heat transfer model for CW laser material processing. *Journal of Applied Physics,* **Vol. 51**(No. 2), 941-947.

Mishra, S. & Debroy, T. (2005). A computational procedure for finding multiple solutions of convective heat transfer equations. *Journal of Physics D: Applied Physics,* **Vol. 38**, 2977-2985.

Oreper, G. M. & Szekely, J. (1987). A comprehensive representation of transient weld pool development in spot welding operations. *Metallurgical and Materials Transactions A,* **Vol. 18A**, 1325-1332.

Pitscheneder, W.; DebRoy, T.; Mundra, K. & Ebner, R. (1996). Role of sulfur and processing variables on the temporal evolution of weld pool geometry during multi-kilowatt laser beam welding of steels. *Welding Journal,* **Vol. 75**(No. 3), 71s-78.

Pitscheneder, W.; Ebner, R.; Hong, T.; Debroy, T.; Mundra, K. & Benes, R. (1997). Experimental and numerical investigations of transport phenomena in conduction mode weld pools. *Proceedings of Fourth International Seminar on Numerical Analysis of Weldability,* pp. 379-395, ISBN, Graz- Seggau, September 1997, Austria.

Price, K.; Storn, R. & Lampinen, J. (2005). Differential Evolution – A Practical Approach to Global Optimization. Springer, Berlin.

Reddy, J. N. & Gartling, D. K. (2000). The Finite Element Method in Heat Tranafer and Fluid Dynamics, CRC Press, Florida.

Sahoo, P.; Debroy, T. & Macmillan, M. J. (1988). Surface tension of binary metal-surface active solute systems under conditions relevant to welding metallurgy. *Metallurgical and Materials Transactions B*, **Vol. 19**, 483-491.

Storn, R. (1997). Differential evolution, a simple and efficient heuristic strategy for global optimization over continuous spaces. *Journal of Global Optimization*, **Vol. 11**, 341-359.

Tanriver, U.; Longobardi, J.; Latham, W. P. & Kar, A. (2000). Effect of absorptivity, shielding gas speed, and contact media on sheet metal laser welding. *Science and Technology of Welding and Joining*, Vol. **5**, 310-316.

Teng, T. L.; Fung, C. P.; Chang, P. H. & Yang, W. C. (2001). Analysis of residual stresses and distortions in T-joint fillet welds. *International Journal of Pressure Vessels and Piping*, **Vol. 78**, 523-538.

Trivedi, A.; Bag, S. & De, A. (2007). Three dimensional transient heat conduction and thermomechanical analysis for laser spot welding using adaptive heat source. *Science and Technology of Welding and Joining*, **Vol. 12**(No. 1), 24-31.

Tzeng, Y. F. (1999). Pulsed Nd:YAG laser seam welding of zinc coated steel. *Welding Journal*, Vol. **78**(No. 7), 238s - 244s.

Tzeng, Y. (2000). Parametric analysis of the pulsed Nd:YAG laser seam-welding process. *Journal of Materials Processing Technology*, **Vol. 102**, 40-47.

Zhao, H.; White, D. R. & DebRoy, T. (1999). Current issues and problems in laser welding of automotive aluminum alloys. *International Materials Reviews*, **Vol. 44**, 238-266.

Zhang, W.; Roy, G. G.; Elmer, J. W. & DebRoy, T. (2003). Modeling of heat transfer and fluid flow during gas tungsten arc spot welding of low carbon steel. *Journal of Applied Physics*, **Vol. 93**(No. 5), 3022-3033.

7

Laser welding process: Characteristics and finite element method simulations

Yannick Deshayes
University of Bordeaux 1-IMS Laboratory
France

1. Context and objectives

Expertise of packaging for optoelectronic components requires the solution of optical, mechanical and electrical problems in the same way. The purpose of this study is to present three-dimensional simulations using finite element method (FEM) of thermomechanical stresses and strains in transmitter Laser modules induced by Nd:YAG crystal Laser welds on main sub-assembly Laser submount. Non-linear FEM computations, taking into account of experimental $\sigma(\varepsilon)$ measured curves, show that Laser welding process can induce high level of strains around the Laser welding zone, bearing the Laser diode, responsible of an optical axis shift and a gradual drop of the optical power in relation with relaxation of accumulated stresses in the sub-assembly (Sherry and al., 1996). Typical stresses are close to 160 MPa with drift about 5 MPa with the dispersion of energy level of laser Nd : YAG beam. The introduction of both material and process dispersion in order to evaluate their impact on product life time distribution has been taking into account. Thermal cycles (-40°C/+85°C VRT) are used to estimate the robustness of the technology assembly. Previous paper demonstrated that Laser submount near laser welding zones is the most sensitive part of optical system (Deshayes and al., 2003).The gradual changes of stresses distribution from the laser welding process and after thermal cycles are estimate using FEM. Experimental analyses were also conducted to correlate simulation results and monitor the output optical power of Laser modules after 500 thermal cycles.

The development of high bandwidth single mode fibre optics communication technologies coupled with the availability of transmitter components for wavelength multiplexing has created a revolution in the transmission technology during the last fifteen years. These performances can be reached by packaging interface and control circuits with the optical chips leading to the concept of high reliable technically-advanced Laser modules. Reduced cost, low consumption, hermetical and highly efficient optical coupling between the Laser diode and the single-mode fibre associated to a mechanical stability are some of the key issues. Moreover, packaging of such systems requires the resolution of optical, thermomechanical and electrical problems.

These problems are often highly interactive and the stability of optoelectronic devices is still an essential factor to ensure high bandwidth data transmission, acceptable bit-error rate and develop reliable solutions. In actual telecommunication applications, photonic systems involve a non direct mechanical alignment between the laser diode and the optical fibre

(Deshayes and al., 2003; Breedis and al., 2001). Generally, one or two lens are used to for this optical alignment. For instance, mechanical stability requires tolerances less than 1 μm to avoid a power change higher than 10 %, which must be consistent during the lifetime of the module and across the temperature range.

For optical alignment, three primary techniques have been developed to align and connect the light-emitter to the optical fibre associated with different package configurations (Jang, 1996; Song and al., 1996) :

- Solder with V-groove,
- Epoxies,
- Nd:YAG Laser welds.

It has been already demonstrated that Nd:YAG Laser welding technique is the most effective method to satisfy performances criteria previously described. Due to inherent advantages, a growing number of communication systems integrators are requesting Laser welded packages for their end-users. However the challenge of containing the solidification shrinkage caused by the light-metal interaction during the welding process, resulting in a weld shift leading to the reduction of coupling efficiency and device throughput stability (Song and al., 1996).

Standard qualification procedures, in particular power drift monitoring, must be conducted to validate the system with respect to tolerances through temperature cycling or storage temperature characterizing the limits and the margins of the technology. Actual standards tend to be 500 cycles in the temperature range -40°C/+85°C without failures (Goudard and al., 2002). These ageing tests are generally realized in order to evaluate all the parameters in relation with failure distribution but more than one hundred modules must be performed during several thousands hours mixing different life test conditions. These results can allow determining the robustness of the technology but due to a high complexity of the package, cannot give accurate information on the failure origin, which is responsible of the optical power drift. To face qualification challenges, new processes are now being proposed focusing on reliability concerns at the early stage of the product development. In this approach, the qualification is considered as a long-term process rather than a final exam at the end of the development (Goudard and al., 2002). Based on environmental and functional specifications, the product development can start with a technical risk analysis phase. This phase aims at pointing out the major risks for a given product design. In this case, physical simulation (thermal and/or mechanical) represents an attractive tool to assess and weigh up the risk criticality (Mcleod and al., 2002).

The purpose of this paper deals with results achieved from nonlinear thermomechanical simulations using finite-element method (FEM) of a direct modulation 1.55 μm Laser module (10 mW) for telecommunication applications. This study completes the thermomechanical studies in laser diodes module emitting at 1550 nm (Mcleod and al., 2002).

This paper will be developed in three main parts:

- description of the methodology to implement in FEM the Nd:YAG Laser welding using electro-thermal analogies,
- calculations of stresses and strains after Laser welding process between the Laser diode platform and the lens holder taking into account of experimental process parameters,

- impact of calculated strains on optical misalignment (angular deviation of the optical axis) with respect to dispersion process.

2. Laser welding model for FEM

2.1 Theory of laser material interaction

a. Spatial structure and coherent

The structure of laser wave is critical for understand the thermal flow during the laser welding process. This part presents the basic structure of laser wave.
The spatial structure of laser wave can be expressed considering the electric field $E(x,y,z)$ by equation (1):

$$E(x,y,z)=E_0\left[\frac{\omega_0}{\omega(z)}\right]\exp-i\left(k\left[z+\frac{r_\perp^2}{2R(z)}\right]-\phi(z)\right)\times\exp\left[\frac{r_\perp^2}{\omega^2(z)}\right] \quad (1)$$

With $r_\perp^2=x^2+y^2$: transversal radius, $E_0=E(x,y,0)$: transversal electric field, $\omega^2(z)=\omega_0^2\left[1+\left(\lambda z/\pi\omega_0^2\right)^2\right]$: the radial extension of the laser beam, $R(z)=z\left[1+\left(\pi\omega_0^2/\lambda z\right)^2\right]$: curvature radius of the laser beam.
The geometry of the laser beam can be represented by the fig 1.

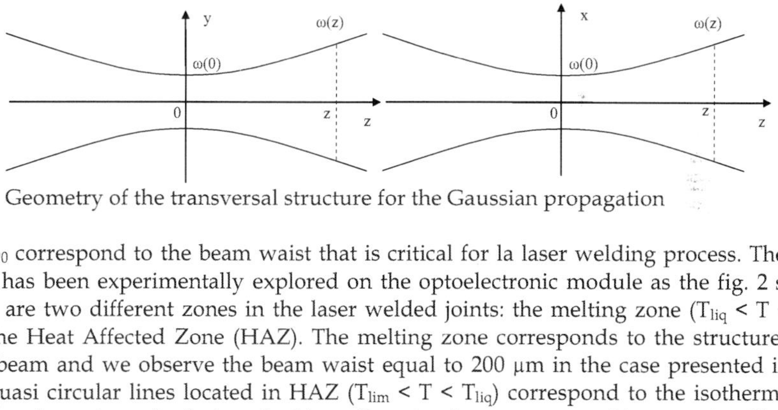

Fig. 1. Geometry of the transversal structure for the Gaussian propagation

The ω_0 correspond to the beam waist that is critical for la laser welding process. The beam waist has been experimentally explored on the optoelectronic module as the fig. 2 shown. There are two different zones in the laser welded joints: the melting zone ($T_{liq} < T < T_{max}$) and the Heat Affected Zone (HAZ). The melting zone corresponds to the structure of the laser beam and we observe the beam waist equal to 200 μm in the case presented in fig.2. The quasi circular lines located in HAZ ($T_{lim} < T < T_{liq}$) correspond to the isothermal line. The laser beam intensity is described by a Gaussian Low as proposed by equation (2):

$$T(r)=T_{lim}+\left(T_{max}-T_{min}\right)\exp\left(-\frac{r_\perp^2}{\omega_0^2}\right) \quad (2)$$

The T_{max} is the maximal temperature estimated at 1823 K, T_{min} = 600 K is the minimal temperature corresponds to the solidification of material and T_{liq} is the limit between liquid-solid phase temperature. In this condition, the material is not liquid but melting with liquid and solid and the Young Modulus is 100 time less than solid material. This condition explain the HAZ until diameter around 800 μm

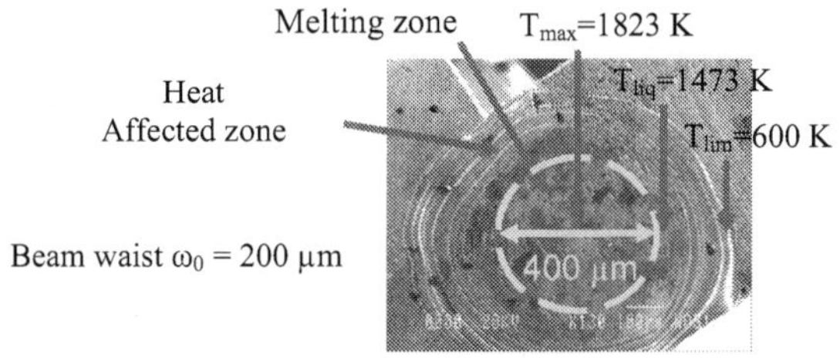

Fig. 2. Geometry of the transversal structure for the Gaussian propagation

The main objective of the FEM model is to simulate the same phenomenon but the great difficulty is to ***apply Gaussian thermal flux on the laser welded zone***.

2.2 Methodology for build FEM model

a. Pattern geometry

Actual graphic performances of the simulator allow, without difficulty, to building a complex geometric model generally coming from the design plans of the considered system. The most important difficulty resides in the dividing of the geometric model in finite elements. It is necessary to follow a procedure of model building very strict allowing to insure, in the final, to obtain an optimum pattern.

b. Finite elements definition

The FEM consist to find an approximate solution of exact solution with field of discretization defined on sub-domain Ωi of global domain Ω. The sub-domain Ω_i have been build considering equation (3).

$$\bigcup_{i=1}^{i=n} \Omega_i = \Omega \text{ and } \Omega_i \cap \Omega_j = \emptyset \;\; \forall i \neq j \tag{3}$$

Fields $\tilde{f}(M,t)$, defined for each sub domain Ω_I is called local field family and is generally defined as polynomial function. The global field family $\tilde{F}(M,t)$, obtained by superposition of local fields, is called interpolation function space of domain Ω.

The field for each sub-domain Ω_I is determined by a finite number of value of field (or value of first order derivation) in arbitrary points in the sub-domain and called nodes. The local field is an interpolation between the values associated to nodes. The sub-domain with its interpolation is called element.

Find a solution by finite element method consists in the determination of local field that can be attributed for each sub-domain for the global field $\tilde{F}(M,t)$ calculated by superposition of local field is close to real solution of the problem.

c. Elements and geometry combination

The elements constitute for software resolution a little volume element for integration or differential equation resolution. In international reference, it exist only three different geometric references: Cartesian, geometrical and cylindrical. In Finite element model, the single element can be associated as an elementary discrete volume $d^3\tau$.

For example, we consider Cartesian representation. In this case, the elementary volume is a parallelepiped element defined by equation (4):

$$d^3\tau = dx \cdot dy \cdot dz \quad (4)$$

and represented by fig. 3.

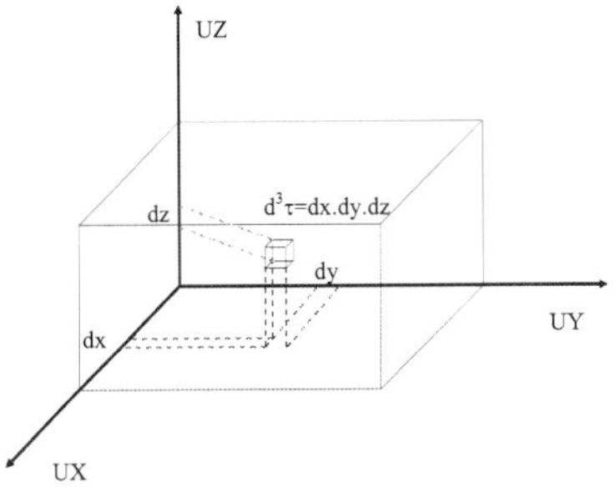

Fig. 3. Cartesian geometric reference and elementary volume

For cylindrical and spherical representation, the finite element can be adjusted to form a single elementary volume allowing to describe respectively cylindrical or spherical system. The accuracy of the simulation results for system considered is, in part, related to both element design and their number.

d. Physical associated to element

- Mechanical elements

  Elements of this type require material's features properties listed below: Young modulus, poison ratio, plastic modulus, coefficient thermal dilatation, material density. This element allows to perform linear or non linear simulation taking into account of non-linearity of materials (plastic deformation, creep,…). The non linearity is also induced by geometry structure as contact of area to simulate pivot link for example.

- Thermal elements

  Elements of this type require material's features properties listed below: thermal conductivity, material density, heat capacity, and enthalpy for material phase change. This element allows to perform linear or non linear simulation taking into account of non-linearity of materials (phase change).

- Coupled field elements

  Coupled field associate in same time the mechanical, thermal, electrical and magnetic simulations. In this case, the material behaviour is considered as linear and all simulations are performed in same time. In the case where it is necessary to perform thermal and mechanical simulations, for example, on the structure with non-linear material properties, the solution is to perform, in the first time the thermal simulation and in the second time to execute the mechanical simulation. The thermal simulation became boundary conditions of the mechanical simulation for this example.

e. Geometry of elements

- Regular elements

Elements are also characterized by their geometry in Cartesian coordinates (x, y, z), cylindrical (ρ, θ, z) and spherical (ρ, θ, φ) for three dimensional simulation. Generally this kind of elements is constituted by 8 or 16 nodes as it indicates in fig. 4 and is named regular elements.

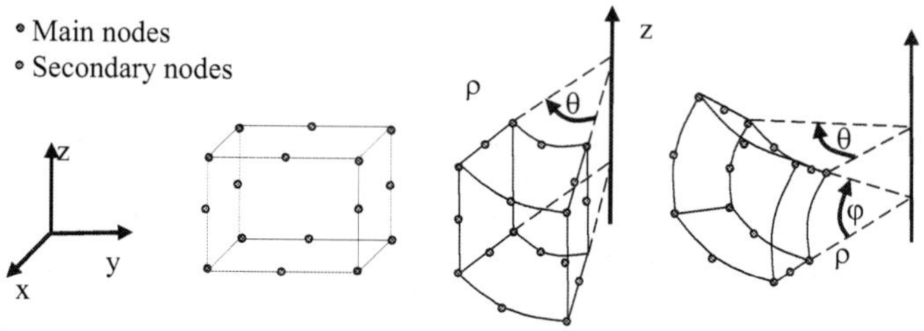

Fig. 4. Regular element

- Tetrahedral elements

The second geometry of elements is tetrahedral useful for mesh an intersection with two different regular geometry, cylindrical and cubic for example. The structure of this element is described by fig. 5 and the element is build with 4 or 9 nodes.

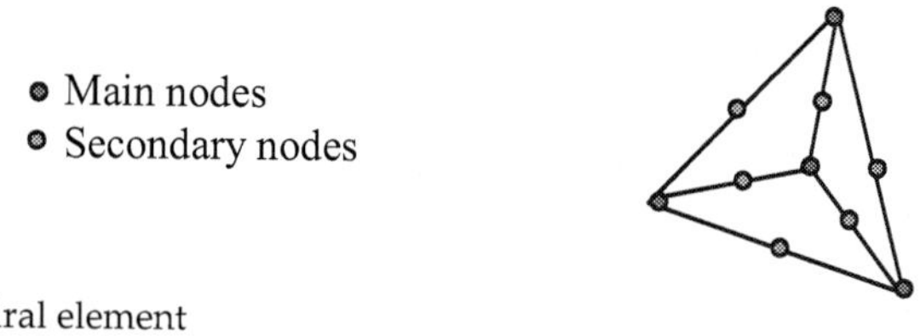

Fig. 5. Tetrahedral element

- Prism elements

The last geometry of elements is prism useful for mesh a centre of cylindrical geometry for example. The structure of this element is described by fig.6 and the element is build with 6 or 15 nodes.

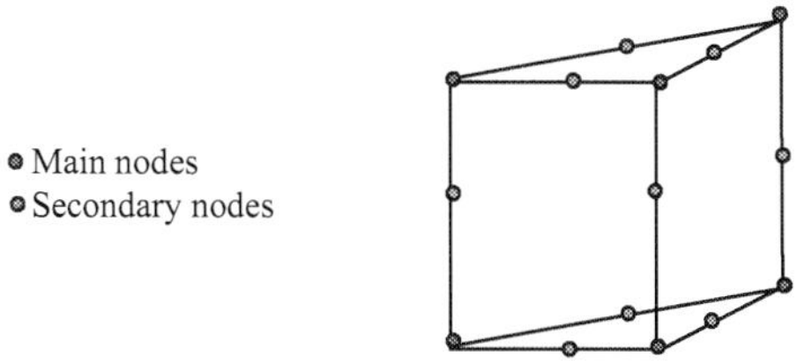

Fig. 6. Prism element

2.3 Building of laser welding process model

To build the laser welding process model, the main idea is to consider an electro-thermal model. The electrical part can give us a Gaussian distribution of the equipotential line. For simple application of electrical condition, we can simulate a Gaussian spatial distribution of equipotential. Using a adapted electrical conductivity, the thermal heat, caused by potential different between two isopotential, has also a Gaussian distribution. Then it is very simple and optimum to elaborate an electrothermal model about laser welding process.
Fig. 7 presents the electrical boundary condition on the cubic volume around the real laser weld volume. The laser beam simulate has the same characteristics than one previously presented in fig.2. The boundary conditions of potential are ground for five face of cube as it shown in fig. 7. On the last face of cube, the punctual potential V_{YAG} is applied in the centre of the face O(0,0,0). The meshing method must build node for this specific point O.

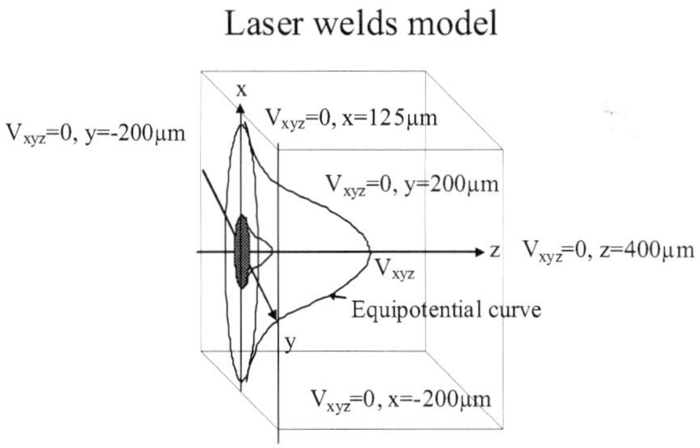

Fig. 7. Laser welding zone in electrical boundary conditions

The V_{LAS} potential is calculated using the thermodynamic relations of enthalpy ΔH based on the heat transfer and phase change. To simulate laser pulse energy E_{LAS} deposited on the surface of the laser welding zone, the V_{LAS} is also a pulse. The time whitening of the pulse Δt is related by the total power P_{LAS} deposited during the laser welding process. The relation between electrical model and thermal one is proposed in equation (5).

$$\Delta H = \frac{V_{LAS}^2}{R}\Delta t = mC_p\Delta T + mL_f \tag{5}$$

With m: the total ma of laser welded zone, C_p : the calorific capacity of material: L_f: the latent heat of material.

3. Modelling setup

3.1 Design of the laser module

Semiconductor Laser package bodies are typically either cylindrical-type or box-type styles. For light wave communication systems, box-type bodies are widely used and in particular Dual-In-Line or Butterfly packages with fibre pigtails. This study is focused on 1.55 μm Butterfly package Laser module and a technological description is presented in fig. 8. The DFB Laser diode (Distributed Feedback Laser diode InP/InGaAsP) emitting at 1.55 μm is soldered with a AuSn solder joint (8 μm) on the Laser submount (AlN), and then the submount is attached to the Laser platform (composed by a submount and 2 columns bearing the lens holder) in Kovar by a SnSb solder joint (8 μm). Lens 1, used to collimate the Laser beam from the Laser diode, and the isolator are welded to a lens holder (Kovar) by means of Nd:YAG Laser welding process.

The Laser platform and the lens holder are also welded corresponding to the sub-assembly 1. This last element is then attached to the thermoelectric cooler and mounted with a SnPbAg solder joint (10 μm) in a Butterfly-type package (Kovar). The sub-assembly 2 is composed of a second lens, used to focalized Laser beam into the fibre core, glued with an adhesive material into a circular ferule (Zirconia/Kovar). Finally, the sub-assembly 2 is Nd:YAG Laser welded to the Butterfly-type package providing a complete hermiticity for the system.

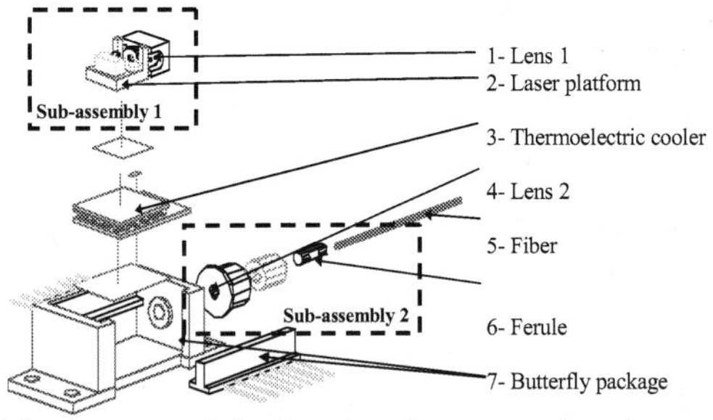

Fig. 8. Description of the Laser module showing the two main sub-assemblies

As an alternative to the common adhesives or solders used in the light-coupling process, Nd:YAG Laser welding offers a number of attractive features such as high weld strength to weld size ratio, minimal heat affected zone, reliability providing some benefits : low heat distortion, non-contact process, repeatability and ability to automate (Sherry and al., 1996). Nevertheless, the main drawback of Laser welding is that the intense energy input, resulting in severe thermal gradients, can contribute to generate strains driving elements out of alignment. Motions in excess of 10 μm can be thus introduced and submicron alignment usually requires some type of motion compensation after the initial welds to hold required tolerances (Hayashi and al., 1996).

3.2 Finite element analysis conditions

In order to calculate levels of stresses and strains after Nd:YAG Laser welding process between the Laser diode platform and the lens holder, FEM simulations are performed. Different models and boundary conditions are defined in this part. Sub-assembly 1 is composed of the Laser platform and the lens holder essentially in Kovar. The model is based on electrical, thermal and mechanical simulation using a multiphysics approach and will allow to extract isothermal contour plots to evaluate magnitude of thermal gradients in the sub-assembly 1. The final goal is to calculate residual stresses and the resultant optical beam axis deviation after this process.

Fig. 9-a presents the global model of sub-assembly 1 with a planar symmetry (Ox, Oy). Fixed points, considered as nodes without any degrees of freedom, represent fixing flanges used in manufacturing process to maintain the lens holder and the Laser platform during Laser welding process. The external loads are listed below:

- Weight is applied on the gravity centre,
- A Clamp forces F_{pres} are applied on the back of the lens holder during Laser welding process to guarantee an adjustment between Laser platform and lens holder,
- Laser heating boundary conditions are shown in fig. 9a and modeled by Joule heating considering the well-known thermal/electrical analogies. The equipotential surface is adjusted to obtain a maximum temperature at 1400 K. We use two different electrical characteristics to traduce a localized heat source as generated by a Nd:YAG Laser beam. This part will be developed in the next section.

Fig. 9 describes an optimized model with 7526 elements and 11803 nodes using three-dimensional tetrahedral multiphysics transfer elements was used for this sub-assembly 1 which is mainly composed of Kovar.

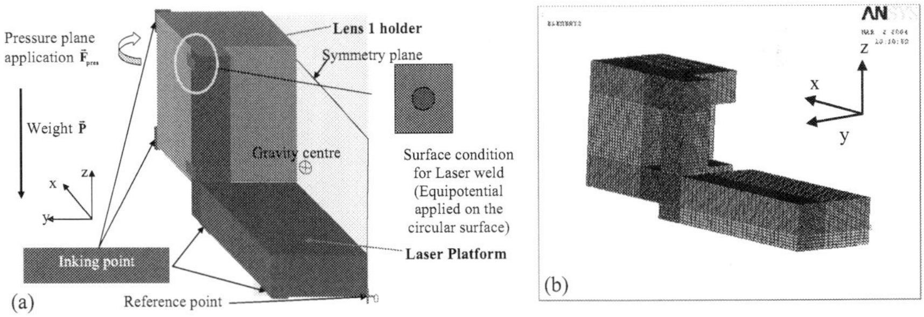

Fig. 9. Sub-assembly 1 simulated design with external loads (a), optimized FEM mesh model (b)

Material properties are assumed to be dependent on temperature (see table 1 below). The time dependence of Laser welding process has conducted us to run time-dependent (transient) simulations.

Fig. 10 summarizes the chronograms of pressure force, voltage of the top volume welded V_{YAG1} and voltage the bottom volume welded V_{YAG2}. On the top of fig. 10, we note the time dependence of the thermal cartography analyses. In this study, the laser heating is modelled by Joule heating taking into account electrical/thermal energy considering that a Laser welded joint can be associated to an electrical resistance calculated from the same area of material (Kovar).

	Kovar		
	27°C	600°C	1200°C
CTE (µm/°C)	5.13	5.86	11.5
Young modulus (GPa)	138	138	138
Yield strength (MPa)	345	245	50
Poisson ratio	0.317	0.317	0.317
Thermal conductivity ($W.m^{-1}.°C^{-1}$)	17.3	17.3	17.3
Heat capacity ($J.kg^{-1}$)	439	439	649
Melting point (°C)		1450	

Table 1. Values of the material physical properties of Kovar using in sub-assembly 1 versus temperature

To evaluate the thermal energy developed in the volume of the welded joint, we considered the relation between the enthalpy variation and the electrical energy (6) :

$$\Delta H = \frac{V^2}{R} \Delta t \qquad (6)$$

with ΔH defines the enthalpy variation, R is equivalent of an electrical resistance of the Laser welded joint volume, V corresponds to the time-dependent applied voltage (V_{YAG1} and V_{YAG2}) and Δt is the YAG Laser pulse duration (2.5 ms).

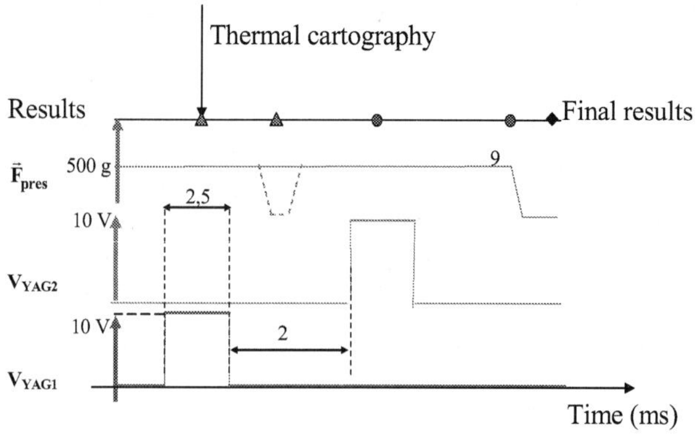

Fig. 10. Time dependence of boundary conditions for Sub-assembly 1

V_{YAG1} and V_{YAG2} correspond respectively to inferior and superior Laser welded joint and are not applied simultaneously as shown in fig. 10.
Deposed energy with a YAG Laser is calculated considering electrical energy dissipated from a resistance on which an applied voltage allows to simulate a thermal energy in the volume of the spot weld with a temperature close to the melting temperature. Our simulations are based on the following expressions giving the relation between electrical energy and thermodynamically conditions (7):

$$\begin{aligned} \frac{V^2}{R}\Delta t &= mC_p\Delta T + L_f \\ \Delta H &= mC_p\Delta T + L_s \end{aligned} \tag{7}$$

Equations (7) give the heating conditions corresponding to the Laser energy quantity deposed on the material with Cp defines as heat capacity (in $J.kg^{-1}.°C^{-1}$) and Lf represents the latent heat of melting given in Joule. Cooling conditions taking into account of latent heat solidification Ls, given in Joule, are resumed by equation (3). It is known that, for Kovar, the heat capacity parameter has temperature dependence and literature allows to extract the value until 1200°C rather than heat latent of solidification which is difficult to obtain (Cheng and al., 1999). This parameter traduces cooling effect which is critical in this case. So proposed simulations are computed at a temperature close to 1473K corresponding to the temperature at which thermomechanical constants are given in table 1. All thermomechanical properties have been used to simulate the Laser welding process and give thermal and mechanical solutions.

4. Results and discussions

4.1 Thrmomechanical simulations

Nd:YAG Laser welding process involves a highly focused Laser beam responsible of a non-uniform temperature distribution on the focus point. Simulated energy deposed allows being close to melting temperature of Kovar material (1473K). Fig. 11 shows the nodal solution contour plot of thermal cartography of Laser platform after first Laser welding process.

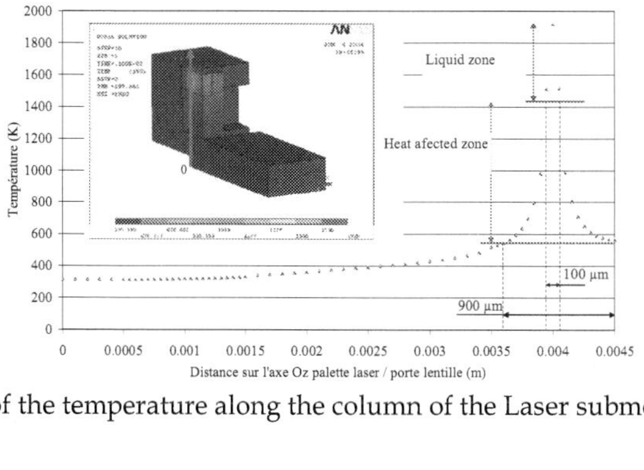

Fig. 11. Variation of the temperature along the column of the Laser submount

The temperature variation along the column of Laser platform can be fitted by a Gaussian law expressed as equation (8):

$$T(r) = T_1 + (T_0 - T_1)\exp\left(-r^2/W^2{}_0\right) \tag{8}$$

With T_0 = 1427K, the maximal temperature of Laser weld, T_1 = 600K the minimal temperature of Laser weld and W_0 the beam waist defined as the minimum radius of the Laser beam.

Experimental and calculated beam waist values are the same and evaluated around 150 µm (Deshayes and al., 2003). The good agreement between experimental and calculated values validate the simulation approach for Laser Nd:YAG welding process.

Fig. 12 compares strains in sub-assembly 1 structure before and after Nd:YAG Laser welding process. Strain occurring in the column is observed and this particular view (deformed and no deformed nodal solution plots) allows highlighting optical beam axis deviation of the lens holder.

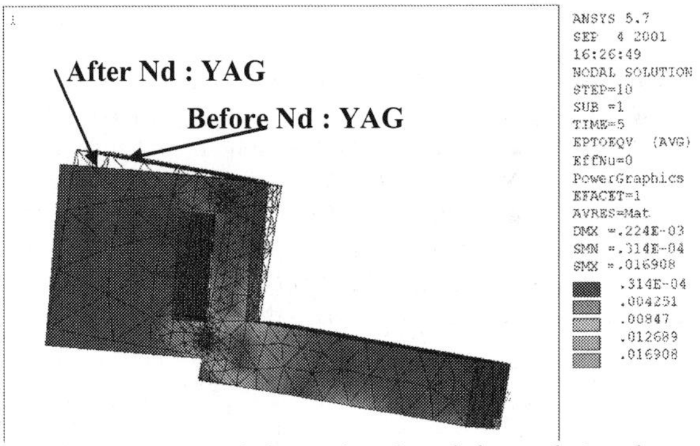

Fig. 12. Residual effective strains deformed and undeformed view located in sub-assembly 1

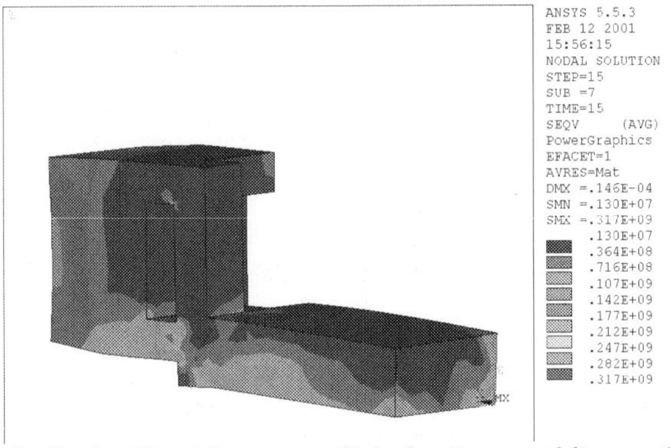

Fig. 13. Residual effective Von Mises stress (Pa) after Laser welding on the sub-assembly 1

Fig. 13 clearly shows the residual effective Von Mises stresses close to 55 MPa located in the column base of the Laser Platform after Laser welding.
Thermal gradients (≈ 1100K) along columns of Laser platform induce maximal displacements close to 2 µm located in the column base after Nd:YAG Laser welding process in manufacturing conditions and maximal strains of 0.05% (fig. 13). These displacements are observed by optical axis angular deviation θ and Δx, Δy and Δz axial deviations. Calculated values of previous parameters are obtained from simulation results.
In fig. 14, two nodes of lens holder representing optical axis have been considered. We also reported Δx, Δy, Δz and θ deviations of optical axis allowing to give its final position after total Laser welding process.
Considering this two nodes, optical axis deviation resulting from residual stress after Laser welds can be evaluated close to an angular deviation of θ = 0.03° and axial deviations of Δx_{max} = 2 µm, Δy = 0 and Δz_{max} = 0.1 µm. Experimental analyses for evaluation of optical coupling drop in 1550 nm Laser modules have reported that optical power losses about 40 % result from two critical values of parameters variation:

- An angular deviation θ of 0.02° between sub-assembly 1 and sub-assembly 2,
- Δy, Δz deviations of 10 µm between Laser diode and Lens holder.

At this step of the manufacturing process, optical coupling between lens holder and Laser diode is correct because Δy, Δz deviations are close to 0.1 µm corresponding to optical coupling losses lower than 0.1%. Thus, an operator could not suspect a possible displacement of the first lens axis after Nd:YAG Laser welds. The assembling step between the pigtail and the sub-assembly 1 requires then a dynamic alignment to find maximal optical coupling. In this case, the operator can adjust a possible optical beam axis deviation without any information about the value of the previous deviation and the level of accumulated stresses trapped in the sub-assembly one.

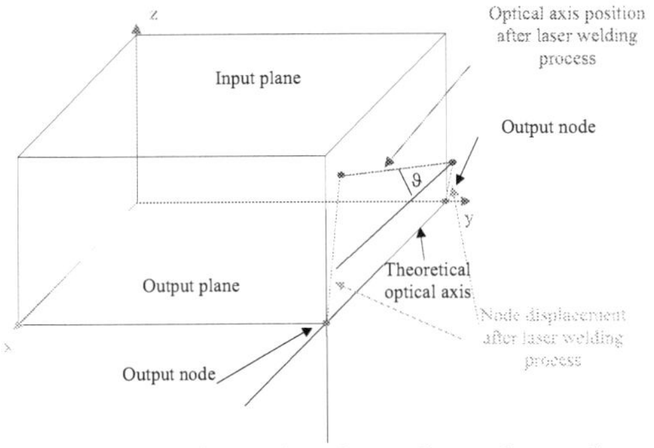

Fig. 14. Residual effective strains deformed and unreformed view located in sub-assembly 1

After Nd:YAG Laser welds, intrinsic and extrinsic stresses can appear. Intrinsic stresses are generally related to only Laser YAG energy deposition on metal surface. The most important accumulated stress is located inside the HAZ caused by huge plastic deformation and very rapid thermal variation in welded joints (Jang, 1996; Panin and al., 1998). Extrinsic

stresses are caused by external loads applied during process and discontinuity of materials on the interface (Inoue and Koguchi, 1997). In our case, the most important external load is represented by pressure strength F_{pres} used to ensure an adjustment between Laser platform and lens holder.
Relaxation of accumulated stresses in the sub-assembly 1 can occur and could be accelerated by defects induced in the welded zone (Inoue and Koguchi, 1997; Hariprasad, Sastry and Jerina). Rapid solidification processing in HAZ leads to a metastable phase formation, solid solution or dispersion strengthened alloys and intermetallics and the whole physical phenomenon is at the origin of defects formation located in welded joints (Hariprasad, Sastry and Jerina; Cheng and Wang, 1996). It has demonstrated that metallic alloys creep fatigue is related to defects rate located in welded joints considered as a metallic alloy zones (Asayama, 2000). In particular, a model based on molecular dynamics calculations, developed by J.D. Vazquez, has discussed on isotropic and anisotropic relaxation phenomenon from simulations of lattice relaxation of metallic alloys considering the sudden appearance of vacancy or an interstitial site in the crystal (Dominguez-Vazquez, 1998). This microscopic relaxation model allows highlighting macroscopic effective displacement of system responsible of relaxation phase. Experimental measurements, using in particular an optical method, have been also conducted to observe strains, stresses and fractures of welded joints at the mesoscale level (Panin and al., 1998). This study has characterized, in bulk material, the accumulated stresses located in HAZ and their evolution after Laser welding process. So our interpretation of gradual optical power drift between the sub-assembly 1 and the pigtail can be explained by relaxation phenomenon and time evolution can be directly related to the number and the location of defects into the welded joints but also in the structure.
Experimental procedure has been established to localize strains and stresses in sub-assembly 1 during the whole step Nd:YAG Laser welding process and evaluation of relaxation phenomenon after thermal cycles.

4.2 Ageing tests analysis

Qualification procedures, in particular power drift measurement, must be conducted to validate the system with respect to tolerances through temperature cycles or storage temperature characterizing the limits and the margins of the technology. Actual standards tend to be 500 cycles in the temperature range -40°C/+85°C with a failure criterion of 10% of optical power drift. The methodology of failure diagnostic for optoelectronics components and modules for telecommunication applications imposed to do ageing tests to validate different assumptions coming from the simulation results. The detailed of this procedure is presented by (Y. Deshayes and al., 2003).
First ageing tests have been made on 1550 nm InGaAsP/InP DFB Laser diodes. After 500 thermal cycles -40°C/+85°C, no failure occurred on Laser diodes. Measurements have been made with a specific test bench with temperature dependence has been developed to monitor P(I), I(V) and L(E). This result demonstrates that optical power drift is only associated to misalignment in relation with thermomechanical aspects. The second ageing test is made on nine different optoelectronic modules in final packaging. Fig. 13 shows variations of ΔE_{ta} (%) defined by :

$$\Delta E_{ta} = \left[\frac{\Delta P_{opt}}{P_{opt}} \right]_{I=100mA} \tag{9}$$

with P_{opt} is initial optical power measurement of the laser module, ΔP_{opt} is the difference between optical power measured after ageing time and initial optical power measurement and I is the current value for optical power measurement.
This experimental procedure has been applied on nine InGaAsP/ InP 1550 nm Laser modules (LM1 to LM9) versus thermal cycles -40°C/+85°C. In fig. 8, evolution of ΔE_{ta} (%) measured at 100 mA from 0 to 500 thermal cycles (-40°C/+85°C) are reported. Experimental and simulation results lead to give failure modes and assumptions on failure location (Deshayes and al., 2003):

- sudden total optical power drop explained by a break located in the optical fibre core,
- gradual optical power drift outside the failure criteria limit in relation with thermomechanical aspect responsible of columns deformation in sub-assembly 1 and related by stresses relaxation phenomenon,
- gradual optical power drift inside the failure criteria demonstrating the relative instability of optical coupling in Laser module especially on sub-assembly 1.

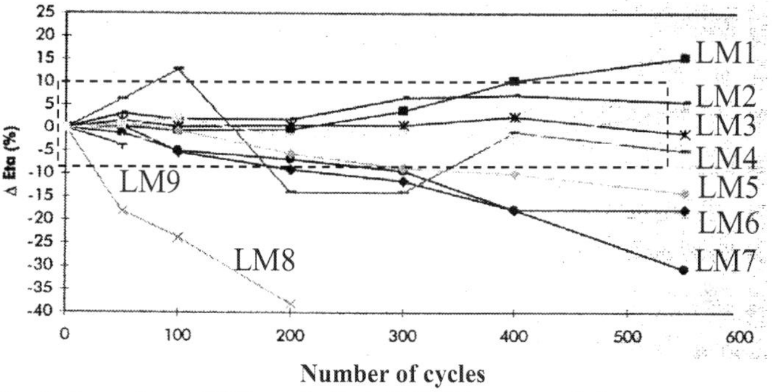

Fig. 15. Ageing test results on 1550 nm InGaAsP/InP Laser module

4.3 Optical misalignment using process dispersion

The new method proposed in the introduction of this paper corresponds to an evolution of optoelectronic qualification practices needing to develop new working methods than the usual "go-no go" qualification tests. The final objective is to define relevant tests performed to define "generic" accelerated test and assess both robustness and reliability of the component. In this case, technological dispersion modelling represents an attractive tool to identify the effect of a critical technological parameter on the optical deviation distribution and reduce time duration of tests. Among these parameters, we can list: material properties, geometric dimensions, welding and solder processes...
Fig. 13 reveals the difference of behaviour between optical modules in term of optical coupling deviations, could be related to manufacturing process dispersion. As we have yet

demonstrated, the most sensitive manufacturing process is Nd:YAG Laser welding associated to clamp forces F_{pres} and Laser heating conditions (E_0). Until now, 3D FEM simulations have been performed considering F_{pres} and E as average constant values called Fpres0 and E0. The range of these last parameters is limited by manufacturing process. The parameter F_{pres} is set from F_{pres0} ±20% and laser Nd:YAG energy from E_0 ±20% according with manufacturer specification (Gibet, 2001).

In the case of clamp force F_{pres} variation limited by F_{pres0} ±20%, less than 10^{-5} degree on angular deviation is observed and stresses stay constant. For this configuration, the impact of clamp force variations on the optical coupling efficiency could be considered as negligible.

The Laser Nd:YAG energy E corresponds to the one absorbed by the welded joint. The amplitude of dispersion can be correlated both to the reflectance of the Laser impact area and thickness of gold deposed on the Kovar mainly composing the sub assembly 1. The absorbance of Laser energy is related to the thickness of gold, water concentration and roughness of the material surface (Watanabe and al., 2004; Martin, Blanchard and Weightman, 2003; Zhang, 2004). The thin film of gold allows to adsorbed infrared 1μm wavelength laser Nd:YAG beam.

Fig. 16 reports variations of optical angular deviation versus energy of the Laser beam. In the same time, we report the maximal stress located in top welded zones. The global study indicates that welding zone is the most critical zone, so FEM simulation has been optimized to precise stresses in welding zone. After specific analyses, we identify that top welding zone is the most critical zone and amplitude of stress is optimized. The energy variation is the experimental data given by manufacturer. It is shown that higher is the energy deposed on the welded zone, higher is the stress level but lower is the optical deviation. This key result is closely correlated with results reported by W.H. Cheng (Jerina; Cheng and Wang, 1996). The displacement is critical because 2/100° induces 40 % of optical power losses and explain the magnitude of ΔE_{ta} (%) shows in fig. 8. The drift of stresses and displacements versus energy E/E_0 is weak and indicates that energy level of Nd: YAG cannot be adjusting to reduce the optical misalignment. So, this key result indicates that the architecture of the system should be optimized to reduce the impact of laser welding process on the optical misalignment.

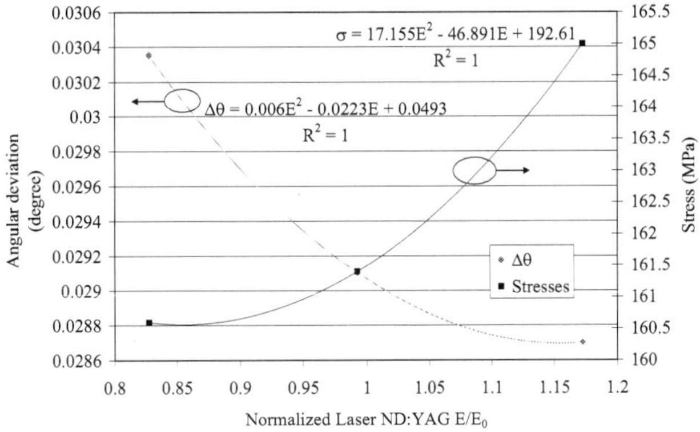

Fig. 16. Optical angular deviation and stresses accumulated versus energy laser Nd:YAG

The different behaviour of different modules shows in fig.16 can be explained by the initial stresses and displacements. This phenomenon is associated to the fact that laser submount and lens support are hyper static system. The methodology presented in this paper conduct manufacturer to modified the design of laser submount taking into account all these results. The new optical module are now qualified using the same standard requirements for telecommunication applications.

5. Conclusion and perspectives

Laser welding process in sub-assembly 1 has been identified as the most potential critical zone and to correlate simulation results using ANSYS software, experimental analyses have been also investigated (Deshayes, Béchou and Danto, 2001).

Calculated optical misalignment in sub-assembly 1 have demonstrated an angular optical beam axis deviation of 0.03° and responsible of a possible first lens axis movement confirming that Laser welding process can induce optical instability of Laser modules and degradation of performances for telecommunication applications. The main solution could be given by a better optimization of the Nd:YAG Laser power density close to $1.5.10^5$ W/cm^2. For this technology, average Nd:YAG Laser power density reaches $2.5.10^5 W/cm^2$ and can generate bulk defects and thermal stresses in welded joints (fig. 17). W.H. Cheng has established that optical losses in Laser modules can relate to the presence of bulk fractures (Jerina; Cheng and Wang, 1996). It has also been highlight that power density is responsible of bulk defects and accumulative stresses. In our case, the presence of bulk defects, observed in fig. 17, could explain random acceleration of time stress relaxation allowing optical power decrease. The time before failure corresponding to ±10% of the optical power drift is directly related to the manufacturing process and to the order of static non determination from a mechanical point of view of the system strongly dependent on the Laser platform and the lens holder design. All conditions are correlated to a mechanical misalignment between Lens axis and pigtail. The major cause of bulk defects formation in the Laser welding process for sub-assembly 1 is due to the excess Laser energy. The other causes are gas bubbles trapped within the weld sections and the heterogeneous nucleation in welded joints (Jerina; Cheng and Wang, 1996).

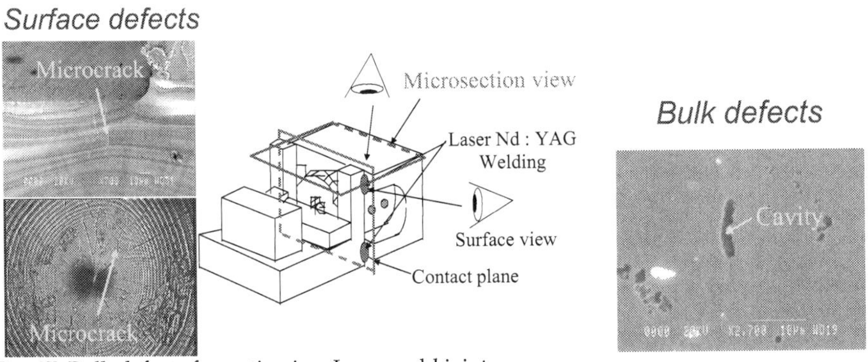

Fig. 17. Bulk defects formation in a Laser weld joint

This chapter reports 3D thermomechanical simulations and experimental tests in order to identify critical zones in a Butterfly-package Laser module showing that three main zones must be carefully analyzed: shape and volume of glue in the ferule, solders and, in particular, Laser welds. Laser welding process is a useful and effective method to ensure hermeticity and secure metal parts but the mechanical distortions due to severe thermal gradients should be controlled within allowance limits. The accumulated stresses are close to 160 MPa in welded zones. The main advantages of this technique are given by precision of alignment close to ±0.2 µm, the whole process fully automated to contain the cycle's time within 60 to 90 seconds. But it has been shown that one of the main inconvenient of the Laser welding process is the excess of deposed Laser energy resulting in high thermal gradients (700 K on 200 µm) and residual stresses (around 160 MPa) in the Laser platform responsible of an optical misalignment and a possible failure in terms of optical power drift requirements. We have demonstrated that FEM simulations, to predict distortion of Laser welding which is very difficult to measure, is very attractive and can be applied to different package configurations.

Such a study is attractive for the definition of more realistic and optimized realistic life cycle profiles, taking advantages of previous methodologies already experienced in the field of microelectronics or military industries.

Experimental failure analyses will be also conducted to validate thermomechanical simulations, focused in particular on Laser welded joints in order to propose assumptions for accumulated strains relaxation phenomenon. In this context, both thermal, electrical and thermomechanical simulations on the package must be realized using an original approach based on multiphysics computations of ANSYS software, in particular for electro-thermal Nd:YAG Laser modelling (Fricke, Keim and Schmidt, 2001). First, a description of the Laser module is given and 3D-FEM models of each sub-assembly are presented taking into account of the different materials characteristics versus temperature and external loads related to manufacturing steps. The last section gives simulation results of the main sub-assemblies of the Laser module concluding on thermomechanical sensitivity of critical zones and the impact on a possible optical axis misalignment.

Our activities are now focused on FEM predictions that could be improved by a detailed knowledge of the effect of bulk defects located in Laser welded joints on stresses relaxation phenomenon and also by a better implementation of heating and cooling conditions in computations. The final objective is to improve packaging design rules and optical misalignment reduction in order to achieve highly reliable bandwidth single mode fibre communication systems.

6. References

Asayama (2000). Creep fatigue evaluation of stainless steel welded joints in FBR class 1 components. *Nuclear Engineering and Design,* 198, 2, (February 2000), pp. 25-40, ISSN: 00295493

Breedis (2001). Monte Carlo tolerance analysis of a passively aligned silicon waferboard package, *Proceeding of Electronic Components and Technology Conference,* pp. 247-254, ISBN: 05695503, United States, 29 May 2001 through 1 June 2001, IEEE, Orlando

Cheng and al. (1999) Thermal stresses in box-type Laser packages, *Optical and Quantum Electronics,* 31, 4 (April 1999), pp. 293-302, ISSN: 03068919

Deshayes, Béchou and Danto (2001). Experimental validation of thermomechanical simulations on 1550 nm Laser modules, *Internal Report, ALCATEL Optronics-IXL*, September 2001.

Deshayes and al. (2003). Three-dimensional FEM simulations of thermomechanical stresses in 1.55 μm laser modules, *Microelectronics Reliability*, 43, 7, (July 2003), pp. 1125-1136. ISSN: 00262714

Dominguez-Vazquez and al. (1998). Relaxation of metals: A model based on MD calculations. *Nuclear Instruments and Methods in Physics Research, Section B: Beam Interactions with Materials and Atoms*, 135, 1-4, (February 1998), pp. 214-218, ISSN: 0168583X

Fricke, E. Keim and J. Schmidt (2001). Numerical weld modeling-method for calculating weld-induced residual stresses, *Numerical engineering and design*, 206, 2-3, (June 2001), pp. 139-150, ISSN: 00295493

Gibet (2001). Procédure de fabrication de têtes optique, Alcatel Optronics - France, Research and development department, Internal report, 2001.

Goudard and al. (2002). New qualification approaches for opto-electronic devices, *52nd Electronic Components and Technology Conference*, pp. 551-557, ISBN: 05695503, United States, 28 May 2002 through 31 May 2002, IEEE, San Diego

Hariprasad, Sastry and Jerina (1996). Deformation behavior of a rapidly solidified fine grained Al-8.5%Fe-1.2%V-1.7%Si alloy, *acta Materialia*, 44, 1, (January 1996), pp. 383-389, ISSN: 13596454

Hayashi and Tsunetsugu (1996). Optical module with MU connector interface using self-alignment technique by solder-bump chip bonding, *Proceedings of the 1996 IEEE 46th Electronic Components & Technology Conference*, pp. 13-19, ISBN: 05695503, United States, 28 May 1996 through 31 May 1996, IEEE, Orlando

Inoue and Koguchi (1997). Relaxation of thermal stresses in dissimilar materials (approach based on stress intensity), *International Journal of Solids and Structures*, 34, 25, (September 1997), pp. 3215-3233, ISSN: 00207683

Jang (1996). Packaging of photonic devices using Laser welding, *Proceedings of SPIE - The International Society for Optical Engineering*, pp. 138-149, ISBN: 0819419745, United States, 25 October 1995 through 26 October 1995, Society of Photo-Optical Instrumentation Engineers, Philadelphia

Martin, Blanchard and Weightman (2003), The effect of surface morphology upon the optical response of Au(1 1 0), *Surface Science*, 532-535, (10 June 2003), pp. 1-7, ISSN: 00396028

McLeod and al. (2002). Packaging of micro-optics component to meet Telcordia standards, *Proceeding of Optical Fiber Communication Conference and Exhibit*, pp. 326-327, United States, 17 March 2002 through 22 March 2002, IEEE, Anaheim

Panin and al. (1998). Relaxation mechanism of rotational type in fracture of weld joints for austenic steels, *Theoretical and Applied Fracture Mechanics*, 29, 2, pp. 99-102, ISSN: 01678442

Sherry and al. (1996). High performance optoelectronic packaging for 2.5 and 10 Gb/s Laser modules, *Proceeding of Electronic Components and Technology Conference*, pp. 620-627, ISBN: 05695503, United States, 28 May 1996 through 31 May 1996, IEEE, Orlando

Song and al. (1996). Laser weldability analysis of high-speed optical transmission device packaging, IEEE Transaction on Component, Packaging and Manufacturing Technology, 19, 4, (November 1996), pp. 758-763, ISSN: 10709894

Watanabe and al. (2004) Optimizing mechanical properties of laser-welded gold alloy through heat treatment, *Dental Materials*, 20, 7, (September 2004), pp. 630-634, ISSN: 01095641

Zhang and al. (2004). Relationship between weld quality and optical emissions in underwater Nd: YAG laser welding, *Optics and Lasers in Engineering*, 41, 5, (May 2004), pp. 717-730, ISSN: 01438166

8

Development of digital laser welding system for car side panels

Hong-Seok Park and Hung-Won Choi
University of Ulsan
South-Korea

1. Introduction

Because of the extremely globalized competition, all manufacturing enterprises not only have to decrease the product development time and reduce manufacturing cost, but also develop new technologies. Automotive enterprises are facing to two market requirements such as the increment in demand for improvement of safety and the reduction of fuel consumption. From the design point of view, the improvement of safety means high strength of car body while the reduction of fuel consumption is treated to light car body. Because car body consists of panels, the improvement of strength can be treated as material property and welding structure. Welding structure is considered as an important factor in case of car body strength. Also, as a part of efforts to lighten car body in automotive enterprises, they try to make car body using new technologies such as TWB(Tailor Welded Blank) (Ku et al., 2004; Zhang, 2006) or hydro forming (Gao et al., 2006; Park et al., 2002; Saito et al., 2006; Suh et al., 2006) which can minimize the overlapping areas of welding. At the same time, they attempt to lighten car body by substituting the existing steel-oriented panel with new materials such as aluminum or magnesium.

The existing spot welding is not anymore appropriate for strength of welding structure and new materials. In order to overcome these problems, laser welding is studied and carried out for car body welding instead of spot welding. Because laser welding has so many advantages such as good accessibility, fast welding speed and good welding quality, the automotive enterprises try to develop and apply laser welding technology. BMW and Volkswagen try to design two layer structures for application of Nd:YAG laser which increase greatly the flexibility of welding process. And AUDI used seam tracking system to perform laser welding without jig/fixture (Emmelmann, 2000; Koerber et al., 2001; Sasabe et al., 2003) (Fig. 1). In case of Korean automotive enterprises, laser welding is still not activated so that just some parts of car body are welded by laser welding (Jung et al., 2002). But they began to recognize the necessity of laser welding and then carry out many experiments and researches for the extensive application. In spite of the high performance of laser welding, it is currently used in only a few area of the theoretically possible application. This is due to the fact that a lot of companies, owing to the complex, time-consuming and cost intensive planning of the laser welding cell, exercise restraint when it comes to entering the field of laser material processing, such inhibitions can be eliminated by providing application specific solution, i.e. a reasonable planning method when planning complex system like a laser welding cell.

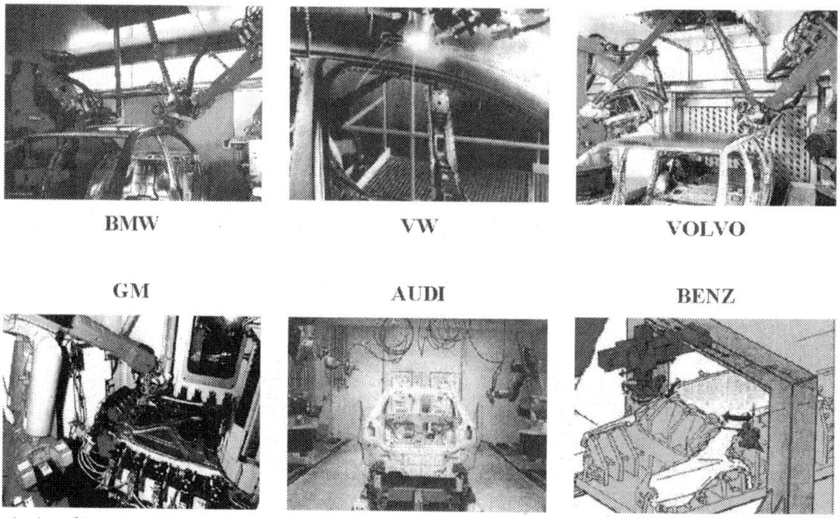

Fig. 1. Application of laser welding in car body assembly

The objective of this paper therefore is to conceive a method of the planning of laser welding cell and its implementation with digital manufacturing.

For the implementation of the laser welding cell as planning object, this means that it should be followed the systematic planning procedure(Fig. 2).

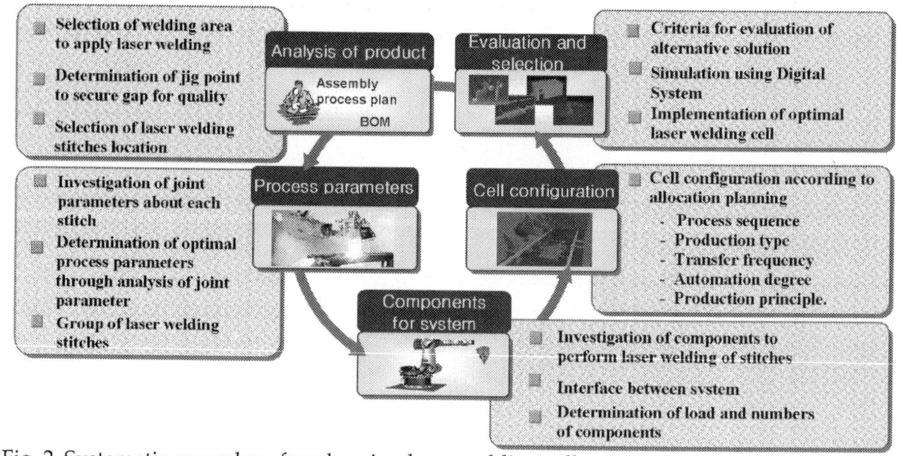

Fig. 2. Systematic procedure for planning laser welding cell

Through the analysis of product as the first step the requirements for executing a welding process and configuring a welding cell are grasped. Based on the these information, the process parameters guaranteeing the welding quality are chosen and grouped for planning the welding sequence and for deriving the needed characteristics of the cell components. To execute the appropriate components are determined through the comparison between the requirement profiles of them and the ability of the commercial products. With the selected components, the cell configurations are generated and evaluated using digital manufacturing.

2. Characteristics of laser welding

2.1 Advantages of laser welding compared spot welding

Most automotive enterprises have assembled car body using spot welding. With this technology, spot guns are big, heavy and have to take lots of direction change to perform the welding task. These problems lead to decrease the flexibility of system and tool accessibility. As a result, the number of cell to perform the welding task increases. However, laser welding using laser beam radiated from optic head can weld, even if accessibility of optic head is allowed at only one side. As laser welding is applied, it offers greatly flexibility of product design and tool accessibility and dramatically decreases welding time than the existing spot welding. In addition, laser welding is expected to improve the welding strength, to prevent from car body deformation as well as to have better quality. Also, we can make the lighter car body through benefits of laser welding such as elimination of redundant reinforcements, minimization of part numbers and overlapping areas of panels.

2.2 Influential factors and process parameters of laser welding

For laser welding, heat conduction welding and deep penetration welding can be distinguished (Dawes, 1992; William, 2001). In the heat conduction welding method, the material melts due to the absorption and thermal conduction of laser beam radiated from optic head. This method has fast welding speed but has low penetration depth because of insufficient thermal energy. The other is deep penetration welding or keyhole welding method, which is normally used for welding car body to ensure reliability of quality and to be easy to exhaust fusion vapor of material. Because of diffused reflection of laser beam in keyhole, welding depth is deep and welding speed is fast.

In order to perform laser welding effectively, process planning should be generated after a examining factors that influence to the laser welding process. The first important factor was gap between panels which was recognized through lots of experiments with the different combination of materials(Fig. 3).

The results of the experiments carried out with the different materials and gaps show that in case of gap greater than 0.2 mm the laser beam cannot penetrate the panels at all combinations. At the first- and second combination, the welding quality was satisfied when welding with the given range of the gap, i.e., $0.0 \leq$ gap ≤ 0.2 mm. In the last two case, the welding failures such as sinking, protrusion, etc. occurred by welding without a gap.

In case of galvanized steel usually used for car body, if there is no exit for evaporated zinc vapor, it may permeate into the inside of welding area because the evaporation point of zinc coated layer is lower than the melting point of steel (1320°C) and could be the main reason of poor welding. Thus, gap between panels should satisfy the gap between 0.1 mm to 0.2

mm and jig was used to keep the suitable gap. As the results of the experiments, a jointing area with more than 4 layers or high reflexibility also produced poor welding because of insufficient thermal energy.

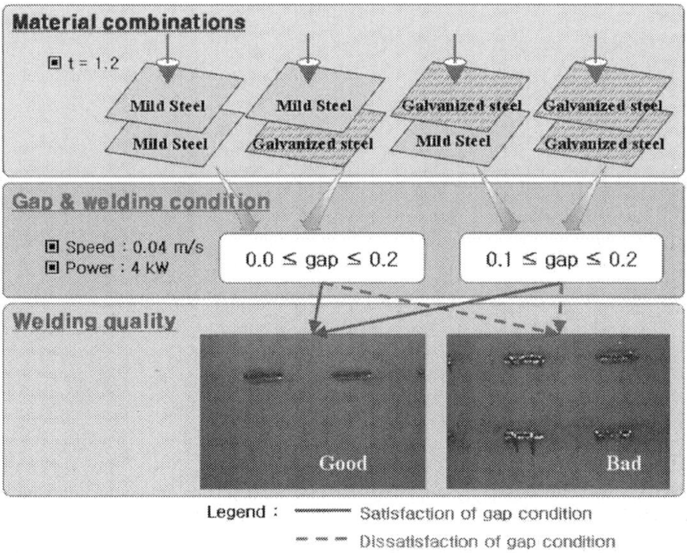

Fig. 3. Welding quality according to material combination and gap condition.

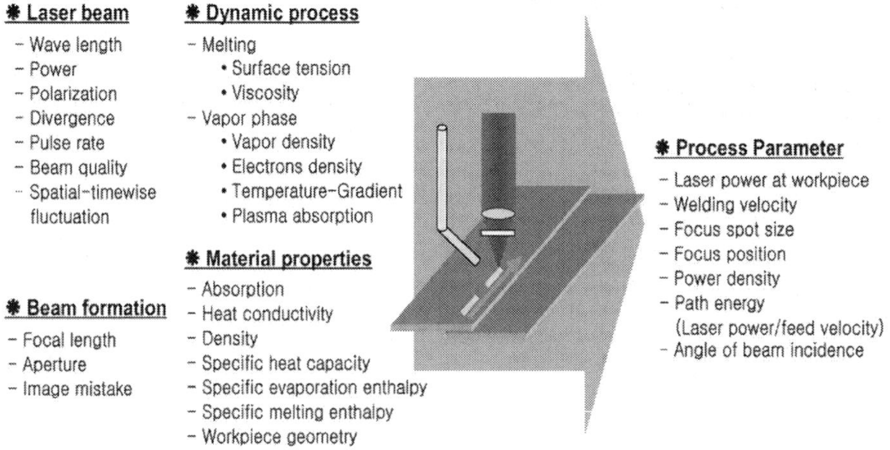

Fig. 4. Important influential variables of laser welding process.

Contrary to spot welding, laser welding has various lengths of welding lines called stitch i.e. the unit length of laser welding path. Fig. 4 presents some influential process variables obtained from laser welding experiments. Because it was impossible to consider all of the influential variables to control welding process, angle of beam incidence, laser power, welding speed, welding depth and stitch path were chosen as the most important process variables for laser welding.

The values of these process variables were determined by the state of the joint variables such as the location and shape of stitch and material. Joint variables were also divided into the geometrical and the technical variables concerning to shape and material of welding area respectively. The geometrical variables include joint type, location of welding area, curvature in the normal direction and tangential direction and stitch path. The technical variables are represented by material, thickness, panel layers, gap, surface condition and surface coating condition. In order to find out the correlation between the process variables and the joint variables, matrix shown at Fig. 5 was developed based on the analysis of the results of the experiments to determine relevant process variables corresponding to joint variables of each stitch.

	Joint variables										
	Geometrical variables					Technical variables					
Process variables	Joint type	Location of welding area	Curvature about normal direction	Curvature about tangential direction	Starting - and ending point of stitch	Material	Thickness	Layer	Gap	Surface condition	Coating
Angle of beam incidence	⦿	⦿									
Laser power						⦿	⦿			⦿	⦿
Welding speed						⦿	⦿			⦿	⦿
Welding depth							⦿	⦿	⦿		
Teaching points			⦿	⦿	⦿						

Fig. 5. Relationship of the process variables and the joint variables.

3.Concept for implementation of laser welding system

3.1 Generation of stitch through product analysis

The conventional assembly method of side panel is spot welding, which is formed by about 250 spot points. In order to change it to laser welding, the possibility of laser welding application and its requirements have to be surveyed through the exact analysis of product. For the flawless generation of welding stitch that is the basic element of laser welding, first of all, the gap between panels must fulfill the condition for quality assurance, i.e, $0.1 \leq$ gap

≤ 0.2 mm. This is accomplished by jig for fixing locations. With the condition that the key points and the inapplicable and inappropriate points for laser welding are performed by spot welding, the determination of jig locations is carried out. The key points are the welding points for forming shape of a module of car body at the beginning stage of welding. The inapplicable points can't be welded with today's laser welding technology. That is, Nd:YAG laser which is popular for welding car body in automobile industry can't penetrate 4 layers welding structure. With regard to the inappropriate points, many jig points are normally needed in corner area for securing of the gap. Some of them can be replaced with spot points to reduce the number of jigs. In Fig. 6, the jig points for the laser welding of side panel are presented considering only the secure of gap.

Since the left part of side panel has a little unevenness, jig point was created with comparatively regular interval. But in case of right rear side due to 4 layers structure or panel combination having severe unevenness, jig points exist close each other shown at Fig. 6.

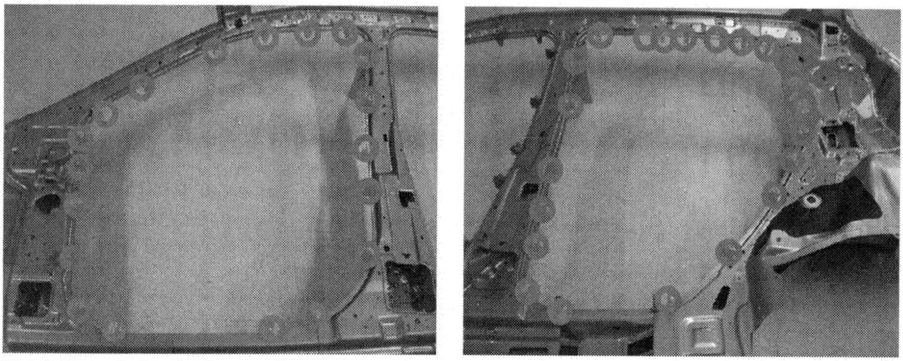

Fig. 6. Jig points of side panel

There are some parts where the application of laser welding is not easy because the existing car body is oriented to spot welding. This leads that it is difficult or impossible to use only laser welding for side panel assembly. Also laser welding results in more unnecessary jigs than that of the existing spot welding.

The side panel has some structures where laser welding is not acceptable such as 4 layers, high reflexibility and so on. For solving these problems, spot welding is applied to the inapplicable points. This contributes to assure quality and reduce whole assembly time by decreasing set up time. After determination of jig points and spot welding points, laser welding stitches are generated between jig points and spot welding points.

The path of stitches is intended to have shape of straight line to make robot teaching easily. The stitch length of 20-30 mm is found on the basis of the experiment results and references. The interval of stitches is set to be greater than 10 mm regarding strength and thermal stress. As a result, 92 stitches are generated on the side panel. Jig points, spot welding points and laser welding stitches generated on the side panel are shown in Fig. 7.

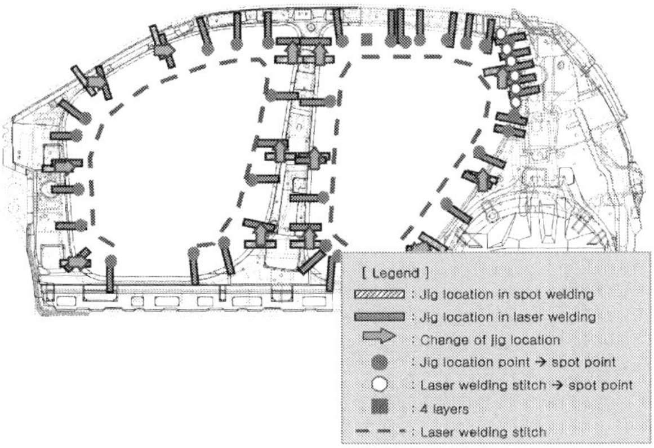

Fig. 7. Determination of jig points, spot welding points and laser welding stitches.

3.2 Derivation of process parameters

In order to ensure the quality in carrying out laser welding for the generated stitches, process parameters should be determined. As explained in section 2, process parameters are determined by analyzing joint variables of each stitch. Joint variables were investigated from the analysis of BOM, process plan, 3D modeling files and measuring data of the side panel in a real car. Fig. 8 shows joint parameters acquired from stitch No. 01. After the investigation of all 92 stitches, stitches having same joint parameters were grouped and 6 groups were generated. According to the data acquired by various welding experiments, process parameters are determined to perform optimal laser welding of each group. The group A including stitch No.1 shown in Fig. 7 is welded in vertical direction (-Z) and velocity of 0.04 m/sec at Nd:YAG laser 4kW, the best welding quality is gained as shown in Fig. 9. The 6 groups of the jointing parameters in Fig. 9 are combined into 3 groups in terms of process parameters.

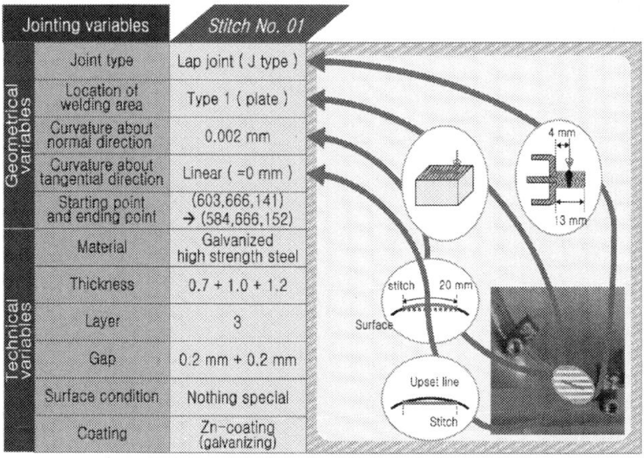

Fig. 8. Joint parameters of welding stitches

	Jointing variables: Geometrical				Jointing variables: Technical							Process variables				
	Joint type	Location of welding area	Curvature about normal direction	Curvature about tangential direction	Material	Thickness	Layer	Gap	Surface condition	Coating	Number of stitches	Angle of beam incidence	Laser power	Welding speed	Number of stitches	
Group 1	J	Type 1	OK	L	a+b+c	2.7	3	OK	N	Zn	62	- Z	4kW	0.04 m/s	72	Group A
Group 2	J	Type 1	OK	L	a+b+d	2.9	3	OK	N	Zn	10	- Z	4kW	0.04 m/s		
Group 3	J	Type 1	OK	L	a+b+e	2.5	3	OK	N	Zn	2	- Z	4kW	0.05 m/s	5	Group B
Group 4	J	Type 1	OK	L	a+f+g	2.4	3	OK	N	Zn	3	- Z	4kW	0.05 m/s		
Group 5	J	Type 1	OK	L	a+f	1.4	2	OK	N	Zn	11	- Z	4kW	0.09 m/s	16	Group C
Group 6	J	Type 1	OK	L	a+e	1.5	2	OK	N	Zn	5	- Z	4kW	0.09 m/s		

Fig. 9. Process parameters of laser welding stitches

3.3 Determination of system components

For the configuration of laser welding cell, optimal components should be determined which can effectively perform the required process. Optic head which is one of important equipments to build laser welding cell is a device to focus laser beam on the welding area. Optic heads are offered in variety according to laser type, focal length and existence or nonexistence of injection device of protection gas or filler wire.

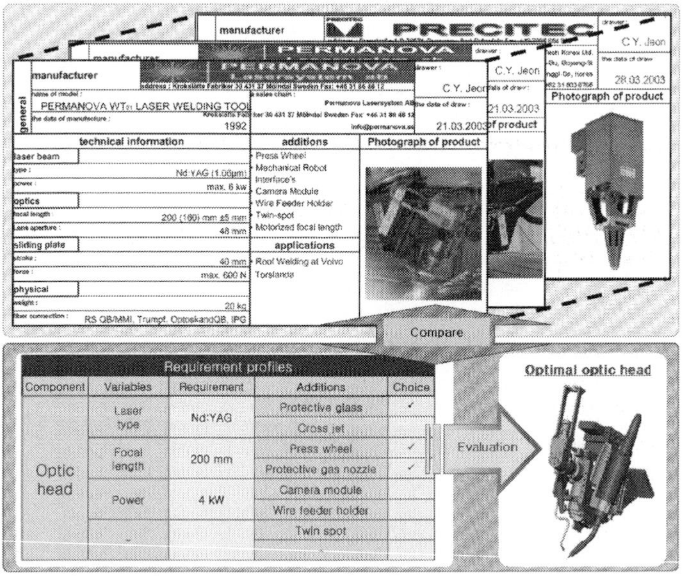

Fig. 10. Selection of optimal components (optic head).

In order to select optimal optic head, the ability of optic head was grasped through the developed form for technical profiles(Fig. 10). The requirement profileswere derived from the analysis of welding problem.

In the following step, evaluation for the best applicable optic head was carried out in the consideration of flexibility, availability, price and compatibility with company specific conditions, e.g. mechanics and information interface, and so on. As the result, the optimal optic head emerged for the definite problem.

4. Digital laser welding system

Digital manufacturing is a technology which uses sophisticated computer to model physical and logical schema and behavior of real manufacturing systems including manufacturing resources, environments and products.

Based on these models, digital manufacturing supports decision making and error checking in the entire manufacturing processes (Lin et al., 2001; Park et al., 2004; Zhai et al., 2005). Implementation of digital cell is started from the modeling of the selected components and the processes(Fig. 11).

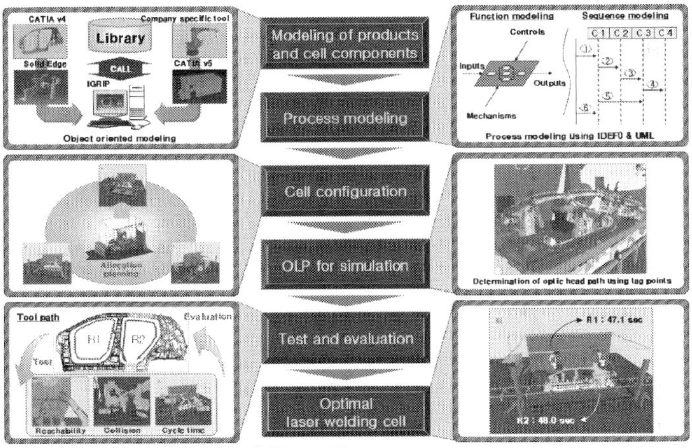

Fig. 11. Procedure for digital laser welding cell

The components are modeled based on object-oriented modeling method by using 3D CAD tools. The models of car body panels, handling devices, transfer devices, jigs and robots modeled from different tools used in virtual environment of the DELMIA's IGRIP tool to layout the workspaces of them and to investigate their kinetic behaviors. For modeling the dynamic components like robot, the kinetic characteristics such as freedom degree, range of motion and angular velocity etc. are assigned to each element of a component according to the design specification of it by using the function of IGRIP "Kinematics", e.g., θ(motion range): $-120° < \theta < 158°$ and ω(angular velocity): $\omega = 140°/s$ for the third axis of the robot used for building the cell. Auxiliary components such as booth, fiber, laser generator and so on, which have variable dimension, are modeled by using CATIA v5 based on dimensional information acquired from field. These 3D models of the components are systematically stored in library by using the function "Catalog" of CATIA v5 and used for new configuration of cells by calling back through a search. After that, the process modeling is performed to define the cell functions and the static and dynamic relationship among the components.

The IDEF0(Integration DEFinition0) model is used for describing those functions and the control mechanism of the cell for executing a welding process. The class diagram and the sequence diagram of UML(Unified Modeling Language) (Martin, 2003; Rosenberg et al., 2001) are used to describe the relationships among the components.
The class diagram defines the static relationships among the components to model the static design view of a system. Laser welding cell consists of two activities, one is a fixing task of components and the other is a welding task. The interfaces and relationships of the components for the cell are grasped with respects to two tasks.
Generally, according to the lapse of time, the events of the components proceed in sequence. The sequence diagram, which represents the messages among the components, is used to describe the dynamic relationship among the components. The sequence diagram shows a set of components and the messages sent and received by those components. The dynamic processes described in the sequence diagram are expressed as the movements of the components by the GSL(Graphic Simulation Language) offered from IGRIP. After modeling the components and the processes, it is needed to allocate the components optimally in satisfaction of constraints given for building the cell, e.g. operation sequence, production organization and type, automation degree, allowed space and investment cost etc. The satisfaction of cycle time and minimization of transfer time for material flow are main points because assembly of car body is the typical type of mass production. The cell is configured so that the quality is ensured by executing welding processes properly in considering the management of tool and jig/fixture, the control regulation planned with the sequence diagram and the connection relationship between pre and post cell. Three types of the cell layout are proposed as the result of allocation planning.
To verify the executing capability of processes within three generated cells and to evaluate the complex behavior of them for selecting the optimal one, simulation should be performed in the reality based and computer generated environment.
The geometry information such as the dimension of the cell and the components and the technology information for welding process and material handling are required for OLP (Off Line Programming). For simulating the detailed welding tasks, the OLP of the robot is executed. To generate kinetically correct and collision free robot cell, the tool paths of the robot are planned with tag points in a virtual environment. The robot program is accomplished based on these information by using GSL (Graphic Simulation Language).
After fully examining the accessibility of optic head to all stitches, the collisions with equipments such as jig etc., the welding parameters for the secure of quality, the applicability of sensor to tracing welding path and the operation sequences for preventing from distortion of parts, the program is generated to execute the events offered from the sequence diagram. Three types of the proposed configuration of cell are evaluated through the simulation of laser welding for the side panel with the generated program. The evaluation regarding to technology is done in terms of cycle time related to productivity, minimization shortest moving distance, automation degree, flexibility, available space and possibility of modular system adapting to change of manufacturing environment and so on. From the organization and economy point of view, the terms of expansion ability, control logic, interface between stations and investment and operating cost are chosen for the evaluation. As the result of evaluation, the cell configuration having the parallel arrangement of two robots is selected as optimal due to high score in the load balancing of each robot(the cycle time of each robot is 47.1 sec and 48 sec) and the minimal area required for implementation. The total cycle time for welding the side panel is also saved 3 sec compared to

two another cell configurations. In addition, the selected cell has the following technical and geometrical parameters as shown in Table 1.

Specifications		Substances
Laser type		Nd: YAG 6KW
Optic head type		BEO D70
Focal length		200 mm
Welding condition	Welding power	4KW
	Stitch size	1mm x 20mm
	Welding speed	40mm/s ~ 90mm/s
	Gap	0.2mm
	Welding direction	-Z (90^0)
	Welding stitches	92 stitches
	Distance between stitches	10 mm
Size of welding cell		6.0 x 7.0x 5.0 M (W x L x H)

Table 1. The specifications of the laser welding cell

5. Conclusion

The laser as economical and flexible tool has established a solid ground in industrial manufacturing area. Specially at welding BIW (Body In White) in automobile industry, the importance of it has been increased due to the technological characteristics such as high process speed, slim seam and good capability of automation and so on.

For application of laser welding technique, welding principle and influential factors were investigated based on the analysis of laser welding processes. With that, the main process parameters were determined according to the material and shape of the welding area through a lot of experiments. The results show the best quality was gained when welding with the gap (0.2 mm), welding speed (0.04~0.09 m/s) and laser power 4 kW. The correlation matrix between the process variables and joint variables was also developed for forming the welding group to carry out the welding process.

The whole stitches for welding side panels were classified to three groups in terms of the process parameter for executing the welding processes efficiently. And the appropriate equipments were selected through the comparison the requirement profiles with the ability of them.

For the configuration of a laser welding cell, a methodology was pursued, which is based on digital manufacturing technique and the application of model analysis concept to validate the cell's behavior.

First, the model was created for analyzing the static and dynamic behavior of the cell by using the class- and sequence diagram of UML. Based on that, a layout planning were executed for structuring the working area. For the motion simulation process, a technology- and machine-oriented offline programming was carried out to optimize the spatial position of tool path in consideration of the welding process as well as the kinetic properties of the selected equipments.

The results of the layout planning and the simulation were the three laser welding cells. Through the final evaluation, the laser welding cell having the specification shown in table 1 proved to be the best suited one for welding the side panels. The decisive criteria were the required cycle time of the robot and the space requirement.

The presented method of implementing the laser welding cell based on digital manufacturing permits substantially improved applying of laser technology and structuring of manufacturing system.

6. References

Emmelmann, C. (2000). *Introduction to Industrial Laser Materials Processing*. Rofinsinar.

Dawes, C. (1992). *Laser Welding*. McGraw-Hill, Inc. New York.

Gao, R.; Pan, L., Tyan, T., Mahadevan, K. & Ghouati, O. (2006). Impact simulation of Hydro-formed front end vehicle structure. *SAE Paper No. 2006-01-0312.*

Jung, C. H.; Lee, H. B., Jang, I. S. & Kwon, T. Y. (2002). Laser welding application for side panel in car body manufacturing. *Proc. Laser Materials Processing Conference*, pp. 1–6.

Kperber, K. & Hesse, T. (2001). *Laser in the Production*. VDI-Berichte 1621. pp. 43–56.

Ku, T. W.; Kang, B. S. & Park, H. J. (2004). Tailored blank design and prediction of weld line movement using the backward tracking scheme of finite element method. *Int. J. Advanced Manufacturing Technology*, Vol.25, No. 1, pp. 17–25.

Lin, M. H. & Fu, L. C. (2001). A virtual factory based approach to on-line simulation and scheduling for FMS and a case study. *J. Intelligent Manufacturing,* Vol.12, No.3, pp. 269–279.

Martin, F (2003). *UML Distilled 3rd edn.* Booch Jacobson Rumbaugh.

Park, H. S.; Choi, H. W., Kang, M. J., Choi, B. W. & Lee, H. B. (2004). Digital laser welding cell for car white body. *IMS Int. Forum 2004 Grobal Challenges in Manufacturing*, pp. 617–625.

Park, J. K.; Lee, S. Y. & Santoro, R. (2002). Laserinduced soot vaporization characteristics in the laminar diffusion flames. *Int. J. Automotive Technology,* Vol.3, No.3, pp. 95–99.

Rosenberg, D. & Scott, K. (2001). *Applying Use Case Driven Pbject Modeling with UML: An Annotated Ecommerce Example.* Addison Wesley.

Saito, T.; Hira, T., Yamasaki, Y., Blumel, K. W., Hartmann, G. & Lenze, F. J. (2006). A study of sheet hydro-forming using high strength steel sheets. *SAE Paper No. 2006-01-0546.*

Sasabe, S.; Eguchi, N., Ema, M. & Matsumoto, T. (2003). Laser welding characteristics of aluminium allys for automotive applications. *Welding Int.,* Vol.17, No.11, pp. 870–878.

Suh, J.; Lee, J. H., Kang, H. S., Park, K. T., Kim, J. S., Lee, M. Y. & Jung, B. H. (2006). Optimal processing and system manufacturing of a laser welded tube for an automobile bumper beam. *Int. J. Automotive Technology,* Vol.7, No.2, pp. 209–216.

William, M. S. (2001). *Laser Material Processing*. 2nd edn. Springer.

Zhai, W.; Yan, J., Ma, D., Jin, Y. & Fan, X. (2005). Production engineering-oriented virtual factory: a planning cell-based approach to manufacturing systems design. *Int. J. Advanced Manufacturing Technology*, Vol.28, No. 9, pp. 957–965.

Zhang, J. (2006). Optimazation of contact forces in tailor welded blanks forming process. *Int. J. Advanced Manufacturing Technology*, pp. 1433–3015.

9

Estimation of composition change in pulsed Nd:YAG laser welding

M. J. Torkamany[1], P. Parvin[2], M. Jandaghi[2] and J. Sabbaghzadeh[1]
[1] *Iranian National Center for Laser Science & Technology, P.O .Box 14665-576, Tehran, Iran*
[2] *Physics Department, Amir Kabir university of technology,P.O.Box15875-4413, Tehran, Iran*

Abstract:A numerical model based on the kinetic theory of gases and the thermodynamic laws is developed for keyhole formation in pulsed laser welding. For a single incoming pulse the spatial profile of the created keyhole was simulated as a function of time using this model. Since undesirable loss of the volatile elements affects on the weld metal composition and properties we have focused in our model to find the process conditions that minimum of these losses take place during pulsed laser welding. The major laser welding process parameters including pulse properties have been examined formerly in this model. The power density and pulse duration were the main investigated variables. The model predicts that loss of alloying elements increase at higher peak powers and longer pulse durations. The model was used for two different kinds of metals, one from the ferrous compounds - Stainless Steel 316- and the other an aluminum alloy- 5754 Aluminum alloy-. By running the model for SS316 it was found that concentrations of the Fe base and Nickel were increased in the weld metal region while concentrations of the chromium and manganese were decreased. Pulsed laser welding of stainless steel 316 in keyhole mode was experimentally studied too. The welding work piece was 2 mm thick SS316 sheet metal. After welding experiments, samples were cut and weld cross sections were analyzed. The concentrations of iron, chromium, nickel and manganese were determined in the weld pool by means of the Proton-Induced X-ray Emission (PIXE) and Energy Dispersive X-ray/Wavelength Dispersive X-ray (EDX/WDX) analysis. It was shown that the composition alteration, predicted by the model due to varying of the laser parameters is in well accordance to the corresponding experimental data.
Al5754 was the second material used for laser welding experiments. Weld metal composition change of this alloy in keyhole mode welding, using a long pulsed Nd:YAG laser was investigated by use of the developed numerical model and supported with experimental measurements. During laser welding process, the significant variables were laser pulse duration and power density. It was predicted in the model and concurred experimentally that, the concentration of magnesium in the weld metal decreases by increasing the laser pulse duration, while the aluminum concentration increases. Moreover, the concentrations of aluminum and magnesium elements, in the weld metal were

determined by laser induced breakdown spectroscopy (LIBS) for different welding conditions.

Keywords: Pulsed laser welding, Nd:YAG Laser, Alloying element losses, Keyhole formation model, Stainless Steel 316, Aluminum alloy 5754, LIBS

1. Introduction

During laser beam welding of many important alloys, vaporization usually takes place from the weld pool surface. Undesirable vaporization of volatile alloying elements changes the weld metal composition relative to the base metal and resultantly the mechanical and metallurgical properties of the weld metal will change too. To realize a quantitative estimation of the weld metal composition, while varying the irradiation parameters, a comprehensive model is required. Several authors have used Longmuir equation for the calculation of the changes in weld metal composition due to various welding processes [1,2].The equation is useful for calculation of relative vaporization rates from different alloying elements in vacuum and it results to a higher absolute rate than actual values[2-5].
Mundra and T.Debroy [2] derived equations for the vaporization rate of various alloying elements in conduction mode laser welding of high-manganese stainless steel with a CW CO_2 laser. The model is based on the coupling of the principles of weld pool transport phenomena and vapor phase gas dynamics. In a similar work developed by X.He and T.Debroy [5] the composition changes of stainless steel were estimated during Nd: YAG laser welding. Although these models are valid for conduction mode welding but, when the laser power density is increased to a level sufficient to evaporate a thin layer of material and the second kind of laser welding mode known as keyhole welding occurs they are not able to evaluate the composition change. By keyhole formation, a deep hole is created inside the weld pool, which is an effective trap for the laser beam [6]. Therefore, creation of keyhole will increase the laser energy coupling to the material. U. Dilthey and co-workers [7] developed a theoretical model based on the diffusion equation to evaluate the composition change of aluminum alloy during laser welding with a continuous wave (CW) CO_2 source. They suggested a quasi-stationary model and considered keyhole as a cylinder, with an invariable radius and depth.

In order to obtain a quantitative understanding of composition change in keyhole welding with pulsed lasers, it is necessary to propose a model that predicts the keyhole formation as well as the corresponding physical phenomena that occurs.

In this work, at first the vaporization rates of SS316 alloying elements such as Fe, Ni, Mn and Cr were determined through a theoretical model based on keyhole welding with pulsed Nd:YAG laser. The influences of laser pulse energy and duration on the composition change of the weld metals were predicted by model and compared with the experimental results obtained from WDX analysis.

Secondly, a LIBS (Laser Induced Breakdown Spectroscopy) analysis as a technique of atomic emission spectroscopy (AES) was used in this work to measure the composition change of the weld metal. The purpose is to determine the elemental composition of the sample. LIBS performs real time composition analysis that can be very superficial. Laser-induced breakdown spectroscopy (LIBS), also sometimes called Laser-induced plasma spectroscopy

(LIPS) has developed rapidly as an analytical technique over the past two decades. The technique employs a low-energy pulsed laser (typically ten to hundreds of mJ per pulse) and a focusing lens to generate plasma that vaporizes a small amount of a sample. A portion of the plasma light is collected and a spectrometer disperses the light emitted by excited atomic and ionic species in the plasma, a detector records the emission signals, and electronics take over to digitize and display the results. The spectra emitted are used to determine the sample's elemental constituents [8]. The analysis is ranging from a simple identification of the atomic constituents to a more detailed determination of relative concentrations or absolute masses [8-13]. LIBS technique is regarded as a superior elemental analysis method including simultaneous multi-element detection capability. In addition, because the laser spark uses focused optical radiation rather than a physical device such as a pair of electrodes to form the plasma, LIBS has several advantages compared with conventional AES-based analytical methods. These advantages are simplicity, rapid and real-time analysis, no need for sample preparation, allowing in situ analysis, detection ability of gaseous samples, as well as liquids and solids, and good sensitivity for halogen elements difficult to monitor with other methods [8, 9]. In general, several solid state lasers and in particular, Q-switched Nd:YAG lasers with nanosecond duration are typically used for LIBS measurements. Other types of lasers, most notably the pulsed $CO2$ laser and the UV excimer lasers have been also employed for LIBS exposure [9].
Here, the composition change of the weld metal due to long pulsed Nd:YAG laser welding of Al5754 alloy was studied using the LIBS method based on ArF excimer laser exposure, in order to determine the trace of element loss in the weld metal after welding process.

2. Keyhole Formation Model

There are several models for prediction of keyhole shape during laser welding [14-16]. The fundamentals of the present model are principally similar to the model that was developed by Semak [16]. Accordingly, in the speeds lower than 1cm/s the profile of keyhole is assumed symmetrical, co-axial with the laser beam. Moreover, keyhole is held open due to balance of the surface tension and the (recoil) ablation pressures. At high speed, keyhole axis is deviated from beam axis such that the recoil pressure exceeds the surface tension, whereby the keyhole wall moves inside the weld pool with velocity equal to summation of the evaporation and melt expulsion velocities [16, 17].
In this work, welding process speed is chosen to be 0.5 cm/s, therefore the keyhole's shape remain be symmetrical and co-axial with the laser beam. The melt expulsion velocity is negligible due to balance in surface tension and recoil pressures. Because of the melt flow and presence of Marangoni effect, the effective thermal conductivity is assumed twice the stationary melt conductivity [18,19,20].
For the case of a Gaussian intensity distribution of the incident laser beam, the value of the local absorbed intensity I_{abs} (i) for each point of the keyhole surface is given by the [16,21]:

$$I_{abs}(i) = A_0 \cos^{q+1}(a(i))\left(I_0 \exp\left(\frac{-x(i)^2}{r_l^2}\right)\exp(-\alpha y(i))\right) \tag{1}$$

Where I_0 is laser intensity at the beam axis, and q denote a modification factor to obtain an angular dependence close to the typical experimental curve that depended on the metal. The parameter A_0 =0.27 ascertains the absorption coefficient for normal incidence in boiling temperature of SS316, the x-axis is parallel to the metal surface, and the y-axis coincides with the beam axis. The i quantity refers to the point number on the keyhole surface, a(i) represent to the angle made by beam and keyhole surface vector (see figure1) and α is the inverse Bremsstrahlung absorption coefficient that can be calculated from the following equation[17]:

$$\alpha(m^{-1}) = \frac{n_e n_i Z^2 e^6 g}{6\sqrt{3} m \varepsilon_0^3 \omega^2 (2\pi)^{1/2}} \frac{1}{(m_e k T_e)^{3/2}} \tag{2}$$

Where Z is the average ionic charge in the plasma, c is the speed of light, m_e is the electron mass, ω is the angular frequency, ε_0 is the permittivity of free space, n_i is the ion density, n_e is the electron density, and g is the mechanical Gaunt factor. Where n_e and T_e are the electron density and temperature of welding plasma respectively. These parameters have been measured for Nd:YAG pulsed laser welding by J. Sabbaghzadeh and his co-workers [22]
Figure1 illustrates the schemes of curve interaction with keyhole surface and the corresponding velocity components.

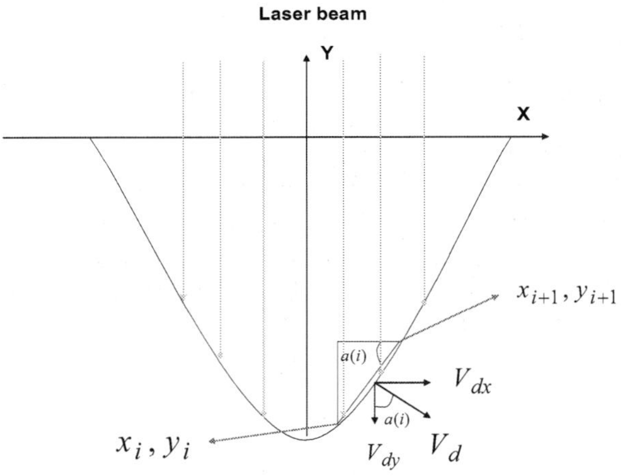

Fig. 1. Schematic illustration of laser interaction with keyhole surface and the corresponding velocity components.

The local energy flux balance can be shown by:

$$I_{abs} = -k\nabla_n T + \rho_m L_v V_{dv} \tag{3}$$

Where k is the heat conductivity, and L_v is the latent heat of vaporization and ρ_m is melted metal density. Temperature gradient on the right-hand side of equation (3) can be estimated to be [16]

$$-\nabla_n T = \frac{T_s - T_m}{\frac{a}{u}} \tag{4}$$

Where a, u, T_s and T_m are heat diffusivity, laser beam translation speed, boiling and melting temperatures respectively.
Substituting equations (1) and (4) in to equation (3) ,the evaporation velocity of the ith point on the keyhole surface is given by:

$$V_{dv}(i) = \left[\frac{A_0 \cos^{q+1}(a(i)) \left(I_0 \exp\left(\frac{-x(i)^2}{r_l^2} \right) \exp(-\alpha y(i)) \right) - (T_s - T_m)\frac{uk}{a}}{\rho_m L_v} \right] \tag{5}$$

Variation of the melt thickness is depended on the mass source and sinks due to melting and evaporation and expulsion velocies [16, 23]. Change of the melt layer can be shown by following equation:

$$\frac{\partial b}{bt} + V_e = -V_{dv} + V_m \tag{6}$$

Where V_e , V_m, V_{dv} are the expulsion, melting and evaporation velocities respectively and b denotes the melt layer thickness. As mentioned above we assume that V_e equals to zero because of negligible process speed, thus variation of the melt thickness is due to the melting and evaporation events.
Since the weld pool profile is strongly affected by the pattern of fluid flow in the weld pool, the convection is significant. There are four different driving forces in the molten weld pool during laser welding causing to convection phenomenon, which affects the pool's shape. These forces are 1-buoyancy or gravity force, 2-surface tension gradient force or Marangoni force, 3-electromagnetic, electromotive force (emf) or Lorentz force and 4-impining or friction force. Lorentz force is absent for gas and laser beam welding [19]. In the weld pool, temperature difference induces a variation in density, thus, the molten metal in the pool boundary is cooler and denser than that on near the center of the weld pool which sinks under the force of gravity. Where as oppositely the molten metal near the center of weld pool is displaced and rise. The circulated velocity is created by gravity force about 1 cm/s. The impinging force is the result of momentum transfer through friction between impinging particles and metal atoms in the molten weld pool. This force induces convection velocity about 1-10 cm/s [24]
Surface tension of liquid depends on the temperature of that liquid. So a temperature gradient causes to a gradient in surface tension. This gradient exerts a force (F_σ) given by:

$$F_\sigma = -\frac{d\sigma}{dT} \nabla T \tag{7}$$

Where σ indicate the surface tension of the molten metal, T is temperature, and ∇T is the temperature gradient at the weld pool surface. In commonly used welding conditions Surface tension gradients induce strong circulation at rates from 10-100 cm/s from the hotter, lower surface tension liquid at the center of the weld pool to the cooler, higher surface tension liquid at the pool edges [18] (figure 2). Finally a dominant Marangoni force is suggested, which results in a wider and shallower weld pool than previous one without the convection.

The Marangoni effect is taken in to account by a simple solution that is considering an artificially higher thermal conductivity for the material in the presence of convection. Effective thermal conductivity in the presence of Marangoni flow is assumed to be at least twice the stationary melt conductivity [18,19,20].

An arbitrary shape of the keyhole wall was assumed in order to start generating the actual keyhole.

The melt thickness is also presumed to be constant during the formation of keyhole. On the other hand, the melting front also moves together with the keyhole wall and a new portion of the metal is melted to replace the evaporation melt, Thus, b is taken to be constant such that $V_m = V_{dv} = V_d$

Where V_d refers to the velocity of the keyhole wall. It is perpendicular to the keyhole surface and its components are given by [16]:

$$V_{dx}(i) = \sin(a(i))V_d(i) \tag{8}$$

$$V_{dy}(i) = -\cos(a(i))V_d(i) \tag{9}$$

Where

$$\sin(a(i)) = \frac{y(i+1)-y(i)}{\left[(x(i+1)-x(i))^2+(y(i+1)-y(i))^2\right]^{\frac{1}{2}}} \tag{10}$$

$$\cos(a(i)) = \frac{x(i+1)-x(i)}{\left[(x(i+1)-x(i))^2+(y(i+1)-y(i))^2\right]^{\frac{1}{2}}} \tag{11}$$

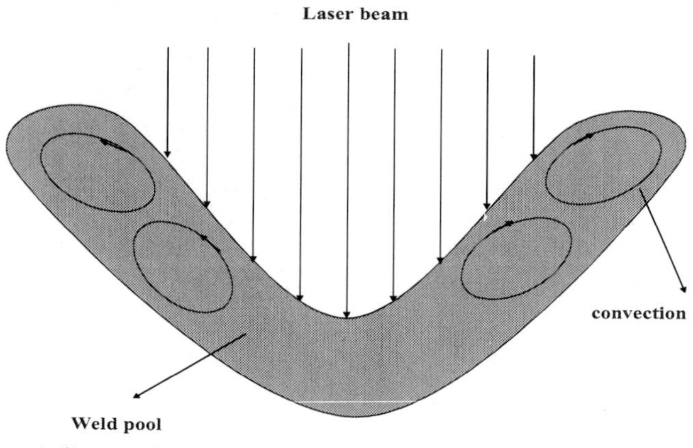

Fig. 2. Marangoni effect inside the weld pool and keyhole pattern.

Subsequently, the change in the position of the ith point on the keyhole surface can be determined as below

$$x_{new} = x(i) + V_{dx}(i)\Delta t \quad (12)$$

$$y_{new} = y(i) + V_{dy}(i)\Delta t \quad (13)$$

Where $x_{new}(i)$ and $y_{new}(i)$ are new coordinates after the time interval Δt. The time interval Δt is selected as nearly 1/1000 of pulse duration. Components of the velocity are shown in figure 1.
The results of first stage of the model containing the shape and depth of the created keyhole for each set of processing parameters was used for the next step of model (vaporization of alloying elements). The computed surface and volume of the keyhole as a function of time are used to determine the vaporization rate as well as the composition change of alloying elements. Results of calculated penetration depth of keyhole are compared with the results obtained from experimental weld profiles of SS316 in figure 3. The thermo physical properties of metal used in the model are presented summarized in table 1.

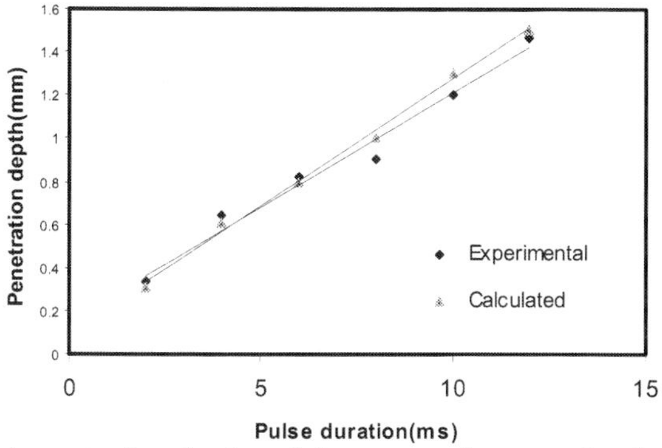

Fig. 3.The calculated penetration depth and the measured penetration for various durations of laser pulses.

PROPERTY	VALUE
Density(kg/m³)	*7200*
Melting point (K)	*1727*
Boiling point (K)	*3100*
Thermal conductivity of solid W/mK	*29*
Effective viscosity(kg/m⁵)	*0.1*
Ratio of specific heats of vapor γ_v	*1.667*
Beam radius (μm)	*200*
Heat of evaporation of (J/kg)	*6.52E6*

Table 1. SS316 Data used for the calculation of vaporization rate and the composition change.

3. Vaporization of Alloying Elements

Vaporization of the alloying element is due to the difference in partial vapor pressure and concentration gradient of each component. The materials vaporization mainly takes place at the keyhole inner wall sheath [2-4].

Pressure and concentration of alloying elements are higher near the weld pool surface in the Knudsen layer than in the bulk shielding gas and in the keyhole bulk (In fact, the pressure of the vapor inside the keyhole is close to the ambient pressure [23]). Partial pressure of each alloying element in the Knudsen layer is related to equilibrium temperature of this layer and can be calculated using the following equation [25]:

$$\log_{10}(p) = A + \frac{B}{T} + C\log_{10}^{T} + DT + ET^2 \quad (14)$$

Where A, B, C, D, and E are constant coefficients which usually differs for the various elements, and T refers to the temperature. For the main elements of table 1 i.e/, Fe, Mn, Ni, and Cr, the corresponding coefficients are listed in table 2 [25].

	A	B	C	D	E
Fe	11.55	−19538	−.6254	$-27E9$	$1908E-13$
Mn	−123.9	−5984	47.07	−.014	$1.5E-4$
Cr	−87.07	−3505	33.65	−.929	$8.3E-7$
Ni	−214.3	−3519	74.94	−.018	$15.1E-7$

Table 2. Constant coefficients for calculation of equilibrium vapor pressure of various alloying elements of SS-316.

Figure 4 illustrates the equilibrium vapor pressure (atm) as a function of temperature (k).

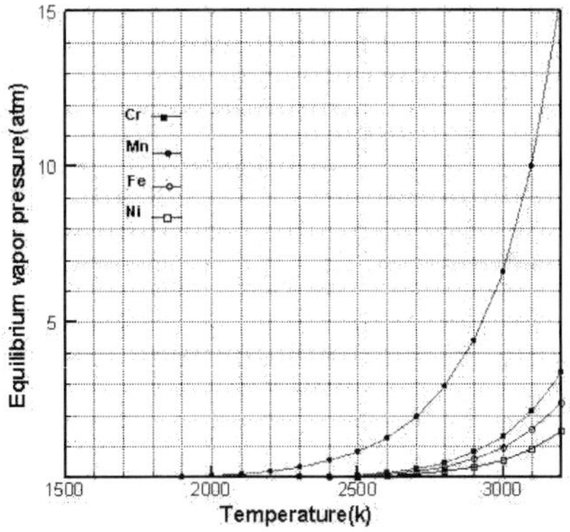

Fig. 4. Equilibrium vapor pressure (atm) of different constituents of SS316 as a function of temperature (K).

Total vapor pressure at the weld pool surface (Knudsen layer) is obtained from summation of the equilibrium vapor pressure of various alloying elements.

$$P_L = \sum_i a_i P_i^0 \tag{15}$$

Where a_i and P_i^0 are the activity and equilibrium vapor pressure of the alloying element i respectively where as i refers to the number of alloying elements.
It is suggested that Knudsen layer is filled only with metal vapor and no shielding gas attends inside. The calculated pressure of the Knudsen layer is subsequently used to obtain the loss of each alloying element due to concentration and pressure gradient vaporization rates.

3.1 Vaporization due to concentration gradient

The vaporization flux due to concentration gradient can be predicted from the kinetic theory of gases [2-5, 26]:

$$J_{c,i} = K_{g,i} M_i \frac{(a_i P_i^0 - P_{i,g})}{RT_e} \tag{16}$$

Where $J_{c,i}$ is the vaporization flux of element i due to concentration gradient, $P_{i,g}$ is vapor pressure of the alloying element i in the keyhole, M_i denotes the molecular weight of the element i, R represents gas constant and $K_{g,i}$ ascertains the mass transfer coefficient of element i . In addition the mass transfer coefficient of element i between the weld pool surface and the shielding gas, outside the keyhole is calculated from the graphical results of Schlunder and Gniclinski for a jet impinging on a flat surface and can be expressed by: [2-5]

$$K_{g,i} = \frac{2Sc^{0.42}\,\mathrm{Re}^{0.5}\,D_{g,i}}{d}\left(1+\frac{\mathrm{Re}^{0.55}}{200}\right)^{0.5}\left[0.483 - 0.108\frac{r}{d} + 7.71\times10^{-3}\left(\frac{r}{d}\right)^2\right] \tag{17}$$

Where d is the diameter of the shielding gas nozzle(Figure 5), r is the radial distance on the weld pool surface, $D_{g,i}$ denotes the diffusivity of the element in shielding gas in $\frac{m^2}{s}$ (see appendix), Re represents the Reynolds number at the nozzle exit and Sh ascertains the Schmit number of the element.
Mass transfer coefficient inside the keyhole is given by:

$$K_{g,i} = \frac{(0.023\,\mathrm{Re}^{0.83}\,Sc^{1/3})D_{i,g}}{D} \tag{18}$$

Where D is mean diameter of the keyhole. Gilliand and Sherwood derived equation (19) for mass transfer between the liquid that flows on the wall of the pipe and the gas current that flows inside the pipe [27].

The longitudinal velocity of the vapor flow inside the keyhole is derived to determine the Schmit number in the keyhole through the following equation [7]:

$$V_Z = \left(\frac{RT}{A}\frac{2\sigma}{P_a a}\right)^{0.5}\left(1+\frac{2(H-Z)}{\sqrt{\pi a}}\right)^{-1} \tag{19}$$

Where Z coordinate origin is taken from the keyhole bottom, A , P_a, a and σ are atom mass of the admixture, external pressure, mean radius and coefficient of surface tension respectively. Mean velocity was written as :

$$<V>=\frac{\int_0^H V_z dz}{H} \tag{20}$$

Were H is keyhole depth.

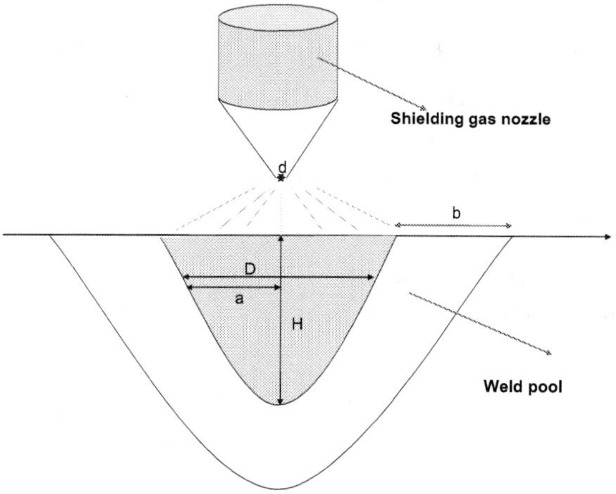

Fig. 5. Schematic illustration of keyhole's geometry and shielding gas nozzle

3.2. Vaporization due to pressure gradient

The vaporization flux due to pressure gradient at the weld pool surface corresponding to a local surface temperature T_s (boiling temperature) is given by [2-5]:

$$J_P = \rho_\upsilon u_n \tag{21}$$

Where u_n and ρ_υ are the mean velocity of particles and density of the vapor at edge of the Knudsen layer. According to kinetic theory of gases the mean velocity of particles can be calculated by equation:

$$u_n = M.s \tag{22}$$

Where M is the Mach number and s is the propagation speed of sound in the gas.
Knudsen layer provokes a rapid change in the density and temperature of the vapor state by its treatment as a gas dynamic discontinuity. In fact, temperature, density, pressure and mean velocity of vapor at the edge of the Knudsen layer can be related to such quantities of vapor on the liquid surface [2-5, 28]. The variations in quantities throughout the Knudsen layer are given by:

$$\frac{T_\upsilon}{T_s} = \left(\sqrt{1 + \pi \left(\frac{\gamma_\upsilon - 1}{\gamma_\upsilon + 1} \frac{m}{2} \right)^2} - \sqrt{\pi} \frac{\gamma_\upsilon - 1}{\gamma_\upsilon + 1} \frac{m}{2} \right)^2 \tag{23}$$

$$\begin{aligned} \frac{\rho_\upsilon}{\rho_s} &= \sqrt{\frac{T_s}{T_\upsilon}} \left(\left(m^2 + \tfrac{1}{2} \right) \exp(m^2)\, erfc(m) - \frac{m}{\sqrt{\pi}} \right) \\ &+ \frac{1}{2} \frac{T_s}{T_\upsilon} \left(1 - \sqrt{\pi}\, m \exp(m^2)\, erfc(m) \right) \end{aligned} \tag{24}$$

Where $m = M\sqrt{\frac{\gamma_\upsilon}{2}}$ and γ_υ is the ratio of specific heat of vapor which is treated as a mono-atomic gas, $T_\upsilon, \rho_\upsilon, T_s, \rho_s$ are temperature and density at the edge of Knudsen layer as well as the temperature and density of the vapor at the liquid surface respectively.
The Mach number is also determined in order to obtain T_υ and ρ_υ . In this model the Mach number is derived from the pressure balance and the mechanical stability of the keyhole. The main forces acting on the keyhole wall are assumed to be the ablation pressure opposed by the surface tension forces. The ablation pressure is given in terms of the density at the edge of the Knudsen layer and the square of the ejected gas mean velocity through equation such that:

$$P_{abl} = m_{Fe} \rho_\upsilon u_n^2 \tag{25}$$

Figure 6 shows the values of the Mach number versus laser power for different welding speeds.

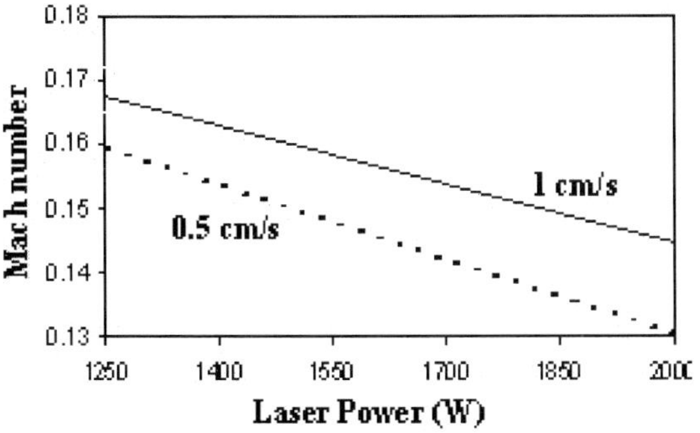

Figure6. Values of the Mach number versus laser power for different welding speeds, a focused Gaussian beam was used with radius 0.2 mm on stainless steel 316.

As a result, by increasing the welding speed, the keyhole becomes narrower and surface tension forces arise at the keyhole boundary, thus ablation pressure must increase in order to fulfill pressure balance. Therefore, the mean velocity of the ejected gas at the edge of the Knudsen layer enhances according to eq (25). Thus, regarding eq (22), the Mach number at higher speeds is grater for same laser power.
Vaporization flux due to pressure gradient of an alloying element, $J_{p,i}$, [2-5] is given by:

$$J_{P,i} = a_i \frac{P_i^0}{P_L} \frac{M_i}{M_\upsilon} J_P \tag{26}$$

Where M_υ is the mean molecular weight of the vapor species in the Knudsen layer that is given by:

$$M_\upsilon = \sum_{i>1}^{n} M_i \frac{a_i P_i^0}{P_L} \tag{27}$$

Therefore, the total vaporization flux, $J_{total,i}$, for an element i is given by:

$$J_{total,i} = J_{P,i} + J_{C,i} \tag{28}$$

If the surface and volume of the weld pool (keyhole profile) are known as functions of time, the total rate of vaporization and the vaporization rate of each alloying element can be obtained using $J_{total,i}$. The vaporization rate of element i, G_i is found by integrating the vapor flux over the entire weld pool surface, and the total vaporization rate of all of elements, G, is given by:

$$G = \sum_{i=1}^{n} G_i = \sum_{i=1}^{n} \iint_s J_{total}\, dx\, dy \tag{29}$$

Where s refers to the weld pool surface. The weight percent of an element i, $(Wt\ \%)_{t,i}$ as function of time is given by:

$$\left(Wt\%\right)_{t,i} = \frac{W_i - G_i t}{W - Gt} \tag{30}$$

Where W and W_i are total mass of the weld metal and mass of species i in the weld metal respectively and t represents time.

4. Laser welding of SS316

4.1. Experimental setup

A long pulsed Nd: YAG laser Model IQL-10 with mean power of 400 W and standard square shape pulse is chosen as the welding laser source to carry out the experiments. The pulse energy of laser was varied from 3 to 18 joules while the pulse duration could be changed from 2 to 12 ms as well. The process speed (5 mm/s) and the pulse repetition rate

(20 HZ) were kept to be invariant during the experiment. Model LA300W-LP Ophir joule meter is used to measure the pulse energies. A focusing optical system composed of three lenses is used with 75 mm focal length and 200 μm minimum spot size. The welding work piece was a 2 mm thick SS316 sheet metal such that the corresponding chemical composition is presented in table 3. 10 lit/min flow of pure Ar gas, blowing from a coaxial nozzle was employed to shield the welded area against the oxidation.

At first, several set of SS316 samples $3cm \times 3cm$ were cut and the welding process was done using laser pulses with 1500 W peak power and 2-12 ms duration. Subsequently, the samples were prepared for the transverse cross section analysis. The samples were etched with a composition of HCL, HCOOH and HNO_3 to develop the weld region. After welding, EDX/WDX and PIXE analysis were performed to study the constituents of the weld metal. Under the same irradiation conditions, the EDX/WDX analysis were carried out using 10 individual points, on the transverse cross section of the weld metal for each sample to determine the concentration of the alloying elements.

Element	*C*	*Mn*	*Cr*	*Ni*	*Si*	*MO*	*P*	*S*	*Al*	*Co*	*Cu*
	Nb	*Ti*	*V*	*Fe*							
Wt%	*.05*	*1.27*	*18.1*	*8.04*	*.49*	*.33*	*.02*	*.003*	*.003*	*.13*	*.32*
	.02	*.002*	*.05*	*balance*							

Table 3. Constituents of 316 stainless steel.

4.2.Comparison of theoretical and experimental results

Average and standard deviation of each set of data was taken for statistical analysis. The Quick calculus software (online calculator for science) was used to determine t-statistic and p-value of the untreated and various laser treated samples to be $p<0.01$. It shows that those are statistically significant and the differences are quite meaningful, as to approve the use of EDX/WDX as a reliable tool for the species identification and subsequent analysis. The results of the modeling indicate that those are consistent with the experiments. Figure 7 depicts the variation of the significant alloying elements including Mn, Cr, Ni and Fe in the weld metal. It is seen that the concentrations of iron and nickel in the weld metal increase whereas the concentrations of the manganese and chromium decrease after the welding process.

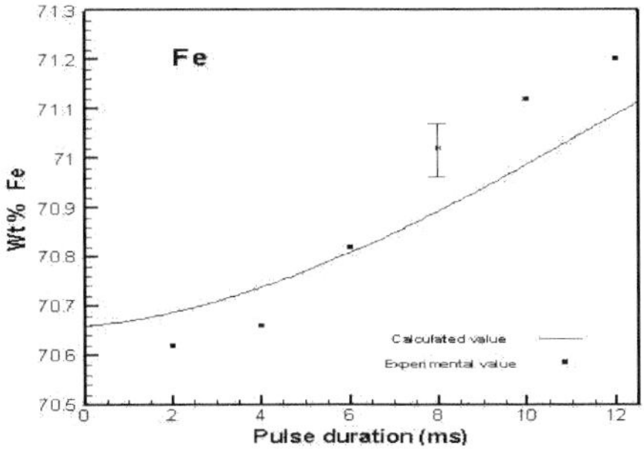

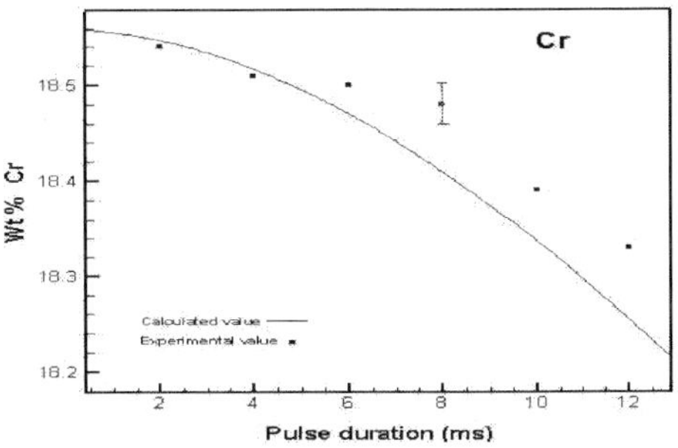

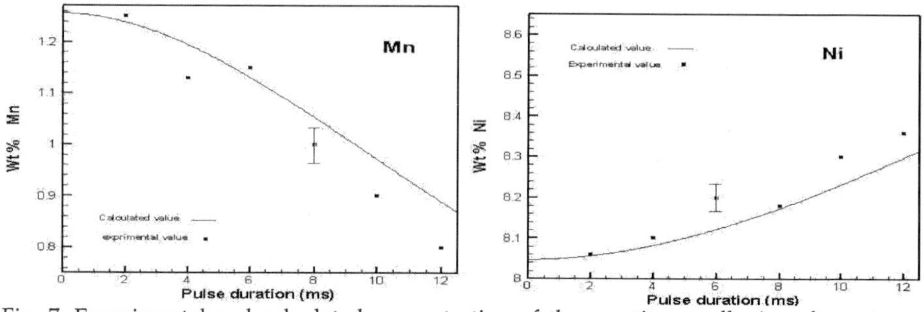

Fig. 7. Experimental and calculated concentration of the constituent alloying elements, as function of pulse duration for a power density of 12 GW/m^2.

Although the total mass of iron and nickel in the weld pool is lower than those before the welding, the total mass of the weld pool has decreased at higher proportion because of the loss of manganese, chromium, iron, and nickel. As a result, the concentrations of iron and nickel in the fusion zone are higher than those in the base metal because of the loss of manganese and chromium.

The geometry of the keyhole including surface to volume of the weld pool were studied using data of section (3.1) in order to calculate the vaporization rate. The keyhole surface temperature was assumed to be the boiling temperature of the base metal due to the two-phase characterizations of the keyhole surface during high power laser irradiation. It is believed that, both pressure and concentration gradients contribute to migrate the alloying elements outside the process area. The results obtained by the model show that the vaporization flux due to the pressure gradient is larger than that of the concentration gradient in the keyhole.

The vaporization mostly occurs not only at the surface of the keyhole but also in the region near the keyhole edge, where the temperature is taken close to the boiling point such that the vaporization out of this region is negligible corresponding to the concentration gradient (diffusion).

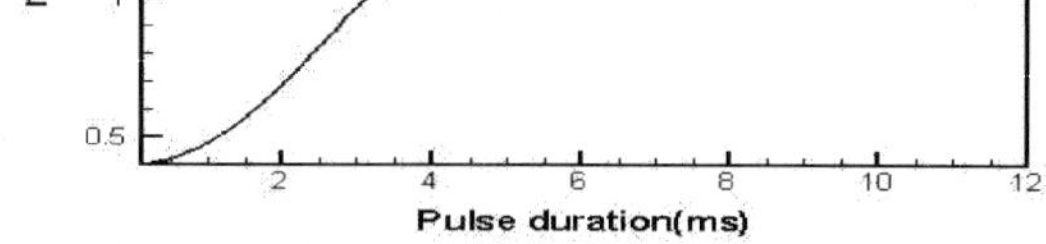

Fig. 8. Ratio of area to volume of the weld pool at end of pulse versus pulse duration for 12 GW/m^2 power Density

According to the theory, the composition changes are mainly due to the volume and surface of the keyhole profile. Model shows that composition changes decrease with increase of the weld pool volume. The reduction of the keyhole surface causes to decrease in elements evaporation from the weld metal thus, a slight composition change happens. In other words, the keyhole area acts as sink and keyhole volume operates as source of alloying elements in the fusion zone. The pulse duration affects on the ratio of area to volume of the keyhole which is illustrated in figure 8.
Figure 9 displays the spectra taken by EDX for the weld and base metal with 4ms pulse duration and power density of 12 GW/m^2.The respected picks of various elements show that the concentrations of Mn and Cr decrease in the weld metal while concentrations of Fe and Ni increase.

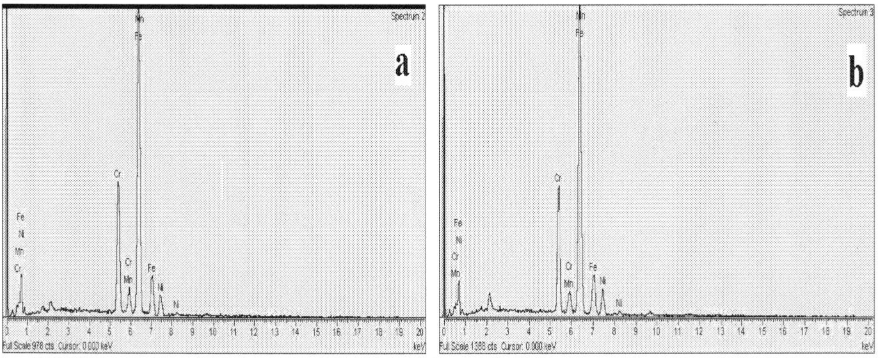

Fig. 9. EDX spectra of (a) base metal, (b) weld metal with power density of 12 GW/m^2 and pulse duration of 4 ms

Figure 10 illustrates the concentration profile of the constituent alloying elements using PIXE analysis. It obviously indicates that the manganese and chromium concentrations in the weld area decreases regarding the chemical composition of base metal.

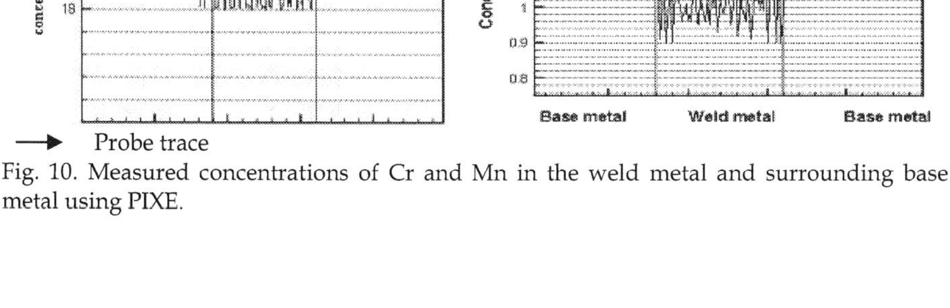

Fig. 10. Measured concentrations of Cr and Mn in the weld metal and surrounding base metal using PIXE.

Similarly, the Mn and Cr densities in the weld metal for power densities of 10-20 GW/m^2 were measured at constant pulse duration (8ms) as shown in figure 11. It indicates that the concentration of manganese and chromium in the weld area decreases linearly versus increasing power density.

The influence of the laser power density on the ratio of the area to volume of the weld pool is depicted in figure 12. It is obvious that the ratio grows linearly with power density, leading to a linear reduction of Mn and Cr with power density mainly because the surface enlargement is dominant relative to volume growth, thus, the composition change significantly increases with laser power density. As the laser power density enhances, both the volume and surface of weld area enlarges, but area increase rate is greater than volume increase rate, thus, composition change in the weld pool increases with the laser power.

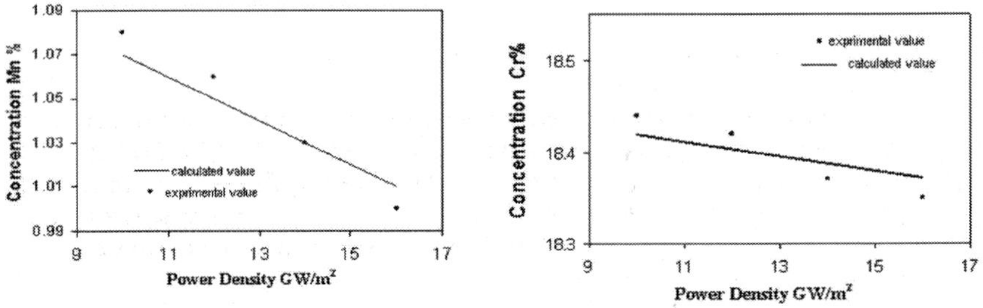

Fig. 11. Concentration of Mn and Cr in weld pool with different power density for 8ms pulse durations.

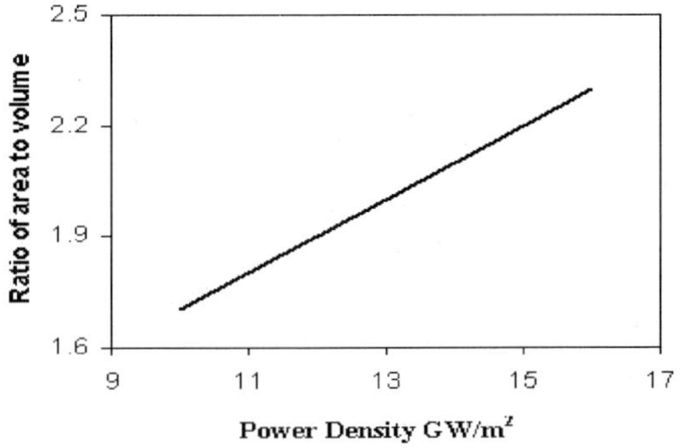

Fig. 12. Ratio of area to volume of the weld pool at end of pulse versus power density (pulse duration=8ms)

5. Laser welding of Al5754

5.1. Laser welding set up

The Specimens of aluminium alloy5754 were irradiated with a long pulsed Nd:YAG laser model IQL-10 with mean power of 400W and standard square shaped pulses. The pulse energy of laser was varied from 4.5 to 10.5 joules while the pulse duration could be changed from 3 to 7ms as well. The process speed (9mm/s) and the pulse repetition rate (10HZ) were kept to be invariant during the experiments. The details of the power measuring equipments, and optical focusing system are the same as mentioned in section 4.1. The welding work piece was a 2mm thick aluminium alloy 5754 sheet metal such that the corresponding chemical composition is presented in table 4. 10lit/min flow of pure Ar gas, blowing from a coaxial nozzle was employed to shield the welding area against the oxidation.

Element	Si	Mn	Cr	Ni	Fe	Mg	Sn	Bi	B	Co	Cu	Ga	Ti	V	Al
Wt%	.12	.25	.08	.006	.37	3.009	.002	.004	.0018	.004	.04	.01	.014	.009	balance

Table 4. Chemical composition of aluminun alloy 5754.

5.2. LIBS Set Up

After welding processes, the concentrations of aluminum and magnesium in the weld and base meal were determined using LIBS analysis. Experimental set-up consists of conducting and focusing optics, laser pulse diagnostics, delay generator, spectrometer, and the processor as shown in Fig.13/1. An excimer laser (Lambda physics LPX200) with 40mJ/pulse, 20ns duration, and pulse repetition rate of 1-10Hz, has been used as the coherent source at 193nm to generate the micro plasma on the target surface. A couple of a quartz lens with 75, and 150mm focal lengths are situated in front of the main laser beam to increase the incident power density irradiating on the Al5754 samples. A semiconductor detector (PIN diode, EG&G, FNT100), a 300-MHz digital storage Tektronix 30145, 2.5Gb/sec oscilloscope, as well as a 400-MHz Tektronix 7844 oscilloscope, and a Coherent TM joule / power meter (Field Master, LM-P10 & LM-P5 100 heads) were used for the relative and absolute measurements of laser power and pulse energy.

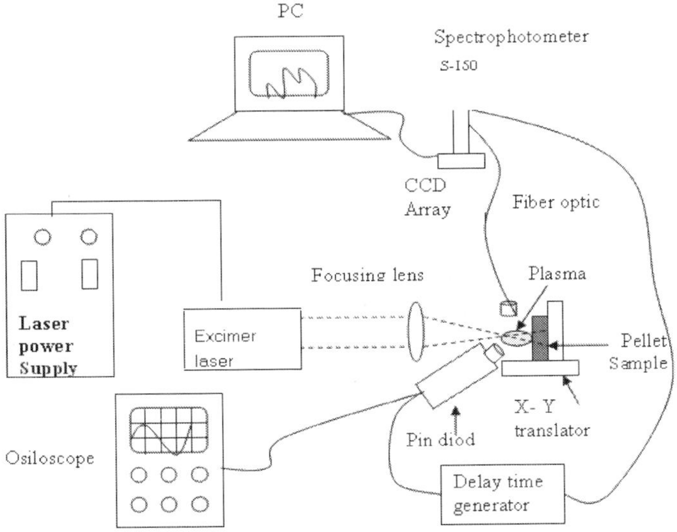

Fig. 13. LIBS set up using excimer laser.

A fiber bundle (UV 600/660 type with SMA-905 fiber connector and 1m length) collected the light emission of the plasma using a quartz lens (25mm diameter, 50mm focal length) placed 80mm away from the sample. The fiber output was coupled to the entrance slit of a compact wide range spectrometer (200-1100nm) model S150 Solar Laser Systems TM (50mm focal length, transmission diffraction grating with 200 grooves/mm with 0.02mm×3.0mm of entrance slit and 0.5nm spectral resolutions). A charge-coupled device (CCD) detector array model Toshiba TCD 1304AP with 3648 number of pixels was used to detect the dispersed light, subsequently. The CCD camera was triggered ~2 µs after the onset of laser shot using a suitable delay generator in order to reduce the continuum Bremsstrahlung radiation.

5.3 Calculation of the relative density of alloying elements using LIBS analysis

Under LTE (local thermodynamic equilibrium) condition, the population of the exited levels for each species follows the Boltzmann distribution [9]:

$$n_i^s = \frac{g_i}{U^s(T)} n^s e^{\frac{-E_i}{kT}} \tag{31}$$

Where n_i^s indicates the population density of the excited level i of species s, n^s is the total number density of the species s in the plasma, A_{ij}, E_i, g_i, K, and $U^s(T)$ are the transition probability, the excitation energy of the level, the statistical weight for upper level, Boltzmann constant and the partition unction of the species at electronic temperature T, respectively. The coefficients A_{ij} and g_i are taken from the National Institute for Standards and Technology (NIST), atomic spectra database [19].

$$U^{s(T)} = \sum_i g_i e^{\frac{-E_i}{kT}} \tag{32}$$

With the detectors typically used in the LIBS measurements, an alternative formula in terms of the integrated line intensity (number of transitions per unit volume per unit time) is preferred by:

$$I_{ij} = n_i^s A_{ij} \tag{33}$$

Now, by considering two lines, λ_{ij} and λ_{mn}, of the same species, characterized by different values of the upper level energy ($E_i \neq E_m$), the relative intensity ratio can be used to calculate the plasma temperature [9]

$$T = \frac{E_i - E_m}{k \ln\left(\frac{I_{mn} g_i A_{ji}}{I_{ij} g_m A_{mn}}\right)} \tag{34}$$

When selecting a line pair, it is advisable to choose two lines as close as possible in wavelength and as far apart as possible in excitation energy. This is to limit the effect of varying spectral response of the apparatus, as well as to minimize the sensitivity to small fluctuations in emission intensity.

To calculate the concentration ratios of the main components in the samples from the intensity ratios, it is desirable to compute the intensity ratio of two different lines from elements A and B. Concentration ratios of the elements A and B can derive from following equation:

$$\frac{n^A}{n^B} = \frac{I_{ij} A_{mn} U^A(T) g_m}{I_{mn} A_{ij} U^B(T) g_i} \exp^{\left(\frac{-(E_m - E_i)}{kT}\right)} \quad (35)$$

5.4. Theoretical modeling

A well-tested 'hydrodynamic' physical model described was used to simulate the keyhole profile as a function of time. The keyhole profile as a function of time was then used to calculate the vaporization rates of alloying elements. The data used for the keyhole formation during pulsed Nd:YAG laser welding of Al5754 [14] are presented in table 5.

PROPERTY	VALUE
Density(kg/m^3)	*2370*
Melting point (K)	*933*
Boiling point (K)	*2792*
Thermal conductivity of solid W/mK	*237*
Effective viscosity(kg/m^5)	*1.1*
Ratio of specific heats of vapor γ_v	*1.667*
Beam radius (μm)	*200*
Heat of evaporation of (J/mol)	*294000*

Table 5. Date used for the keyhole formation during pulsed Nd:YAG laser welding of Al5754.

Close to the liquid phase, inside the keyhole on its walls, vaporization of material takes place within a sheath. The so-called Knudsen layer is governed by the transition from a non-equilibrium state at the keyhole wall to an equilibrium state a few mean free paths distant from the liquid boundary. Vaporization of the alloying elements is due to the difference in partial vapor pressure and concentration gradient of each component. Pressure and concentration of alloying elements are higher near the weld pool surface in the Knudsen layer than in the bulk shielding gas and in the keyhole bulk [2-5,28] (In fact, the pressure of the vapor inside the keyhole is close to the ambient pressure [23]).

Partial pressure of each alloying element in the Knudsen layer is related to equilibrium temperature of this layer and can be calculated using the equation (14).Where A, B, C, D, and E are constant coefficients, which usually differs for the various elements, and T refers to the temperature. For the main elements of table 4 i.e. Al, and Mg, the corresponding coefficients are listed in table 6 [14].

	A	B	C	D	E
Al	9.98	-1.38E+4	-.345	1.13E-11	-1.21E-15
Mg	-72.65	-4.201E+3	30.26	-1.55E-2	2.64E-6

Table 6. Constant coefficients required for calculation of equilibrium vapor pressure of Al and Mg.

Total vapor pressure at the weld pool surface (Knudsen layer) is obtained from summation of the equilibrium vapor pressure of various alloying elements (equation 15).
It is suggested that Knudsen layer is filled only with metal vapor and no shielding gas attends inside. The calculated pressure of the Knudsen layer is subsequently used to obtain the loss of each alloying element due to concentration and pressure gradient vaporization rates.

5.5.Comparison of theoretical and experimental results

Several sets of aluminium alloy 5754 samples with 5cm×5cm area were cut and the welding process was done using laser pulses with 1500W peak power and 3-7ms durations. After welding, LIBS analysis was performed to study the constituents of the weld metal.
Emission spectra were taken from weld metal in atmospheric air employing a standard LIBS arrangement shown in figure13, using a UV ArF laser (40mj, 10Hz, 20ns, 193nm) whose beam was focused through a quartz lens on the weld region of the solid sample. The collected light was launched into an optical fiber bundle coupled to the spectrometer, and recorded with an intensified charge coupled device (CCD) array.
In LIBS, the pulsed laser is initially focused on the target surface. The energies involved are in the GWcm^{-2} range for the nanosecond duration, which are high enough to result in plasma formation through the subsequent process of vaporization, atomization and ionization in a single step. It leads to a high-temperature dense plasma that can be spectrally resolved and detected which contains the characteristic peaks with significant information on the nature and concentration of the elements. The integrated emission of the individual spectral lines, in principle, is associated with the number density of the corresponding emitting species in the plume. However, this, in turn, can be correlated with the element concentration in the sample, via the various calibration methods leading to the quantitative analysis. In general, it is desirable to increase the signal-to noise ratio (S/N) by scaling the strength of the signal for precision measurements of trace elements. The plasma characteristics are influenced by the laser-target interaction mechanisms including various laser properties such as wavelength, power density, laser pulse duration and energy.
LIBS analysis can be done using IR or UV laser exposure. The selective evaporation is the main mechanism, which becomes much more pronounced during IR laser irradiation of the target. The effect could be suppressed to a notable extent by UV laser exposure mainly due to the fact that absorption coefficient for the UV wavelength is considerably higher than IR

[29]. Here, the emission spectra from the weld metal were taken using a UV excimer laser. It was shown that there was a good linear relation between the intensity of emission atomic spectral lines and the contents of alloying elements in samples. In fact, selective evaporation is not a dominant mechanism in UV laser. Therefore the plume concentration is nearly similar to that in the target during UV photo ablation. While using IR laser irradiation, the evaporation of the volatile elements alters the plume concentration from what is found in the target base. Figure 14 illustrates the typical LIBS spectrum obtained from weld metal. The selected lines are Al emission line at 396.1nm and the Mg characteristic line at 285.2nm that are shown more clearly in figure 14(b) for welding condition with various pulse durations. In fact, a proper delay time is required to eliminate the continuum spectrum as the background noise that mostly arises from the ion recombination with free electrons. The strong emission characteristic lines due to dominant neutral Al and Mg trace were used to determine the relative density ($\frac{n_{Al}}{n_{Mg}}$) subsequently.

Reliability was examined through mean and relative standard deviation (%RSD) of each set of experimental data taken from LIBS repeated measurements.

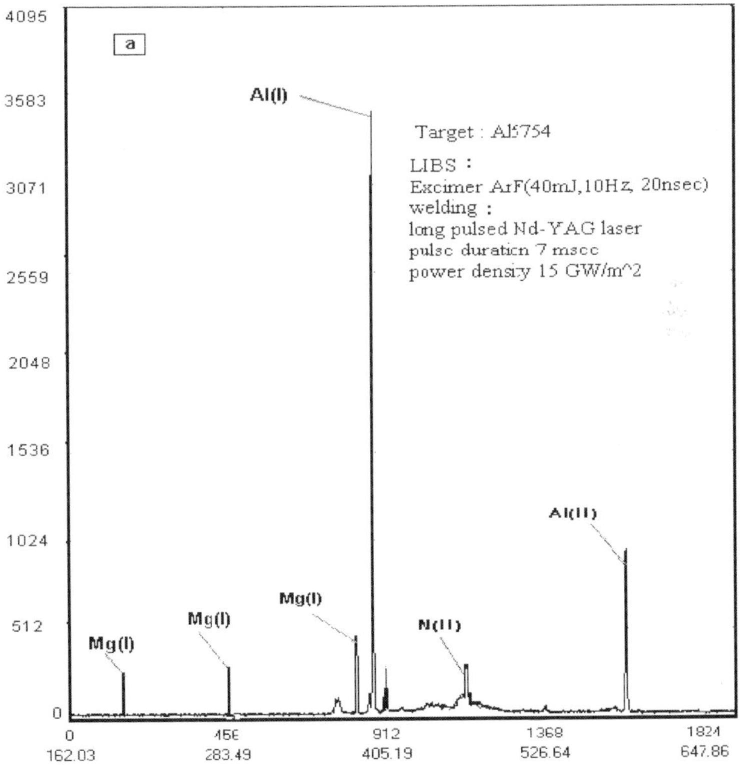

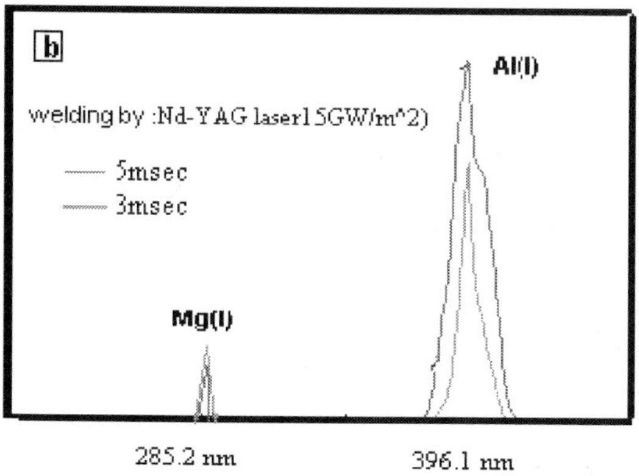

Fig. 14. (a)A typical spectrum of weld metal Al5754 using a long pulse Nd:YAG with 7ms pulse duration and 15Gw/m² power density (b) the emission characteristic lines of Al (396.1nm) and Mg (285.2 nm) for laser welding with different pulse duration (3and 5msec)

Mean Intensities of Al and Mg emission lines for various conditions of the laser welding process (3-7msec pulse duration) are revealed in table 7.

Pulse duration of laser welding process	*3msec*	*4msec*	*5msec*	*6msec*	*7msec*
mean Intensity (285.2 nm, MgI)	*211*	*202*	*197*	*193*	*190*
%RSD (285.2 nm, MgI)	*2*	*1.3*	*1.5*	*2.2*	*1.9*
mean Intensity (396.1 nm, AlI)	*2563*	*2993*	*3021*	*3205*	*3421*
%RSD (396.1 nm, AlI)	*1.9*	*1.8*	*2.1*	*2.2*	*1.9*

Table 7. Mean Intensities of Al and Mg emission lines in LIBS analysis for 3-7msec pulse durations of laser welding

The generated plasma in the laser ablation process is assumed to be in local thermodynamic equilibrium (LTE). The LTE condition is given by [30]:

$$N_e \geq 1.6 \times 10^{12} T^{1/2} \Delta E^3 \tag{36}$$

Where N_e ascertains the electron density, T (K) denotes the plasma temperature, and $\Delta E(eV)$is the largest energy transition for which the condition holds. Electron density is known as an important plasma parameter, which gives indications about the thermal equilibrium. A common method for spectroscopic determination of N_e is based on the Stark effect of the atomic or ionic lines whereas for typical LIBS, the contribution of ion broadening could be negligible. Therefore, the Stark broadening $\Delta\lambda_s$ of the neutral line expressed as the FWHM in nanometers is simplified as [13, 31]:

$$\Delta\lambda_s = 2W\left(\frac{N_e}{10^{16}}\right) \tag{37}$$

Where W is the electron impact parameter.

Therefore, the line width $\Delta\lambda_s$ corresponding to the typical characteristic NII line was determined to estimate the electron density. The experiment satisfies equation (36) to emphasize the validity of LTE condition. Thus, Boltzmann equation is used to relate the population of an excited level to the number density of the species within the plasma. The typical electron density and temperature were determined to be~ $10^{18}cm^{-3}$ and ~10^4K respectively. It is notable to mention that emission line of NII at 500.5nm is an intense line with the stark broadening of about 5nm that is one order of magnitude greater than the optical resolution of the spectrometer (0.5nm) and concludes to an acceptable accuracy in determination of N_e.

Because the transitions are element specific and quantized or of a specific wavelength, a given species has the highest probability of reabsorbing a photon emitted by a member of the same species. Because of the high density of atoms in the micro plasma and its characteristically high temperature and electron density gradients, cool atoms, residing mostly in the ground state, will populate the outer layer of the plasma. The central core of the plasma will contain a higher density of excited atoms. As these atoms decay to the ground state, the emitted photons corresponding to resonance transitions will have a high probability of being absorbed by the cooler atoms in the outer layers, thereby reducing the observed intensity of the emission line. As the concentration of the atoms in the target sample increases, the number of cooler atoms in the outer layer increases and self-absorption becomes evident [8]. Consequently in quantitative laser induced breakdown spectroscopy it is essential to account for the effect of self-absorption on the emission lines intensity.

The self-absorption coefficient (SA) is defined as the ratio of the measured height peak to the value of the line peak in absence of self-absorption. It is clear that, in the presence of Self absorption, the intensity of the line at its maximum (i.e. for $\lambda = \lambda_0$) is lower than in optically thin condition, according to the following relation [32]:

$$\frac{I(\lambda_0)}{I_0(\lambda_0)} = SA \tag{38}$$

$I_0(\lambda_0)$ represents the line profile assuming negligible self-absorption. In turn, the knowledge of the coefficient (SA) allows correcting the peak line intensity. (SA) is equal to one if the line is optically thin, while it decreases to zero as the line becomes optically thick. Self-absorption coefficient is easily derived, using equation (39).

$$(SA) = \left(\frac{10^{16}\Delta\lambda}{2WN_e}\right)^{-0.5} \tag{39}$$

Where $\Delta\lambda$ is line width that is directly obtained from the spectrum analysis and W parameter of the regarded line is obtained from relevant literatures [32, 33].

Self-absorption coefficients SA were evaluated for emission characteristic line of neutral Al at (396.1nm) and neutral Mg at (285.2nm) in order to correct relative density of Al to Mg. Figure 15 depicts the ratio of relative concentration of alloying elements in the weld metal as a function of the welding laser pulse duration. It is seen that the ratio of aluminium to magnesium concentrations linearly increases in terms of pulse duration. In other words, magnesium concentration in the weld metal decreases, whereas the aluminium concentration increases simultaneously. It indicates that Mg loss significantly increases with longer pulses.

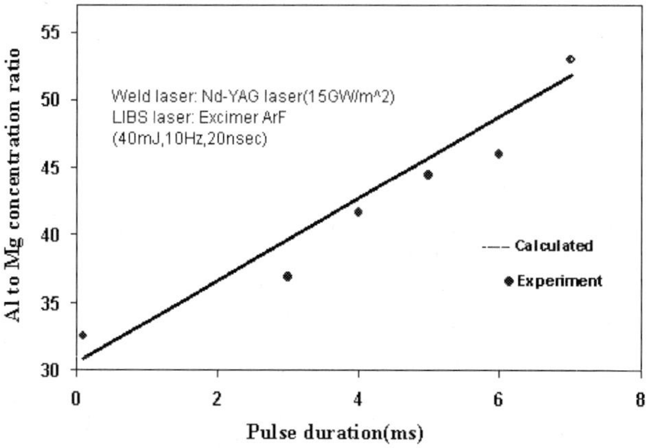

Fig. 15. Ratio of aluminium to magnesium concentrations as a function of pulse duration in Nd:YAG laser welding.

The geometry of the keyhole, i.e. surface to volume of the weld pool, is required to estimate the vaporization rate that was investigated in our previous work exhaustively. Therefore, surface and volume of the keyhole are essentially taken into account as a couple of significant parameters for the element loss measurement. The keyhole area acts as sink and its volume functions as a source of alloying elements within the fusion zone. The reduction of the keyhole surface causes to decrease of the element evaporation leading to smaller loss of element such that, a slight change in the composition occurs. In fact, the pulse duration strongly affects on the ratio of area to volume of the keyhole as displayed in figure 16.

Computational results indicate that ratio of area to volume increases with pulse duration of a single shot accompanied by an increase of the vaporization rate. The element loss becomes more significant during long pulsed welding accordingly.

The keyhole surface temperature was assumed to be kept at the boiling temperature of the base metal due to two-phase characterization of the keyhole surface during high power laser irradiation. The model shows that the vaporization flux due to the pressure gradient is larger than the vaporization flux due to the concentration gradient in the keyhole.

The influence of the laser power density on the ratio of keyhole area to volume as well as the ratio of aluminum to magnesium concentrations in the weld pool are shown in figure17(a, b).

Figure 17(a) illustrates that the ratio of keyhole area to volume is kept nearly invariant for a wide range of power densities to indicate that is not very sensitive to variation of laser power. In addition, figure 17(b) displays that, the ratio of the relative concentrations of magnesium and aluminum within the weld metal are independent of the laser power density. This fact was inferred from the model and confirmed by the experimental data obtained from LIBS analysis.

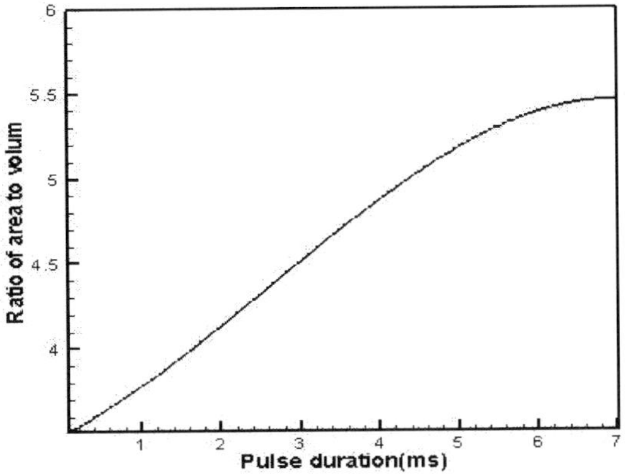

Fig.16. Ratio of keyhole area to volume of the weld pool at the end of a single pulse for various pulse durations at $15GW/m^2$ power density

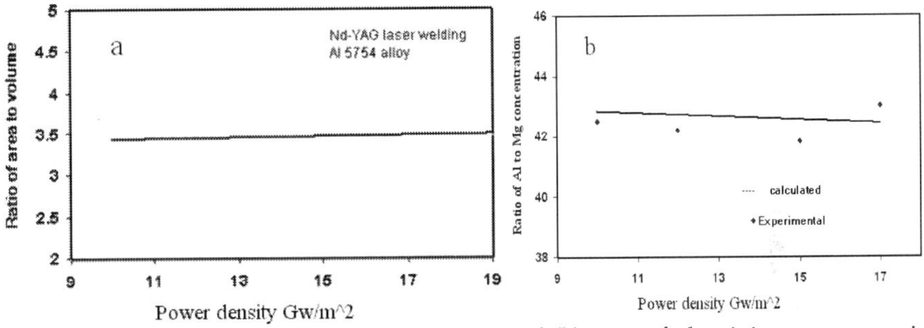

Fig. 17. (a) The ratio of keyhole area to volume, and (b) ratio of aluminium to magnesium concentrations versus laser power density.

6. Conclusion

Here, we have shown that the alloying elements are controlled in the weld metal by changing the laser parameters in the keyhole welding of SS316 and Al5754 using a long pulsed Nd:YAG laser. Several experiments were performed and a theoretical model was developed for the determination of significant alloying element losses such as Mn, Cr, Ni, and Fe in SS316 and Al and Mg in Al5754. Despite laser welding is a complicated process, here, the effect of laser parameters (mainly for various pulse duration at constant power density, as well as the different power densities at the invariant pulse duration.) were investigated on the composition alteration of the weld metal.

Based on the analysis and modeling, we have shown that in SS316 welding the Mn, Cr concentrations reduce within the weld metal however, those of Fe, Ni increase

simultaneously, mainly due to the higher equilibrium pressure of Mn and Cr respect to Fe and Ni according to figure 4 [34].
In fact, a couple of competitive mechanisms are involved including keyhole shape (surface to volume ratio) and the diffusion time of the migrated elements. The concentrations of alloying elements are nonlinear in terms of the laser pulse duration due to the nonlinearity of surface to volume ratio versus the pulse duration. It was found that the keyhole shape is significant for shorter pulse duration; however, the diffusion time becomes dominant at longer pulses according to figures 8 and 16.
Moreover, when power density varies from 10 GW/m^2 to 20 GW/m^2 while the laser pulse duration is kept unchanged, then the element loss increases linearly mainly due to the linear correlation of surface to volume ratio with peak power density.
Elemental change of Al5754 alloy after laser welding was extensively investigated using LIBS technique for element tracing in the weld metal [35]. ArF laser was employed to create micro plasma over the weld region. The LIBS analysis includes the significant finding as below:
i) Mg loss linearly increases with increasing the pulse duration of the laser welding.
ii) The variation of Mg trace is negligible while varying the laser power density.
Moreover, the ratio of keyhole area to volume strongly depends on the pulse duration which is in good agreement to the above conclusion (i).
Finally the keyhole geometry obtained from model remains invariant with the laser power density of pulsed Nd:YAG laser source which is in accordance with the above conclusion (ii). Eventually in order to increase the welding depth, it is suggested to increase the laser power densities rather than using longer pulse durations to assure of minimum Mg loss.

Appendix

The mass diffusivity of an element a in the shielding gas b at temperature T is given by [2]

$$D_{ab} = \frac{1.8583 \times 10^{-7} \sqrt{\frac{M_b + M_a}{M_b M_a}}}{\sigma_{ab}^2 \Omega\left(\frac{KT}{\varepsilon}\right)} T^{\frac{3}{2}}$$

$$\sigma_{ab} = \frac{\sigma_a + \sigma_b}{2}$$

Where σ the collision diameter in angstroms, M is molecular weight, and Ω is the slowly varying function of the parameter $\frac{KT}{\varepsilon}$ which is given by:

$$\Omega = \left(44.45\left(\frac{K_B T}{(\varepsilon_a \varepsilon_b)^{\frac{1}{2}}}\right)^{-4.909} + 1.911\left(\frac{K_B T}{(\varepsilon_a \varepsilon_b)^{\frac{1}{2}}}\right)^{-1.575}\right)^{0.1}$$

Where ε refers to the intermolecular force parameter.

7. References

1. A. Block-Bolten and T. W. Eagar: Metallurgical Transaction B, 15B, 461, 1984
2. K. Mundra and T. Debroy: Metallurgical Transaction B, 24B, 145, 1993
3. H. Zhao and T. Debroy: Metallurgical Transaction B, 32B, 163, 2001
4. P.A.A. Khan, T. Debroy, and S.A. David: Metallurgical Transaction B, 67, pp.1s-7s, 1988
5. X. He and T. Debroy: J. Phys. D: Applied Physics, 37, 4547, 2004
6. M.J. Torkamany, M.J. Hamedi, F. Malek, and J. Sabbaghzadeh: J. Phys. D: Applied Physics, 39, 4563, 2006
7. U. Dilthey, A. Goumeniouk, V. Lopota, G. Turichin and E. Valdaitseva: J. Phys. D: Applied Physics, 34, 81, 2001
8. David A. Cremers, Leon J. Radziemski .Handbook of Laser-Induced Breakdown Spectroscopy, 2006 (John Wiley&Sons,Ltd)
9. Anderzej W. Miziolek, Vincenzo Palleschi, Israel Schechter. Laser-Induced Breakdown Spectroscopy, 2006 (CAMBRIDGE University press)
10. D. A. Rusak, B .C .Castle, B .W .Smith, J .D. Winfordner; Crit. Rev. Anal.Chem., Vol. 27, pp 257, 1997
11. P. Lucena and J. J. Laserna; Spectrochim. Acta B, Vol.56, pp. 1120, 2001
12. L. Barrette and S. Turmel; Spectrochim. Acta B, Vol. 56, pp. 715, 2001
13. S. Z. Shoursheini, P. Parvin, B. Sajad, M. A. Bassam; Applied spectroscopy, Vol. 63, P.423-9, 2009
14. Jae Y. Lee, Sung H. Ko, Dave F. Farson and Choong D. Yoo; J. Phys. D: Applied Physics, 35, 1570, 2002
15. Xi Chenl and Hai-Xing Wang; J. Phys. D: Applied Physics, 36, 1634, 2003
16. A Matsunawa and V Semak; J. Phys. D: Applied Physics, 30, 798, 1997
17. W.W. Duley: laser welding (New York: Wiley), 1998
18. Conny Lampa, Alexander F. H. Kaplan, John Powell, and Claes Magnusson; J. Phys. D: Applied Physics, 30, 1293, 1997
19. T. Zacharia, S.A. David, J.M. Vitek and T. Debroy; Welding Journal, 12, 499, 1989
20. W. Sudnik, W. Erofeev and D. Radaj; J. Phys. D: Applied Physics, 29, 2811, 1996
21. E. Amara and A. Bendib; J. Phys. D: Applied Physics, 35, 272, 2002
22. J. Sabbaghzadeh, S. Dadras, and M.J. Torkamany; Journal of Physics D: Applied Physics, 40, 1047, 2007
23. V. Vladimir, W.D. Semak, B. Bragg, Damkroger and S Kempka; J. Phys. D: Applied Physics, 32, L61-L64, 1999
24. W. Robert, Jr. Messler: "Principles of welding"; (WILY-VCH Publishing Co.), 2004
25. C.L.Yaws, Handbook of vapor pressures (Gulf Publishing Co., Houston), 1994
26. S. Dusham and J.M. Laferty; Scientific foundations of vacuum technique, 2nd edition, John wily, New York, PP.691-737, 1962
27. E.U. Schlunder and V. Gniclinski; Chem. Eng. Technology, 39, 578, 1967
28. O. Solana and J. L. Ocana; J. Phys. D: Applied Physics, 30, 1300, 1997
29. H. Kurniawan, A. N. Chumakov, Tjung Jie Lie, M. O. Tjia, M. Ueda, and K. Kagawae; Journal of Applied Spectroscopy, Vol. 71, pp.5-9, 2004
30. A.W. Miziolek, V. Palleschi, I. Schechter, *Laser Induced Plasma Spectroscopy,* 2006, (Cambridge: Cambridge University Press) chapter 3, 122pp
31. H. R. Griem, *Spectral line broadening by plasma* (Academic Press, 1974), Appendix. 4, pp. 320.

32. A.M. El Sherbini, Th.M. El Sherbini, H. Hegazy, G. Cristoforetti , S. Legnaioli, V. Palleschi, L. Pardini, A. Salvetti, E. Tognoni; Spectrochimica Acta Part B, Vol. 60, pp. 1573 - 1579, 2005
33. H.R. Griem, Plasma Spectroscopy, Mc Graw Hill, New York, 1964.
34. M. Jandaghi, P. Parvin, M. J. Torkamany, J. Sabbaghzadeh; Journal of Physics D: Applied Physics, Vol. 41(23) 235503 (9pp) (2008)
35. M. Jandaghi, P. Parvin, M. J. Torkamany, J. Sabbaghzadeh; Journal of Physics D: Applied Physics, Vol. 42(20) 205301 (8pp) (2009)

10

Laser welding: techniques of real time sensing and control development

Xiaodong Na
Cummins Inc.
USA

1. Background

As shown in Figure 1, Laser Welding is a non-contact fusion process with various lasers applying to materials. Laser welding accomplishes the welding work through laser beam. With laser beam, energy is concentrated and used directly on the small welding area. Consequently, the welding zone is very narrow and hardly distorted due to little heat influence. Compared to traditional processes, Laser Welding is of potential. Its non-contact, localized, and narrow heat zone can create high quality result. Common re-working and after-work procedure are no more required, which saves cost and labour. Till now, Laser welding as been widely applied in various fields including automotive, microelectronics, aerospace, etc.

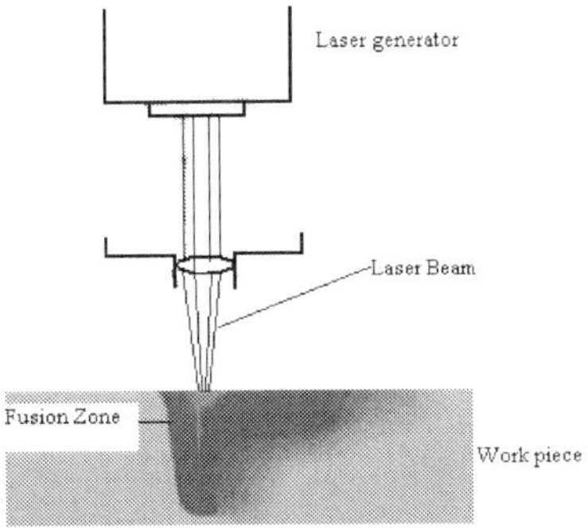

Fig. 1. Simple laser welding process

Common types of lasers applied to welding include CO2 gas laser, Solid state laser (YAG type), and Diode laser welding. CO2 laser uses a mixture of high purity carbon dioxide with helium and nitrogen as the medium, infrared of 10.6 micro-meters. Argon or helium is additionally used to prevent oxidation. YAG laser takes advantage of a solid bar of yttrium aluminium garnet doped with neodymium as the medium, whose infrared is only 1.06 micro-meters. Diode laser is mostly based on the conversion between high electrical to optical powers (Migliore 1998, Sun 1999, Sun 2002, Pedrotti 1993, Williams 1997).

Despite the quality performance in Laser Welding, the going concerns centres on any possible compromise of human and environmental health and safety. Indeed, these considerations have been challenging engineers to develop advanced automatic manufacturing process without any need of human involvements. However, successful development of automation system is beyond challenging because first of all, no exact model has been developed to describe the process and even it does, the model is much more complicate for control design; second of all, intelligent welding system requires appropriate and real time measurement working with specific developed control algorithm so that the process is robust and adaptive.
The major focus of this chapter will be on the real time sensing and control methods to the laser welding such that a practical automation system can be developed and implemented for heavy manufacturing and industry.

2. Overview of Laser Welding

Laser welding is an advanced fusion joining process that applies the energy converted from a laser beam to melt and joint metal pieces together. Laser beams can be either continuous or pulsed. Continuous laser systems are mostly used for very deep welding, whereas pulse lasers are used to weld very thin materials together. Depending on how the laser light is generated, Laser can be categorized into solid state lasers and gas lasers. Solid state lasers use solid media, such as synthetic ruby and crystal, to form the laser beam, such as Nd:YAG laser and Diode laser. Gas lasers use gaseous media, such as helium, nitrogen and carbon dioxide to form the laser beam, such as CO2 laser. Solid state lasers operate on much shorter wavelength than gas lasers, but they have much lower power outputs.
As shown in Figure 2, the advantage of laser welding is remarkable, e.g. low distortion, high speed and small heat affected zone. This is mostly because laser welding is applying a beam of light that is monochromatic, collimated and of sufficient power density. With adjustment power density, very high values of irradiance and much localized heating can be easily achieved. Because the light is collimated and monochromatic, the heat-affected zone can be very small without need of post processing, especially in the case of spot welding with extremely small weld diameter. System set up and configuration is also relatively easier and there is no contact of any material with the work piece. The disadvantage of laser welding is its cost and possibly limited capability. The initial capital cost of laser machine is usually very high. Depending on the laser system capacity, the depth of penetration in laser welding is also limited. Careful process monitoring and control is also required to avoid material vaporization due to high temperature around the weld.

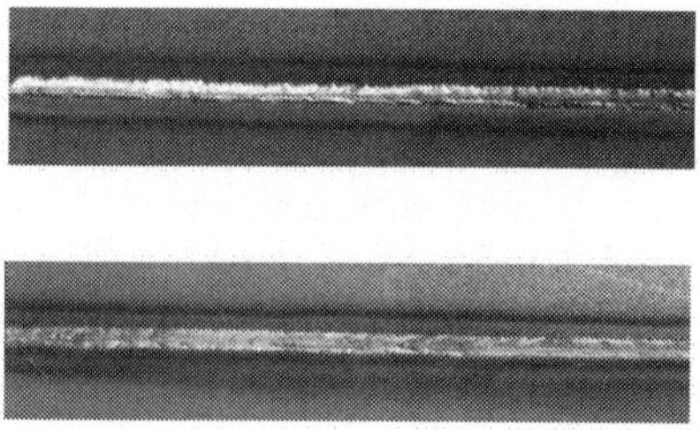

Fig. 2. Standard diode laser (1KW) welding results (9.5mm/s), 1.5mm thickness steel

By far Laser welding has been benefiting as many industries as possible from its advantages. Its applications vary with power-generation capability. Low-power applications are mostly seen in the instrumentation and electronics industries, while higher-power applications exist in the automotive, shipbuilding and aerospace industries. One potential disadvantage limiting its application is the cost, the more power of the laser provides, the higher cost it requires. For each application, the trade-off always involves with the capital cost of laser systems and the future economic returns.

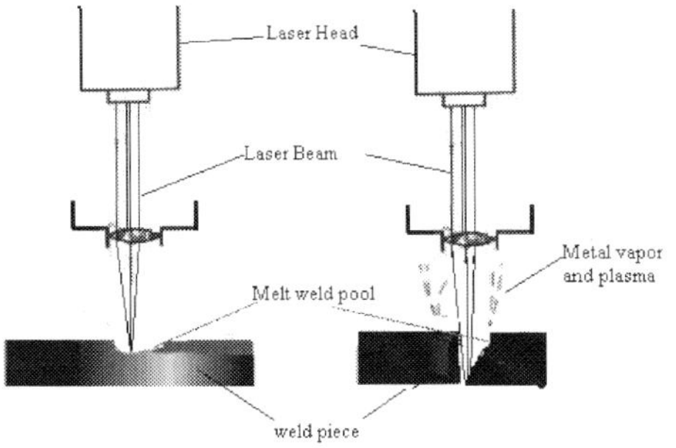

Fig. 3a. Conductivity based Laser Welding; 3b: Penetration based Laser Welding

Figure 3 presents standard system configuration for laser welding. As introduced earlier, fundamentally laser welding is through heat distribution process. Accordingly any factors that affect the laser power, welding speed and material complexity can impact the whole melting and jointing process. As shown in Figure 3a, heat distribution has the most significant impacts on the welding performance. For a laser with low power density, mostly heat converted from optical energy is completed through a conductive distribution. When laser power density is as big as KW level, heating the spot after laser focus transferred to the surface can boil and even vaporize the metal; accordingly a hole can be formed and filled with ionized metallic gas. The hole is also frequently referred as key-hole. The advantage of the cylindrical keyhole is that with key-hole formation, more effective heat energy will be absorbed and significantly boost welding process, especially by penetration, as shown in Fig. 3b. As a result, not only is the welding speed going to be much faster, but also the weld seam depth to width ratio much bigger. In addition, the heat-affected-zone can be relatively smaller, which is the most critical factor to welding quality.

3. Importance of Welding Automation

As introduced above, although laser welding highly advantageous, its process is potentially hazardous. For example, because of the heat and melting, particular fume, toxic noises and irradiation will be generated and exhausted to the working environment. Although with special care and human maintenance, these hazards can be reduced significantly, the risk of human error to some extent exposes operators and those around them to latent risks. Accordingly, it is always necessary to develop automatic control laser welding processes with limited or even without any need of human interference. Automation as a result offers a means of removing the operator from the process, reducing application-related hazards, and more importantly improving the control of the welding environment. This is particularly beneficial for those heavy-duty manufacturing systems.

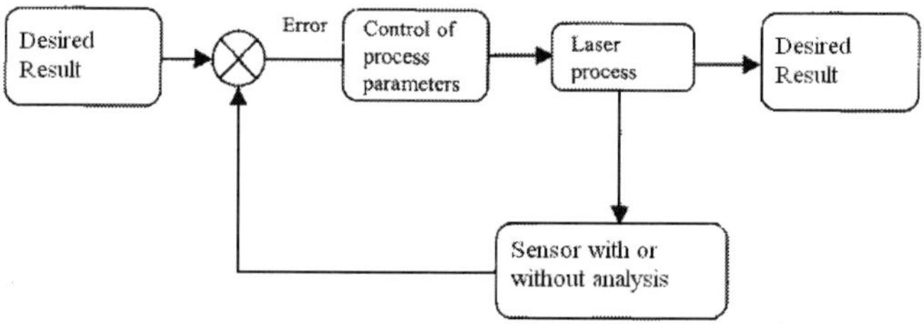

Fig. 4. standard automatic control laser welding process

Figure 4 represents a standard automatic laser welding process. Certainly any successful automation requires suitable sensor system for process signal acquisition and processing and practical control development for self adjustment. This way, control development and sensing system can constitute complete close-loop automation and intelligently make possible corrections online based on the current system state. For example, increase laser power when it is low; decrease travelling speed when it is high. The problem is that to perform online adjustment in general requires sufficient knowing knowledge about how the welding conditions such as laser power and speed impact the welding performance, i.e. weld pool geometry. In other words, the difficulty is how to implement a suitable relation to evaluate the process. Apparently successful control of the power and the speed during the welding process is the most significant step for the automation implementation. Since laser welding takes advantage of the heat energy to melt and joint the weld pieces, critical parameters that impact the energy should be taken care of at first. Power and travel speed are always considered the influential factors before designing laser welding automation. An easiest way is, taking the power and travel speed as input, the welding quality parameter as output, laser welding process could be achieved as a tracking control.

4. Real time sensing: a prerequisite for automation

As shown in Fig. 4, sensor system acquires and processes signals regarding the welding process such that any successful decision can be made, without which automation can not be complete. Sensing system development proves necessary to reduce any impacts on human health, detect any weld detect in real time and reduce overall operational cost, although it can be challenging because laser welding is dynamic, complex and uncertain. When the focused laser beam is applied to a material piece, spot of surface will be melted and around the fusion zone, energy is emitted in various forms. Each signal might carry information describing the characteristics of laser welding process. Figure 5 shows a selection of detectable emissions, such as optical, acoustic, infrared, vibration, and so on. Because of the significance of penetration, most of the sensing system is aimed at detecting either in-complete or over-penetration. In addition, the geometrical parameters of the keyhole and melt pool to some extent represent the welding quality. Accordingly analysis based on these measurable signals can help understand characteristics of the welding process. So far, various studies have been done to monitor the laser welding process. Some focused on the emission signals such as acoustic, infrared, ultraviolet, plasma, and so on (Shao 2005, Ostendorf 2003, Ono 1992, Li 2002, Farson 1999, Steen 1986, Gu 1996). Others aimed to the weld pool images acquired with CCD cameras (Beersiek 2001, Zhang 1996, Na 2009, Beersiek 1999).

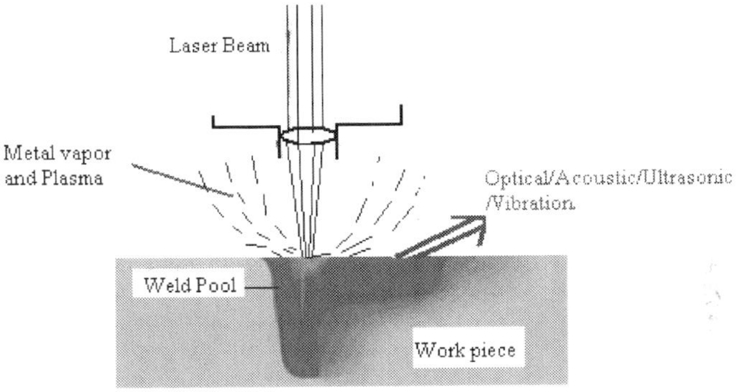

Fig. 5. Signal emissions and sensors in laser welding

5. Acoustic Signal sensing system

A lot of studies have been made on acoustic sensing for laser welding process (Sun 2002, Shao 2005, Ostendorf 2003, Ono 1992, Li 2002, Farson 1999, Steen 1986, Gu 1996). Others aimed to the weld pool images acquired with CCD cameras (Beersiek 2001, Zhang 1996, Na 2009, Beersiek 1999). Though it is promising, its application might be compromised by the complexity and noises in the laser welding. As shown in Fig. 6, in general the sensor is used to convert the measured sounds (e.g. weld pool surface) into electrical variable. The signal frequency can be up to 1MHz, from the human audible to ultrasonic range.

Because a normal audible signal falls into the range of 20Hz to 20 kHz, to monitor acoustic signal variation, a regular microphone can be used in applications. As presented in (Shao 2005, Farson 1999) it is quite possible to take advantage of micro-phone acquired acoustic signal to correlate the acoustic signal and welding performance. For example, spectrum analysis could indicate welding progress by analyzing any spike like fluctuations. A model was built describing the spectrum characteristics of the weld pool under various conditions. It has been demonstrated that there is a possible relation between key-hole oscillation and the frequency (Kroos 1993)

Ultrasonic signals were specifically investigated in search of a pattern or relationship describing the emission and the welding performance, e.g. weld pool geometry. A piezoelectric sensor is installed on the back of laser beam to capture any acoustic mirror signal generated by the back-reflected laser signal (Li 2002). Doing so was considering the dynamic vibration of the weld pool surface cause fluctuation of the reflected laser beam during welding process. It has been demonstrated that the signal strength varies with the distance from the weld pool. Moreover, in the case of deep penetration, i.e. generating a key-hole, a signal spike was noticed whenever key-hole was about to complete. Similar results were also shown in (Steen 1986) . In a study presented in (Gu 1996), FFT (Fast Fourier

Transform) was applied to obtain the frequency response between 20 kHz and 0.5 MHz to investigate resonant relationship during last welding. It has been proved that with manipulating frequency components, it was possible to isolate welding process from over-penetration or partial penetration. Similar studies are also described in (Miller 2002, Klein 2002)

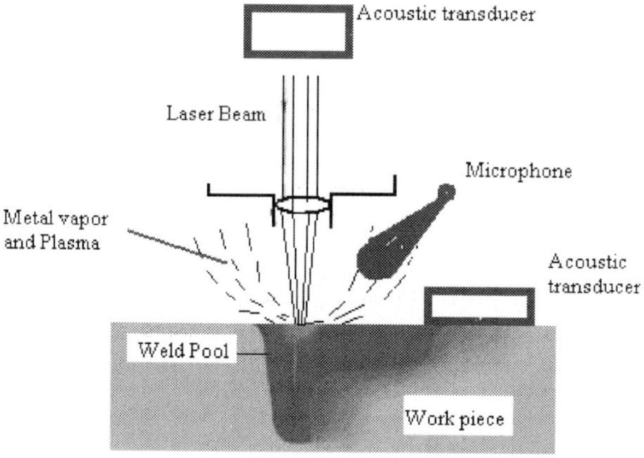

Fig. 6. Acoustic sensing system set up

With aid of acoustic emission signal analysis, reliable and applicable measurement and control of laser welding becomes possible. The disadvantage of these methods is its limit while differentiate from partial penetration and full penetration. The sensing system is also very easily influenced by instable noise.

6. Optical Signal sensing system

As shown in Figure 7, Optical sensor is mostly based on the detection of plasma plume emission and the thermal radiation of the weld pool. Fundamentally this is considering the associative relation between the degree of penetration and the emission intensity measurement. Various studies have been done on optical signal sensor design (Ostendorf 2004, Tonshoff 1998, Park 1999, Park 2002, Sforza 2002). In (Tonshoff 1998), a silicon photodiode assisted with a preamplifier is developed to detect dynamic plasma intensity fluctuation during laser welding. Similar study is also done in (Park 2002) based on Ultra-violet photodiodes and Infrared photodiode to measure the emission from the plasma and metal vapor in the CO2 laser welding searching for a relationship describing the heat distribution and the emissions. A pattern to correlate the laser energy under conditions such as optimal heat input, slightly low heat input and low heat input with welding quality such as complete and partial joining is also developed. Similar experiments were implemented in (Park 1999, Park 2002) to develop a relationship between the plasma and spatter and bead shape according to the welding variables, based on a multiple regression analysis and neural network to estimate the penetration depth and width of the weld bead. In (Sforza

2002, Zhang 2004), IR and UV signals were captured and analyzed to detect the IR and UV waveband of the optical emissions induced in the underwater laser welding and searched for a relationship between the optical signals and the weld quality with various shielding conditions.

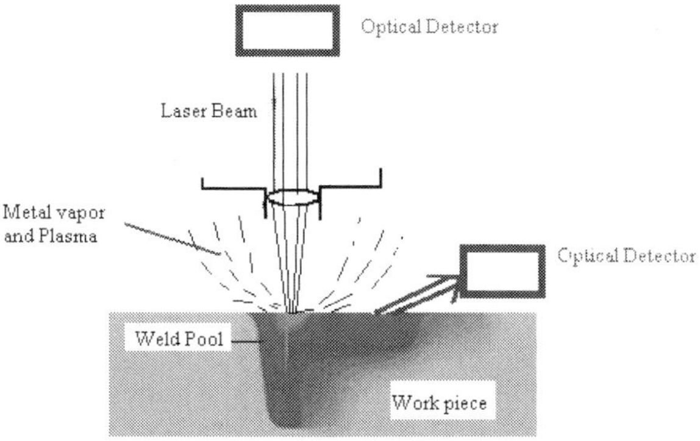

Fig. 7. Optical Signal Sensors

Similar to those acoustic sensing system, optical sensor is easier to implement and of low cost. However, its accuracy is compromised by the system noises.

7. Vision based Keyhole Sensing with CCD/CMOS Camera

During the process of laser welding, high energy of laser beam is focused onto a single location and a keyhole is created. As shown in Figure 8, to ensure successful welding and avoid effects like burnt-through, keyhole depth should be controlled not too much beyond the height of the material. The advantage of Keyhole sensing is the possible small heated zone, which results in better after-work quality.

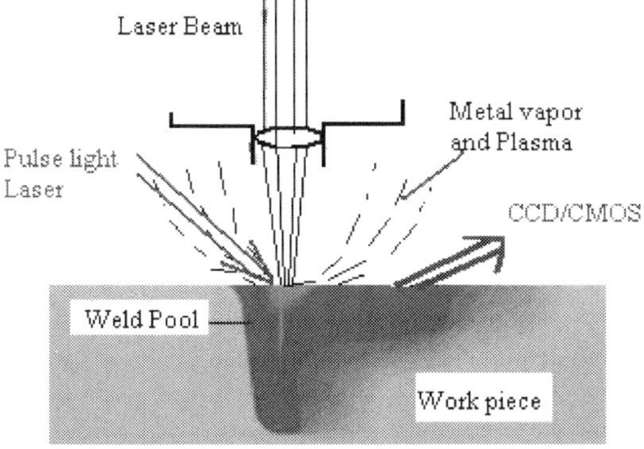

Fig. 8. Vision Sensor

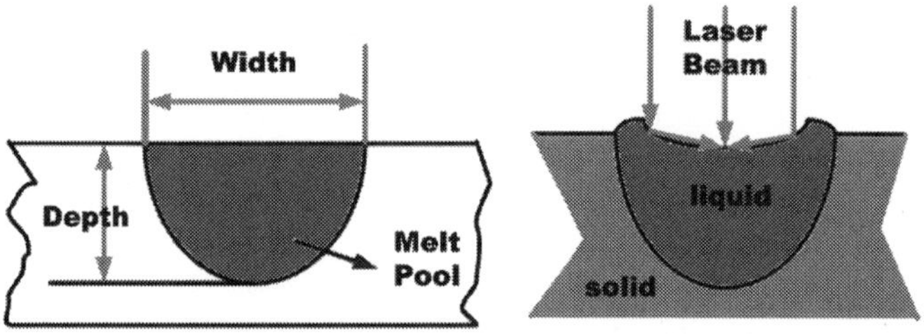

Fig. 9. Weld pool (1a) Laser Welding (1b)

As shown in Figure 9, the full knowledge of the weld pool geometry includes the length, the width, and the depth. Weld pool images were acquired with CCD/CMOS cameras and studied to investigate the relations among various parameters along with laser welding process. Various Researches has been done related to welding pool measurement based on the vision sensing tool such as monochrome camera and optic sensors. Through the sensors, welding related information is achieved online or offline so that control loop can adjust the welding process parameters. Considering the role of penetration to welding process, many study started with penetration measurement and control. Successful 2D or 3D measurement can provide sufficient geometric weld pool information and consequently the control performance (Farson 1999, Lankalpall 1996, Becker 1995). In 0, a camera system was developed to measure the surface geometric size of the weld pool. In (Na 2009, Zhang 1996,Beersiek 2000), a system for process monitoring of laser beam welding based on a CMOS-camera was presented. The system observed the welding process online and coaxial to the laser beam. It was used to investigate the geometrical parameters of the keyhole.

8. Other sensing techniques

Besides acoustic and optical signals, temperature distribution can also be examined. In (Jeon 1998), the surface temperature variation in the laser brazing of a pin-to-hole joint is studied using an infrared radiation sensor. In (Bertrand 2000), surface temperature in Nd:YAG continuous laser welding is monitored to identify the variation of brightness temperature whenever certain welding defects occur. In (Lim 1998) an infrared sensor is used to study laser spot welding. In (Li 1996) a special plasma charge sensor (PCS) based on plasma density and ionization variation is implemented to measure weld penetration and detect weld defects. Some other researches are also involved with direct mode estimation (Morgan 1990, Sun 1993) to assist better control development.

9. Vision sensor based system identification and control

Figure 10 presents a standard diode laser welding system, developed in the Welding Research Lab at the University of Kentucky. The laser is current-driven and the output energy is theoretically proportional to its input current. According to the manufacturing configuration, the wavelength of the laser light is 850nm and the power can reach to 1 kW at most, which corresponds to the current at the level of 58mA. The workstation holding the work-piece is driven by a servo motor. Both the driving current of the laser and the welding speed are controlled by a digital computer. In addition, there are two other parameters, namely the laser focus distance and incident angle. For simplification purposes, the authors adjusted the laser focus distance and incident angle at 89mm and 42 degree respectively according to manufacturing settings. As shown in Figure 11, the vision sensor measures the top surface width for model identification and control design. A close seam tracking is finally possible.

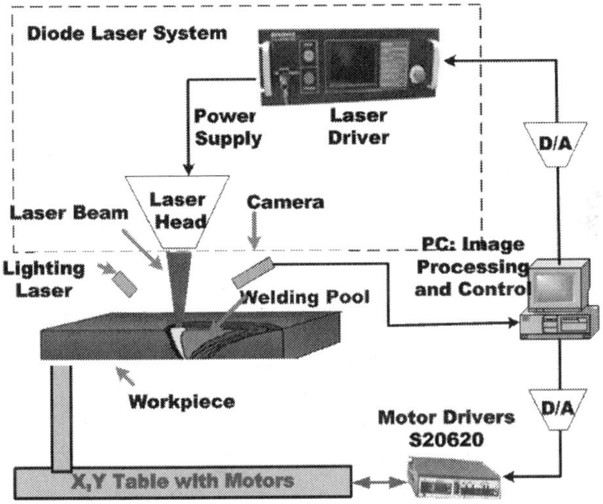

Fig. 10. Standard vision sensor based automatic diode laser welding system

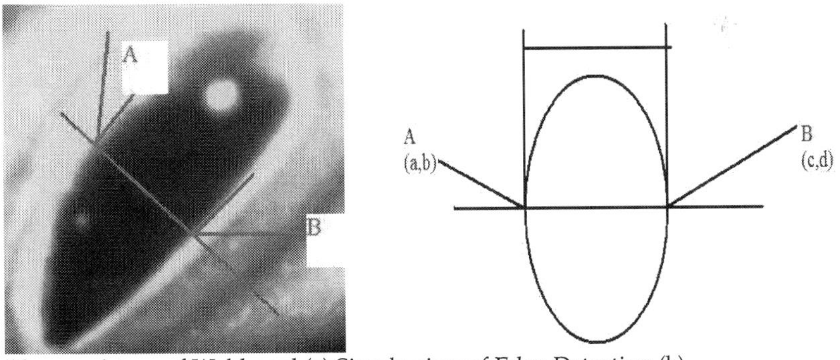

Fig. 11. raw picture of Weld pool (a) Simple view of Edge Detection (b)

10. Nonlinear Hammerstein identification (Na 2009)

As introduced above, Laser welding is a complicated thermodynamic and physicochemical process, which involves material melting, evaporating, plasma forming, keyhole occurrence and so on. The weld shape is determined by many parameters in the thermodynamic and physicochemical process, such as the relative speed between the laser and the work-piece. Thus, the laser welding should be treated as a nonlinear process. Traditionally, a linear model is often used to approximate a nonlinear system. However, the linear model can only describe the local dynamics in the vicinity of the set point and the control system designed by using the linear approximated model has small operation region. To understand/describe the global behavior of the laser welding process and to obtain a large operation region for the control system, a nonlinear model is preferred. However, the process is so complicated that it is hardly possible to build the model solely through physicochemical analysis. Therefore, nonlinear identification, which is based on input-output data, is needed.

Fig. 12. Continuous Hammerstein structure

Figure 12 is the continuous Hammerstein structure used for identifying a practical model to the laser welding process. The structure consists of nonlinearity in series with a linear dynamics, where u and y are the input and the output respectively. The linear dynamics can in general be represented by a linear transfer function:

$$G(s)=\frac{b_0 s^m + b_1 s^{m-1} + \cdots + b_m}{a_0 s^n + a_1 s^{n-1} + \cdots + a_n} = \frac{B(s)}{A(s)}, \tag{1}$$

where $\{a_i\}, i = 0,1,\cdots n$ and $\{b_i\}, i = 0,1,\cdots m$ are parameters of the linear dynamics; and the nonlinearity $f(u(t))$ can be considered as a nonlinear function of the system input and the input of the linear dynamics. Further, as will be seen later, the linear dynamics of the diode laser process can be sufficiently approximated without zeros, in other words, a class of nonlinear dynamic systems whose linear dynamics is of minimum phase.
Then in the Laplace domain:

$$y(s) = G(s)L(f(u(t))), \tag{2}$$

where L is the operator of Laplace transform.
Without loss of generality, denote

$$a_n = 1 \tag{3}$$

$$K(u(t)) = L^{-1}(B(s)L(f(u(t)))), \tag{4}$$

where L^{-1} is the operator of inverse Laplace transform.
Then with substitution, (2) can be rewritten as:

$$a_0 y^{(n)}(t) + a_1 y^{(n-1)}(t) + \cdots y(t) = K(u(t)) \tag{5}$$

Rewrite (5) into matrix form:

$$y(t) = \phi^T(t)\theta \tag{6}$$

$$\phi^T(t) = \begin{bmatrix} -y^{(n)}(t) & \cdots & -y^{(1)}(t) & 1 \end{bmatrix} \tag{7}$$

$$\theta^T = \begin{bmatrix} a_0 & \cdots & a_{n-1} & K(u(t)) \end{bmatrix} \tag{8}$$

Accordingly this Nonlinear Hammerstein identification becomes to estimate unknown parameters in (8). Among these parameters only $K(u(t))$ will vary with the inputs.
Suppose $f(u(t))$ is a polynomial:

$$f(u(t)) = C_0 + C_1 u(t) + \cdots + C_p u^p(t); \tag{9}$$

And further consider step input $u(t) = U$ in Fig. 7. Therefore, $f(u(t))$ is also a step signal whose Laplace transform can be obtained by $L(f(u(t))) = f(U)/s$; therefore, the linear dynamics (1) can be simplified to:

$$G(s) = \frac{b_m}{a_0 s^n + a_1 s^{n-1} + \cdots + 1} \tag{10}$$

Furthermore, the system can be can be simplified as:

$$a_0 y^{(n)}(t) + a_1 y^{(n-1)}(t) + \cdots y(t) = K(u(t)) \tag{11}$$

where

$$K(u) = b_m f(U) \tag{12}$$

and b_m is a constant.
As a result, the proposed identification for a specific Hammerstein structure with (10) as the linear dynamics and (12) as the nonlinear static function in general can be achieved in two steps:
Step 1: Identify A(s) under different step input $\{U_j\}, j = 1, \cdots, M$ (13)

Due to the use of a step input U_j, the nonlinearity function,

$$K(u) = K(U_j) = b_m f(U_j), \tag{14}$$

becomes a constant parameter and can be identified by:

$$\theta^T(U_j) = \begin{bmatrix} a_0 & \cdots & a_{n-1} & K(U_j) \end{bmatrix}. \tag{15}$$

Hence, both the format of A(s) and the value of $b_m f(U_j)$ can be identified.

Step 2: Determine the nonlinear function $f(u)$ and b_m from the steady-state responses $Y(U_j)$ under step inputs, $\{U_j\}, j = 1,\cdots,M$.
In this case, the nonlinearity function can be identified by:

$$K(U_j) = \phi^T(U_j)\theta, \tag{16}$$

where
$$\phi^T(U_j) = \begin{bmatrix} 1 & U_j & U_j^2 & \cdots & U_j^p \end{bmatrix} \tag{17}$$

and
$$\theta^T = \begin{bmatrix} b_m C_0 & b_m C_1 & b_m C_2 & \cdots & b_m C^p \end{bmatrix}. \tag{18}$$

Then the parameters in $b_m f(U_j)$ can be estimated by such as linear least square algorithm:

$$\theta = (\Phi^T\Phi)^{-1}\Phi^T[Y(U_1) \quad \cdots \quad Y(U_M)]^T \tag{19}$$

where $\Phi^T = [\phi^T(U_1) \quad \cdots \quad \phi^T(U_M)]^T$.

11. Nonlinear Constructive Control Design (Liu 2002, Liu, 2003, Liu 2004)

We consider the system that can be represented by:

$$y^n(y) = f(y, \dot{y}, \cdots, y^{n-1}) + \theta u + g(u) \qquad (20),$$

where y, u are the output and the control respectively; $y^i, i = 1,2,\cdots n$ is the ith derivative of y; and $f(y, \dot{y}, \cdots, y^{n-1}) \leq \theta \bar{f}(y, \dot{y}, \cdots, y^{n-1})$, where $\theta > 0$
$g(u) \leq c\|g(u)\|$, where $c > 0$

Let
$$\begin{aligned} x_1 &= y \\ &\cdots \\ x_n &= y^{n-1} \\ v &= u \\ z &= v \end{aligned}$$
, where v is the virtual control input for the system

Then we have:

$$\begin{aligned} \dot{x}_i &= x_{i+1} \\ \dot{x}_n &= f(x) + u, \\ 1 &\leq i \leq n-1 \end{aligned} \tag{21}$$

where $x = [x_1 \quad x_2 \quad \cdots \quad x_n]^T$

During the identification, $g(u)$ is approximated by a polynomial. In fact, it can be other complex formats. However, we assume that $g(u)$ is continuous and bounded
We further assume all the states of the system are available for feedback and define:

$$
\begin{aligned}
e_1 &= x_1 - x_d \\
&\cdots \\
e_{n-1} &= x_n - x_d^{\,n-2} \\
e_n &= x_n - x_d^{\,n-1}
\end{aligned}
\tag{22}
$$

With simple substitution, we can have:

$$
\begin{aligned}
\dot{e} &= Ae + b\left(f(e + \bar{x}_d, z) + \theta v - x_d^{\,n} + g(e + \bar{x}_d)\right) \\
z &= v
\end{aligned}
\tag{23}
$$

where $A = \begin{bmatrix} 0 & 1 & 0 & 0 \\ \vdots & \vdots & \vdots & \vdots \\ 0 & 0 & 0 & 1 \\ 0 & 0 & 0 & 0 \end{bmatrix}$, $b = \begin{bmatrix} 0 \\ \vdots \\ 0 \\ 1 \end{bmatrix}$, and $\bar{x}_d = \lfloor x_d \quad \cdots \quad x_d^{\,n-1} \quad x_d^{\,n} \rfloor$.

Define $A_m = A - bK$, where K is chosen so that the matrix $A_m = A - bK$ is Hurwitz. Then,

$$
\begin{aligned}
\dot{e} &= A_m e + b\left(f(e + \bar{x}_d, z) + \theta v - x_d^{\,n} + g(e + \bar{x}_d)\right) \\
z &= v
\end{aligned}
\tag{24}
$$

Further, we define a dynamic signal described by:

$$
\begin{aligned}
\dot{r} &= -c_0 r + r_m(e, \bar{x}_d) \\
r(0) &> 0 \\
c_0 &> 0 \\
r_m &= \|e + \bar{x}_d\|^2 \gamma_0 \|e + \bar{x}_d\| + d_0
\end{aligned}
\tag{25}
$$

From [40], signal r has the property of :

$$
V \le r(t) + D(t) \tag{26}
$$

Then the robust adaptive controller can be designed by:

$$
\begin{aligned}
v &= -\beta e^T Pb\left(\bar{f}(e + \bar{x}_d, z) + \|e + \bar{x}_d\|^2 + \|z\|^2 + \left(\alpha^{-1}(2r)\right)^2 + (Ke)^2 + 1\right) \\
\dot{\beta} &= \beta_m(e, z, r, \bar{x}_d) - \Gamma\sigma\beta \\
\beta_m &= \Gamma\left(e^T Pb\right)^2\left(\bar{f}(e + \bar{x}_d, z) + \|e + \bar{x}_d\|^2 + \|z\|^2 + \left(\alpha^{-1}(2r)\right)^2 + (Ke)^2 + 1\right) \\
\Gamma &> 0 \\
\sigma &> 0
\end{aligned}
\tag{27}
$$

$\alpha^{-1}(\bullet)$ is the inverse of function of $\alpha(\bullet)$ and P is the solution of

$$
\begin{aligned}
&PA_m + A_m^{\ T} = -Q \\
&Q = Q^T > 0
\end{aligned}
\tag{28}
$$

Define the Lyapunov candidate as:

$V = e^T Pe + \theta\Gamma^{-1}\left(\beta - \widehat{\beta}\right)^2$, where $\widehat{\beta}$ is a positive constant and the desired value of β

$$
\begin{aligned}
\dot{V} &= -\left(A_m e + b\left(f\left(e+\bar{x}_d, z\right) + \theta v - x_d^{\ n} + g\left(e+\bar{x}_d\right)\right)\right)^T Pe + e^T P\left(A_m e + b\left(f\left(e+\bar{x}_d, z\right) + \theta v - x_d^{\ n} + g\left(e+\bar{x}_d\right)\right)\right) + V_1 \\
&\le -e^T Qe + 2\bar{\beta}\theta\left(\left(e^T Pb\right)\left(Ke\right) - \frac{1}{2\widehat{\beta}\theta}\right)^2 - 2\bar{\beta}\theta\left(\left(e^T Pb\right)\bar{f}\left(e+\bar{x}_d, z\right) - \frac{\theta}{2\widehat{\beta}\theta}\right)^2 + \frac{\theta^2}{2\widehat{\beta}\theta} - 2\bar{\beta}\theta\left(\left(e^T Pb\right)\left(e+\bar{x}_d\right) - \frac{c_1}{2\widehat{\beta}\theta}\right)^2 \\
&+ \frac{c_1^{\ 2}}{2\widehat{\beta}\theta} - 2\bar{\beta}\theta\left(\left(e^T Pb\right)\left(z\right) - \frac{c_2}{2\widehat{\beta}\theta}\right)^2 + \frac{c_2^{\ 2}}{2\widehat{\beta}\theta} - 2\bar{\beta}\theta\left(\left(e^T Pb\right)\left(\alpha^{-1}(2r)\right) - \frac{c_3}{2\widehat{\beta}\theta}\right)^2 + \frac{c_3^{\ 2}}{2\widehat{\beta}\theta} - 2\bar{\beta}\theta\left(\left(e^T Pb\right) - \frac{c_4}{2\widehat{\beta}\theta}\right)^2 \\
&+ \frac{c_4^{\ 2}}{2\widehat{\beta}\theta} - \theta\sigma\beta^2 + \theta\sigma\widehat{\beta}^2 - \theta\sigma\left(\beta - \widehat{\beta}\right)^2
\end{aligned}
\tag{29}
$$

where,

$$
V_1 = 2\theta\Gamma^{-1}\left(\beta - \widehat{\beta}\right)\left(\Gamma\left(e^T Pb\right)^2\left(\bar{f}\left(e+\bar{x}_d, z\right) + \left\|e+\bar{x}_d\right\|^2 + \left\|z\right\|^2 + \left(\alpha^{-1}(2r)\right)^2 + (Ke)^2 + 1\right) - \Gamma\sigma\beta\right)
$$

$$
c_4 = \sup\left(\left\|y_r^{\ n}\right\| + c_3\left\|g\left(e+\bar{x}_d\right)\right\| + \alpha^{-1}\left(D(t)\right)\right)
$$

Accordingly,

$$
\begin{aligned}
&\dot{V} \le -e^T Qe - \theta\sigma\left(\beta - \widehat{\beta}\right)^2 + M \\
&M = \frac{1}{2\beta\theta}\left(c_1^{\ 2} + c_2^{\ 2} + c_3^{\ 2} + c_4^{\ 2} + \theta^2 + 1\right) + \sigma\theta\widehat{\beta}^2
\end{aligned}
\tag{30}
$$

As a result, the Lyapunov function V will decrease monotonically, which means that (e, β) are bounded. The system is accordingly bounded asymptotically stable

12. Related to the Diode Laser Processing System Without $\dot{u}$

As shown in the identification, our laser welding system can be represented by:

$$
\begin{aligned}
\ddot{y}(t) &= \theta_1 y + \theta_2\dot{y}(t) + \theta_3 + \theta_4 u(t) + \theta_5 u^2(t) \\
&+ \theta_6 u^3(t) + \theta_7 u^4(t)
\end{aligned}
\tag{31}
$$

Define the state function as:

$$
\begin{aligned}
x_1(t) &= y(t) \\
x_2(t) &= \dot{y}(t) \\
z(t) &= u(t) \\
z(t) &= v
\end{aligned}
\tag{32}
$$

Then the system can be represented by:

$$\begin{aligned} \dot{x}_1(t) &= x_2(t) \\ \dot{x}_2(t) &= \theta_1 x_1(t) + \theta_2 x_2(t) + \theta_3 + \theta_4 z(t) + \theta_5 z^2(t) + \theta_6 z(t)^3 + \theta_7 z(t)^4 + \theta_8 v \end{aligned} \tag{33}$$

For simplicity, we let the nonlinear function be:

$$f(\cdot) = \theta_1 x_1(t) + \theta_2 x_2(t) + \theta_4 z(t) + \theta_5 z^2(t) + \theta_6 z(t)^3 + \theta_7 z(t)^4 \tag{34}$$

Accordingly, the amplitude limit function can be written as:

$$f(\cdot) \le \left\| \theta_1 x_1(t) + \theta_2 x_2(t) + \theta_4 z(t) + \theta_5 z^2(t) + \theta_6 z(t)^3 + \theta_7 z(t)^4 \right\| \tag{35}$$

$f(\cdot) \le \theta\left(\sqrt{x_1^2} + \sqrt{x_2^2} + \sqrt{z^2} + z^2 + \sqrt{z^6}\right)$, where θ can be unknown

Thus:

$$\bar{f}(\cdot) = \sqrt{x_1^2} + \sqrt{x_2^2} + \sqrt{z^2} + z^2 + \sqrt{z^6} \tag{36}$$

Equation 36 gives the boundary of the nonlinearity function.
Let the tracking signal $y_r = \sin(t)$, the standard sinusoidal signal with amplitude 1
Then the error signal can be written by:

$$\begin{aligned} e_1(t) &= x_1(t) - y_r(t) \\ e_2(t) &= x_2(t) - \dot{y}_r(t) \end{aligned} \tag{37}$$

or,

$$\begin{aligned} e_1(t) &= e_2(t) \\ \dot{e}_2(t) &= \theta_1 + \theta_8 v + \theta_4 z + \theta_5 z^2 + \theta_6 z^3 + \theta_7 z^4 - \ddot{y}_r(t) \\ z(t) &= v \end{aligned} \tag{38}$$

or,

$$\dot{e}(t) = \begin{bmatrix} 0 & 1 \\ 0 & 0 \end{bmatrix} e + \begin{bmatrix} 0 \\ 1 \end{bmatrix} \left[f(e + y_r, z) - \ddot{y}_r(t) + \theta_5 v + \theta_3 \right] \tag{39}$$

let

$$A = \begin{bmatrix} 0 & 1 \\ 0 & 0 \end{bmatrix} \text{ and } b = \begin{bmatrix} 0 \\ 1 \end{bmatrix} \tag{40}$$

Then equation 9.3 can be rewritten by:

$$\dot{e}(t) = Ae + b\left[f(e + y_r, z) - \ddot{y}_r(t) + \theta_5 v + \theta_3 \right] \tag{41}$$

With simpler substitution, the error matrix is:

$$\dot{e}(t)=\begin{bmatrix}0 & 1\\ 0 & 0\end{bmatrix}e+\begin{bmatrix}0\\ 1\end{bmatrix}\left[f(e+y_r,z)-\ddot{y}_r(t)+\theta_5 v+\theta_3\right] \tag{42}$$

To design the controller, we reformat the equation as:

$$\dot{e}(t)=A_m e+b\left[Ke+f(e+y_r,z)-\ddot{y}_r(k-3)+\theta_5 v+\theta_3\right] \tag{43}$$

, where $A_m = A - bK$ and K is chosen so that A_m is Hurwitz.

$$\dot{e}(t)=\begin{bmatrix}0 & 1\\ -2 & -4\end{bmatrix}e+\begin{bmatrix}0\\ 1\end{bmatrix}\left[\begin{bmatrix}2 & 4\end{bmatrix}\begin{bmatrix}e_1\\ e_2\end{bmatrix}+f(e+y_r,z)-\ddot{y}_r(t)+\theta_5 v+\theta_3\right] \tag{44}$$

For matrix:

$$A_m=\begin{bmatrix}0 & 1\\ 0 & 0\end{bmatrix}-\begin{bmatrix}0\\ 1\end{bmatrix}\begin{bmatrix}k_1 & k_2\end{bmatrix} \tag{45}$$

$$A_m=\begin{bmatrix}0 & 1\\ -k_1 & -k_2\end{bmatrix} \tag{46}$$

The constant matrix $K=\begin{bmatrix}k_1 & k_2\end{bmatrix}$ is chosen so that the roots of the characteristic equation have negative real parts.
We then can design the robust adaptive controller

$$v=-\beta e^T Pb\left\{\left[f(e+y_r,z)\right]^2+\left\|e+y_r\right\|^2+\left\|z\right\|^2+\left[\alpha^{-1}(2r)\right]^2+(Ke)^2+1\right\}, \tag{47}$$

where β is the adaptive parameter of the controller and $\alpha_1{}^{-1}$ is the inverse function and is a function of class K_∞. For now, we assume $\alpha_1(\cdot)=\left\|\cdot\right\|^2$
P is Lyapunov matrix under the condition of:

$$PA_m + A_m{}^T = -Q\text{, where } Q=Q^T>0 \tag{48}$$

$$\dot{\beta}=\beta_m(e,z,r,y_r)-\Gamma\sigma\beta \tag{49}$$

$$\beta_m=\Gamma\left(e^T Pb\right)^2\left\{\left[f(e+y_r,z)\right]^2+\left\|e+y_r\right\|^2+\left\|z\right\|^2+\left[\alpha^{-1}(2r)\right]^2+(Ke)^2+1\right\} \tag{50}$$

For the design constants we assume they are known and satisfy the condition of

$$T>0,\sigma>0 \tag{51}$$

Because in our experiments, only the position signal is detected, we further implement high gain observer for the purpose of output feedback.

Let the error signal be:

$$\dot{\hat{e}}_1 = \hat{e}_2 + \left(\frac{\sigma_1}{\varepsilon}\right)(e_1 - \hat{e}_1), \qquad (52)$$
$$\dot{\hat{e}}_2 = \left(\frac{\sigma_2}{\varepsilon^2}\right)(e_1 - \hat{e}_1)$$

where $\varepsilon > 0$ is a small constant, $\sigma_i > 0, i = 1,2$ are chosen so that $A_n = A - K_\sigma C$ is a Hurwitz matrix, and

$$K_\sigma = [\sigma_1 \quad \sigma_2],\ C = [1 \quad 0] \qquad (53)$$

Accordingly for the matrix:

$$A_n = \begin{bmatrix} 0 & 1 \\ 0 & 0 \end{bmatrix} - [\sigma_1 \quad \sigma_2]\begin{bmatrix} 1 \\ 0 \end{bmatrix} \qquad (54)$$

$$A_n = \begin{bmatrix} -\sigma_1 & 1 \\ -\sigma_2 & 0 \end{bmatrix} \qquad (55)$$

The positive constants matrix $K_\sigma = [\sigma_1 \quad \sigma_2]$ is chosen so that the characteristic equation $s^2 + \sigma_1 s + \sigma_2 = 0$ has the roots with negative real parts.

To eliminate peaking in the implementation of the observer, we define

$$\hat{e}_1 = \frac{q_1}{\varepsilon} \qquad (56)$$
$$\hat{e}_2 = q_2$$

Thus:

$$\varepsilon\dot{q}_1 = q_2 + \sigma_1(e_1 - q_1) \qquad (57)$$
$$\varepsilon\dot{q}_2 = \sigma_2(e_1 - q_1)$$

In order to prevent the peaking from entering the control system, we saturate the control signal and adaptive controller outside of their domains of interests. In our experiments,

$$M_v = 50 \qquad (58)$$
$$M_\beta = 30$$

With these constants, which are larger than or equal to the upper bound of those signals $v(e, r, \bar{y}_r, \beta), \beta_m(e, r, \bar{y}_r)$, we can denote the following equations:

$$v^S(e, r, \bar{y}_r, \beta) = M_v sat\left(\frac{v(e, r, \bar{y}_r, \beta)}{M_v}\right) \qquad (59)$$

$$\beta_m^{\ S}(e,r,\bar{y}_r) = M_\beta sat\left(\frac{\beta_m(e,r,\bar{y}_r)}{M_\beta}\right) \tag{60}$$

$$sat(\ )\text{: saturation function} \tag{61}$$

Thus the robust adaptive output controller can be obtained by replacing $v(e,r,\bar{y}_r,\beta)$ and $\beta_m(e,r,\bar{y}_r)$ with $v^S(e,r,\bar{y}_r,\beta)$ and $\beta_m^{\ S}(e,r,\bar{y}_r,\beta)$

Simulation results:

We choose the constants:

$$\sigma = 0.00001, K = \begin{bmatrix} 2 & 4 \end{bmatrix}, P = \begin{bmatrix} 1.375 & 0.25 \\ 0.25 & 0.1875 \end{bmatrix}, Q = \begin{bmatrix} 1 & 0 \\ 0 & 1 \end{bmatrix}.$$

The simulation track two different input signal respectively: Both simulations are tested with adding 0.5sin(t) as disturbance.

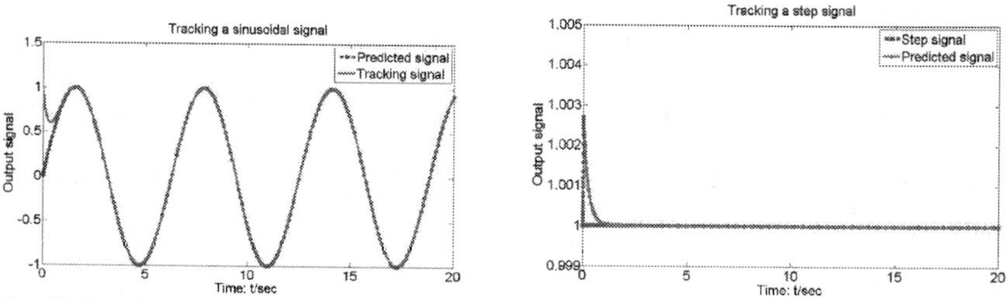

Fig. 13. Tracking sinusoidal and step signals

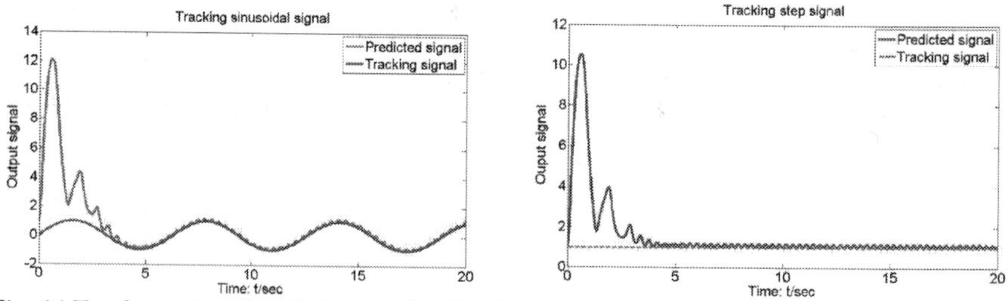

Fig. 14 Tracking step signal: Output feedback

Apparently the state feedback tracks better with either Sinusoidal or Step signal although he observer design somewhat caused a fluctuation with very small amplitude. This might require further tuning on the constants

13. References

Beersiek, J. (1999), On-line monitoring of keyhole instabilities during laser beam welding, ICALEO (R) 99: Section D, 49-58

Beersiek, J. (2000), "A CMOS Camera as a Tool for Process Analysis Not Only for Laser Beam," ICALEO'01 92&93: 1185-1193, 2000

Beersiek, J. (2001), "A CMOS camera as a tool for process analysis not only for laser beam welding," 20th ICALEO Vols 92&93, 1185-1193, 2001

Becker, S. and. Beersiek, J. (2000), "Semi-Automatic 3D Model Extraction from Uncalibrated 2D Camera Views," SPIE Visual Data Exploration and Analysis II 2410: 447-461, 1995

Bertrand, P, Smurov, I and Grevey, D. (2000), Application of near infrared pyrometry for continuous Nd:YAG laser welding of stainless steel, Applied surface science 168 (2000) 182-185, 2000

Farson, D. (1999), "Laser welding penetration monitoring with multiple emissions signal measurements," Journal of Laser Applications 11(2): 47-53, 1999

Farson, D, and Kim, K, R (1999), "Generation of optical and acoustic emissions in laser weld plues," Jurnal of applied physics, 85(1999), 1329-1336, 1999

Gu, H, P and Duley, W. W. (1996), "Resonant acoustic emission during laser welding of metals," J Phys. D: Appl. Phys. 29 (1996) 550-555.

Jeon, M, Kim, W, Han, G and Na, S. (1998), "A study on heat flow and temperature monitoring in the laser brazing of a pin-to-plate joint," Journal of materials processing technology, 82 (1998) 53-60

Klein, M and Bodenhamer, T, Laser ultrasonics technology report, http://www.industrial-laser.com

Kroos, J., Gratzke, U., Vicanek, M. and Simon, G., (1993) "Dynamic behaviour of the keyhole in laser welding," Journal of Physics, D: Applied Physics, 26, 1993, 481-486

Lankalpall, K. N. T., J. F, and Gartner, M. (1996), "A model for Estimating Penetration Depth of Laser Welding Processes," Journal of Physics: Apply Physics 29(7): 1831-1841, 1996

Li, L, Brookfield, D, J. and Steen W., M. (1996), "Plasma charge sensor for in-process, non-contact monitoring of the laser welding process," Meas, Sci, Technol. 7 (1996), 615-626.

Li, L. (2002), "A comparative study of ultrasound emission characteristics in laser processing," Applied surface science, 186 (2002): 604-610

Lim, D, Cho, Y, Gweon, D. (1998), "A robust in-process monitoring of pulsed laser spot welding using a point infrared sensor," Proceedings of the Institution of Mechanical Engineers, Part B, Engineering Manufacture, Vol. 212, Issue 3, 241-250, 1998.

Liu, Y. S. and Li, X. Y. (2002), "Decentralized robust adaptive control of nonlinear systems with unmodeled dynamics," IEEE Transactions on Automatic Control 47(5): 848-856, 2002

Liu, Y. S., and Li, X. Y. (2003), "Robust adaptive control of nonlinear systems represented by input-output models," IEEE Transactions on Automatic Control 48(6): 1041-1045, 2003

Liu, Y. S., and Li, X. Y. (2004), "Robust adaptive control of nonlinear systems with unmodeled dynamis," IEE Proceedings-Control Theory and Applications 15151(1): 83-88, 2004

Migliore, L. (1998), "Welding with Lasers," Industrial Laser Review, 1998

Miller, M, Mi, B, Kita, A and Ume, C. (2002), "Development of automated real-time data acquisition system for robotic weld quality monitoring, Mechatronics, 12 (2002), 1259-1269

Morgan, S. A. F., M D T; McLean, M A; Hand, D P; Haran, F M; Su, D; Steen, W M; Jones, J D C. (1997), "Real-time process control in CO2 laser welding and direct casting: focus and temperature," ICALEO '97: Laser Materials Processing. San Diego, California; USA. 83: G290-G299, 1990

Na, X. D., Zhang, Y. M., Liu, Y. S. and Walcott, B. (2009), "Nonlinear Identification of Laser Welding Process," IEEE Transactions on Control Systems Technology, Digital Object Identifier 10.1108/TCST.2009.2026163

Park, H and Rhee, S. (1999) "Estimation of weld bead size in CO2 laser welding by using multiple regression and neural network," Journal of Laser Applications 11(3): 143-150, 1999

Park, H and Rhee, S. (1999), "Analysis of mechanism of plasma and spatter in CO2 laser welding of galvanized steel," 31 (1999) 119-126

Park, Y, W, Park, H, Rhee, S and Kang, M. (2002), "Real time estimation of CO2 laser weld quality for automotive industry," Optics & laser Technology 34 (2002) 135-142.

Pedrotti, F. L. a. P., L.S. (1993), "Introduction to Optics," New Jersey, Prentice Hall, 1993

Ono, M, Nakada, K and Kosuge, S, An investigation on CO2 laser-induced plasma, J Jpn Weld Soc, 10(1992): 239-245

Ostendorf, A, Specker, W, Stallmach, M and Zeadan, J. (2003), "3D-MID and process monitoring for micro joining applications," Proceedings of the society of photo-optical instrumentation engineers (SPIE) 4977, 508-517, 2003

Ostendorf, A, Kulik, C, Stallmach, M and Zeadan, J. (2004), "Basic investigations for controlling the laser spot welding process when packaging 3-dimensional molded interconnect devices," Proceedings of the society of photo-optical instrumentation engineers (SPIE) 5339, 441-447, 2004.

Ostendorf, A. and Temme, T. (2004), "Laser spot welding of electronic micro parts," Proceedings of the society of photo-optical instrumentation engineers (SPIE) 5662, 306-312, 2004.

Sforza, P and Blasiis, D. (2002), "On-line optical monitoring system for arc welding," NDT&E international 35 (2002) 37-43

Shao, J and Yan, Y. (2005), "Review of techniques for On-line Monitoring and Inspection of Laser Welding," Journal of Physics: Conference Series 15 (2005), pp. 101-107, 2005

Steen, W, M and Weerasinghe, W, M, Monitoring of laser material processes, SPIE Proc. 650(1986), 160-166.

Sun, A., Kannatey-Asibu, Jr. E. (1999), "Sensor systems for real-time monitoring of laser weld quality," Journal of Laser Applications 11(4): 153-168, 1999

Sun, A., Kannatey-Asibu, Jr. E. (2002), "Monitoring of Laser Weld Penetration Using Sensor Fusion," Journal of Laser Applications 14(2): 114-121, 2002

Sun, Z, Salminen, A. S. and Moisio, T. J. I. (1993), "Quality improvement of laser beam welds by plasma control," Journal of Materials Science Letters, Vol. 12, No. 14, Springer Netherlands, January, 1993

Tonshoff H, K, Ostendorf A, Guttler R, Specker W. (1998), "Online monitoring and closed-loop control of laser welding processes," 12th international symposium for electromachining(ISEM), Aachen, Germany, May 11-13, 1998, 1405 603-612, 1998.

Williams, C. (1997), "CO Laser Processing-An Overview," Aircraft Engineering and Aerospace Technology 69(1): 43-52, 1997

Zhang, X, Chen, W, Ashida, E and Matsuda, F. (2004), "Relationship between weld quality and optical emissions in underwater Nd:YAG laser welding," Opticas and lasers in engineering 41(2004), 717-730

Zhang, Y. M., Kovacevic, R. and Li, L. (1996), "Characterization and real-time measurement of geometrical appearance of the weld pool," Int. J. Mach. Tools Manufact. Vol. 36 No. 7, 799-816, 1996